LIPID AND BIOPOLYMER MONOLAYERS AT LIQUID INTERFACES

LIPID AND BIOPOLYMER MONOLAYERS AT LIQUID INTERFACES

K. S. Birdi
The Technical University of Denmark
Lyngby, Denmark

PLENUM PRESS • NEW YORK AND LONDON

0 3463527

CHEMISTRY

Library of Congress Cataloging in Publication Data

Birdi, K. S., 1934–
  Lipid and biopolymer monolayers at liquid interfaces.

  Bibliography: p.
  Includes index.
  1. Monomolecular films. 2. Lipids. 3. Biopolymers. I. Title.
  QD506.B53   1988   541.3'453   88-23172
  ISBN 0-306-42870-9

© 1989 Plenum Press, New York
A Division of Plenum Publishing Corporation
233 Spring Street, New York, N.Y. 10013

Printed in the United States of America

PREFACE

During the past few decades, much research has been reported on the formation of insoluble monomolecular films of lipids and biopolymers (synthetic polymers and proteins) on the surface of water or at the oil–water interface. This interest arises from the fact that monomolecular film studies have been found to provide much useful information on a molecular scale, information that is useful for understanding many industrial and biological phenomena in chemical, agricultural, pharmaceutical, medical, and food science applications. For instance, information obtained from lipid monolayer studies has been useful in determining the forces that are known to stabilize emulsions and biological cell membranes.

The current texts on surface chemistry generally devote a single chapter to the characteristics of spread monolayers of lipids and biopolymers on liquids, and a researcher may have to review several hundred references to determine the procedures needed to investigate or analyze a particular phenomenon. Furthermore, there is an urgent need at this stage for a text that discusses the state of the art regarding the surface phenomena exhibited by lipids and biopolymers, as they are relevant to a wide variety of surface and interfacial processes.

The purpose of this book is to bring the reader up to date with the most recent experimental and theoretical developments in this field. Considerable effort has been devoted to providing a presentation that describes the interactions among the molecules in monolayers, which are of importance for research in chemistry, physics, biochemistry, biophysics, and geophysics. It is my intention to lead the researcher through the vast literature in such a way that he or she will be able to pursue a particular investigation with suitable guidance. Investigations undertaken by the intended readership for this book may be industrial (e.g., emulsions, foams, dispersions, enhanced oil recovery) or biological (e.g., lipid-phase transitions, vesicles, liposomes, bilayers, membranes, ion transport through membranes) in nature.

The treatment in different parts of the text is based on the fundamental concepts of the classical thermodynamics of surfaces (interfaces), i.e., the

Gibbs adsorption theory. A large amount of experimental data from the literature has been included and critically examined from a theoretical viewpoint. This approach is based upon past analyses, which should provide an exciting incentive for new developments in surface chemistry and monomolecular films.

It is now well established that the monolayer is a useful model membrane system, and its usefulness has led to extensive investigations on monolayers of various lipids and biopolymers of biological interest. These systems are described in detail, and the relevant discussions should be useful in understanding the physical forces that are responsible for the stability and function of cell membranes.

Lipid-phase transition phenomena in cell membranes and in particular in monolayers are discussed in detail. The significance of lipid-phase transitions in other systems, such as emulsions, is also pointed out.

The adsorption of proteins (biopolymers) at water surfaces is discussed. This is a surface phenomenon that has to date received little attention, even though many important phenomena in the food and pharmaceutical industries are dependent on it.

The presence of charge on biopolymers (e.g., proteins) and its relationship to the transport of ions across biological cell membranes (in relation to ionophores) is also described.

This book is intended for professional chemists and other scientists with an interest in the surface chemistry of lipids and biopolymers. The book can also be useful as a graduate-level textbook, under appropriate guidance. Wherever necessary, procedures are delineated to guide the reader through a series of typical monolayer experiments. Both experimental and analytical procedures are stressed, to provide the necessary background information. This approach should be useful for both the newcomer to the field and those pursuing new directions in this exciting and expanding area of research.

I would like to thank my colleagues at the Physical–Chemical Institute for help in various stages of the writing of this book. I would especially like to thank Mr. G. Zilo and Mrs. J. Klausen for their technical help. I am indebted to Drs. V. Gevod and R. Sanchez for stimulating discussions and help during the preparation of the book. The author thanks the Danish Natural Science and Technical Research Councils and The Nordic Ministry for research support.

Finally, I thank my wife and my son for their patience during those periods when, as an author, I needed special treatment.

K. S. Birdi

Lyngby, Denmark

CONTENTS

CHAPTER 4. LIPID MONOLAYERS AT LIQUID INTERFACES   57

1

INTRODUCTION

Many natural phenomena attracted the interest of early man. One such phenomenon was the presence of oil slicks on water. The ancient Egyptians were known to have observed the effect of oil in reduction of friction.

All lipid and protein molecules exhibit surface-active properties. Such surface activity is associated with molecules that possess a dual character—that is, one part is hydrophobic ("water-fearing") and the other part hydrophilic ("water-loving"); such molecules are also called "amphiphiles." The primary requirement for an amphiphile to be surface-active is that the balance between the hydrophobic and the hydrophilic parts be such that the molecule is at a lower free energy when it is adsorbed at an interface (e.g., air–water, liquid–liquid, or solid–liquid) than when it is within the continuous aqueous bulk phase.

The system to be described herein mainly involves measurements undertaken on a monomolecular layer of the surface of a liquid (Fig. 1.1a). The

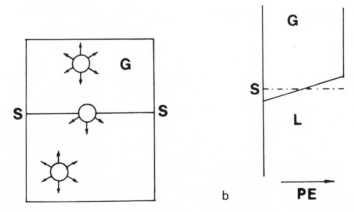

FIGURE 1.1. (a) Schematic diagram of molecules at the surface of a liquid, S—S (the arrows indicate the interactions among adjacent molecules). (b) Schematic diagram of the variation of potential energy (PE) as expected when a molecule moves from the bulk liquid (L) phase to the gas (G) phase.

1

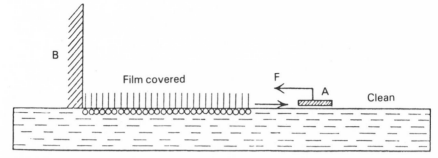

FIGURE 1.2. Schematic diagram of a monolayer apparatus, the so-called "Langmuir film balance" (B). The polar heads (O) are placed within the water phase, while the nonpolar hydrocarbon chains (vertical lines, | | |) are oriented upward from the water phase. The float (A) separates the monolayer-covered area from the clean water surface and allows one to measure the surface pressure, π (F). Redrawn from Adam (1968) with changes.

variation of the potential energy as a molecule moves from the bulk liquid phase to the gas phase is also depicted (Fig. 1.1b)

It was actually Langmuir (1917) who reported the method whereby such studies could be carried out using the apparatus shown schematically in Fig. 1.2.

This apparatus, later called the "Langmuir film balance," allows one to measure the two-dimensional pressure, π, exerted on the surface of water by the monomolecular film of an amphiphile with the necessary physical properties. These investigations have been successful in providing strong support for the ideas of molecular orientation at liquid surfaces and the short-range molecular forces. These investigations further formed the basis for determining why different sorts of molecules form various kinds of films, particularly as regards the interfacial forces.

The literature covers an extensive area of application of monolayer studies and their application to a wide variety of systems. Almost every textbook on surface chemistry includes a chapter or more related to monomolecular film investigations. Some of the more recent and important books are Harkins (1952); Wolf (1959); Davies and Rideal (1963); Gaines (1966); Defay and Prigogine (1966); Adam (1968); Adamson (1982); Chattoraj and Birdi (1984).

1.1. LIPID AND BIOPOLYMER (PROTEIN AND OTHER POLYMER) MONOLAYERS AT LIQUID INTERFACES

To understand the role of lipids and proteins (both of which behave as amphiphiles) in various systems at a more detailed molecular level, it is both convenient and necessary to conduct investigations by using a simple model system, rather than to study complex industrial phenomena (e.g., foams, emulsions, microemulsions) or even more complex biological systems (cell membrane structure and function).

The most important biological system is the cell membrane, which consists of a bilayer of lipid molecules with protein molecules incorporated into this structure. Both these essential components of membranes are known to form a very stable monomolecular film when adsorbed or spread on the surface of water, owing to the fact that the hydrophobic-hydrophilic energy balance is suitably proportionate. In fact, the bilayer lipid structure of membranes was first predicted from such monolayer studies.

It has been found during the last few decades that much useful information about the structure and function of membranes at the molecular level can be obtained in the laboratory by using the monomolecular film method as the model system.

The lipid bilayer structure found in cell membranes is also similar to other bilayer lipid structures, e.g., those found in soap bubbles, foams, and liquid crystals. The latter systems are indeed of great industrial importance for processes such as detergency, emulsifying, microemulsifying, and improved oil recovery.

The liquid crystalline phase transition as observed in membranes can be conveniently investigated by using model systems, e.g., monolayers, vesicles, or bilayers (Fig. 1.3). Since both vesicles and bilayers can be formed from *monolayers*, as described herein, it is important to recognize the importance of investigations carried out on the latter systems. The energetics can be easily estimated from such model systems.

Furthermore, monolayer studies of lipids and proteins are of much value, since the understanding of these systems and the thermodynamics of monolayer systems have been extensively described and rigorous theoretical analyses have been reported.

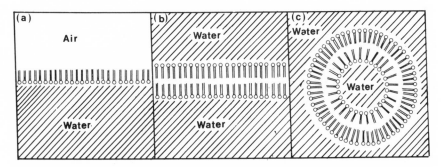

FIGURE 1.3. Schematic representation of model membrane structures of lipid molecules: alkyl part (‖ ‖ ‖) and polar part (O) of the lipid molecule. (a) Monolayer structure; (b) bilayer (bilipid membrane); (c) vesicle (liposome). Redrawn from Noll (1976) with changes.

2

THERMODYNAMICS OF LIQUID SURFACES

It is necessary to discuss the physics of liquid surfaces before describing lipid and protein monolayers at liquid surfaces. This treatment is presented merely as an introduction; for detailed analyses, the reader is referred to more advanced textbooks on this subject (Defay *et al.*, 1966; Adamson, 1982; Chattoraj and Birdi, 1984).

The boundary between two phases, each homogeneous, is not to be regarded as a simple geometric dividing line that separates the two homogeneous phases. It should be considered, rather, as a lamina or film of some thickness.

2.1. SURFACE THERMODYNAMICS

Surface Energy. The state of surface energy is best described by the following classic example (Chattoraj and Birdi, 1984): Consider the area of a liquid film that is stretched in a wire frame by an increment $\gamma\, dA$, whereby the surface energy changes by $\gamma\, dA$ (Fig. 2.1). Under this process, the opposing force is f. From the dimension in Fig. 2.1, we find

$$f\, dx = \gamma\, dA \tag{2.1}$$

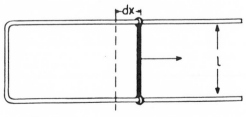

FIGURE 2.1. Surface film of a liquid. From Chattoraj and Birdi (1984).

5

or

$$\gamma = f(dx/dA)$$
$$= (f/2)\ell \qquad (2.2)$$

where dx is the change in displacement and ℓ is the length of the thin film. The quantity γ represents the *force per unit length of surface* (mN/m = dyne/cm), and this force is defined as surface tension or interfacial tension. Surface tension, γ, is the differential change of free energy with change of surface area at constant temperature, pressure, and composition.

The surface tension, γ, and the mechanical equilibrium at interfaces have been described in detail in the literature (Chattoraj and Birdi, 1984). The surface has been considered as a hypothetical stretched membrane, this membrane being termed the surface tension. In a real system undergoing an infinitesimal process, it can be written

$$dW = p\,dV + p\,dV' - \gamma\,dA \qquad (2.3)$$

where dW is the work done by the system when changes in volume, dV and dV', occur; p and p' are pressures in the two phases α and β, respectively, at equilibrium; and dA is the change in interfacial area. The sign of the interfacial work is designated negative by convention (Chattoraj and Birdi, 1984).

The fundamental property of liquid surfaces is that they tend to contract to the smallest possible area. This property is observed in the spherical form of small drops of liquid, in the tension exerted by soap films as they tend to become less extended, and in many other properties of liquid surfaces. In the absence of gravity effects, these curved surfaces are described by the Laplace equation, which relates the mechanical forces as (Chattoraj and Birdi, 1984)

$$p - p' = \gamma(1/r_1 + 1/r_2) \qquad (2.4)$$
$$= 2(\gamma/r) \qquad (2.5)$$

where r_1 and r_2 are the radii of curvature and r is the radius of curvature for a spherical interface. It is a geometric fact that surfaces for which equations (2.4) and (2.5) hold are surfaces of minimum area. These equations thus give

$$dW = p\,d(V + V') - \gamma\,dA \qquad (2.6)$$
$$= p\,dV' - \gamma\,dA \qquad (2.7)$$

since $p = p'$ for a plane surface and V' is the total volume of the system.

2.2. SURFACE FREE ENERGY
(SURFACE TENSION) OF LIQUIDS

The surface-phase (denoted as σ) Helmholtz free energy, F^σ, is given as (Chattoraj and Birdi, 1984)

$$\gamma = (dF^\sigma / dA)_{T,V^\sigma,n_i^\sigma} \tag{2.8}$$

where γ represents the rate of change of the Helmholtz free energy in the σ phase with change of the surface area A.

When the volume of the σ phase is assumed to be negligible, as described by Gibbs (Chattoraj and Birdi, 1984), then one obtains

$$\gamma = (dG^\sigma / dA)_{T,p,n_i} \tag{2.9}$$

2.2.1. SURFACE TENSION OF LIQUIDS

Molecular phenomena at the surface separating a liquid and its saturated vapor, or a liquid and the walls of its containing vessel, are appreciably more complex than those that obtain inside the homogeneous liquid, and it is difficult to say much of a rigorous qualitative nature concerning them. The essential difficulty is that whereas from the macroscopic standpoint there is always a well-defined surface of separation between the two phases, on the microscopic scale there is only a surface zone, in crossing which the structure of the fluid undergoes progressive modification. It is in this surface zone that the dynamic equilibrium between the molecules of the vapor and those of the liquid is established. Owing to the attractive forces exerted by the molecules of the liquid proper on one another, only the fast-moving molecules can penetrate the layer and escape into the vapor; in the process, they lose kinetic energy and, on the average, attain the same velocity as the molecules in the vapor.

Further, the number of molecules escaping from the liquid cannot, on average, exceed the number entering from the much less dense vapor. From the statistical point of view, the density of the fluid is the most important variable in the surface area; it does not suffer an abrupt change, of course, but varies continuously in passing through the surface zone from its value in the liquid to its generally much lower value in the vapor (a decrease by a factor of approximately 1000). In consequence, it is possible to specify only rather arbitrarily where the liquid phase ends and the gaseous phase begins. It is convenient for some purposes to define the interface as a certain layer of constant density, within the surface zone, such that if each of the two phases remains homogeneous to the surface, the total number of molecules in each phase will be the same (Chattoraj and Birdi, 1984).

The work required to increase the area of a surface is that required to bring additional molecules from the interior to the surface. This work must be done against the attraction of surrounding molecules. Since cohesive forces (as described in Chapter 4) fall off very steeply with distance, one can consider as a first approximation interactions between adjacent molecules only. There is strong evidence that the change of density from liquid to vapor is exceedingly abrupt, translational layers being generally only one or two molecules thick. Perhaps the clearest evidence is that derived from the nature of light reflected from the surface of liquids. According to Fresnel's law of reflection, if the transition between air and a medium of refractive index n is absolutely abrupt, the light is completely plane polarized, if the angle of incidence is the Brucetarian angle. But if the transition is gradual, the light is elliptically polarized. It was found (Raman and Ramdas, 1927) that there is still some small amount of residual ellipticity in the cleanest surfaces of water and also that such surfaces scatter light to some extent.

CORRESPONDING STATES THEORY

To understand the physics of liquid surfaces, it is thus important to be able to describe the interfacial forces as a function of temperature and pressure. The magnitude of γ decreases almost linearly with temperature within a narrow range (Defay, 1934):

$$\gamma_t = \gamma_0(1 - \alpha t) \tag{2.10}$$

where γ_0 is a constant. It was found that coefficient α is approximately equal to the rate of decrease of density, ρ, with rise of temperature;

$$\rho_t = \rho_0(1 - \alpha t) \tag{2.11}$$

Different values of constant α were found for different liquids. Furthermore, the value of α was related to T_c (Bakker, 1914).

The following equation relates the surface tension of a liquid to the density of the liquid, ρ_l, and its vapor, ρ_v (McLeod, 1923):

$$\gamma/(\rho_l - \rho_v]^4 = C \tag{2.12}$$

where the value of the constant C is nonvariable only for organic liquids and is not constant for liquid metals. At the critical temperature, T_c, a liquid and its vapor are identical, and the surface tension, γ, and total surface energy, like the energy of vaporization, must be zero. At temperatures below the boiling point, which is $\frac{2}{3}T_c$, the total surface energy and the energy of evaporation are nearly constant. The variation in surface tension, γ, with temperature is given in Fig. 2.2 for different liquids.

These data clearly show that the variation of γ with temperature is a very characteristic physical property. This observation becomes even more important when it is considered that the sensitivity of γ measurements can be as high as approximately ± 0.001 dyne/cm ($=$ mN/m) (as described in detail below).

It is well known that the corresponding states theory can provide much useful information about the thermodynamics and transport properties of fluids. For example, the most useful two-parameter empirical expression that relates the surface tension, γ, to the critical temperature, T_c, is given as

$$\gamma = k_0 (1 - T/T_c)^{k_1} \qquad (2.13)$$

where k_0 and k_1 are constants. Van der Waals derived this equation and showed that $k_1 = \frac{3}{2}$, although experiments indicated that $k_1 \approx 1.23$. Guggenheim (1945) has suggested that $k_1 = \frac{11}{9}$. However, for many liquids, the value of k_1 lies between $\frac{6}{5}$ and $\frac{5}{4}$.

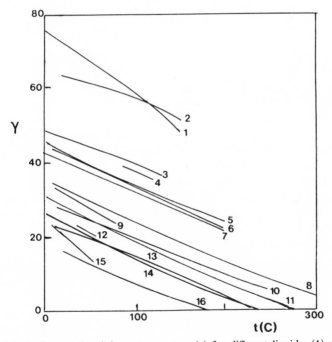

FIGURE 2.2. Surface tension (γ) vs. temperature (t) for different liquids: (1) water; (2) glycerol; (3) glycol; (4) acetamide; (5) nitrobenzene; (6) aniline; (7) phenol; (8) chlorobenzene; (9) ethylene dichloride; (10) acetic acid; (11) benzene; (12) ethylidene dichloride; (13) ethyl alcohol; (14) ethyl acetate; (15) ammonia; (16) diethyl ether.

Van der Waals also found k_0 to be proportional to $T_c^{1/3} P_c^{2/3}$. Equation (2.13), when fitted to the surface tension, γ, of liquid CH_4, has been found to give the following relationship:

$$\gamma = 40.52(1 - T/190.55)^{1.287} \qquad (2.14)$$

where $T_c = 190.55°K$. This equation has been found to fit the γ data for liquid methane from 91 to 190°K, with an accuracy of ±0.5 mN/m. In a recent study, the γ vs. T data on n-alkanes, from n-pentane to n-hexadecane, were analyzed. The constants k_0 (between 52 and 58) and k_1 (between 1.2 and 1.5) were found to be dependent on the number of carbon atoms, n_C, and since T_c is also found to be dependent on n_C, the expression for all the different alkanes that individually were fitted to equation (2.13) gave rise to a general equation where γ was a function of n_C and T, as follows (Birdi et al., to be published):

$$\gamma = \text{function of } T, n_C \qquad (2.15)$$

$$= (41.41 + 2.731 n_C - 0.192 n_C^2 + 0.00503 n_C^3)$$

$$\times \{1 - T/[273 + (-99.86 + 145.4 \ln n_C) + 17.05(\ln n_C)^2]\}^{k_1} \quad (2.16)$$

where $k_1 = 0.9968 + 0.0487 n_C - 0.00282 n_C^2 + 0.000084 n_C^3$.

The estimated values from equation (2.16) for γ of different n-alkanes were found to agree with the measured data within a few percent: γ for n-$C_{18}H_{38}$, at 100°C, was 21.6 mN/m, from both measured and calculated values. This agreement shows that the surface tension data on n-alkanes fit the corresponding state equation very satisfactorily (see Table 2.1). The physical analysis of constants k_0 and k_1 needs to be investigated at this stage.

It is worth mentioning that the equation for the data on γ vs. T, for polar (and associating) molecules like water and alcohols, when analyzed by equation (2.13), gives magnitudes of k_0 and k_1 that are significantly different from those found for nonpolar molecules such as alkanes. This difference requires further analyses so that the relationship between γ and T_c may be more completely understood.

The variation of γ for water with temperature ($t/C°$) is given as (Cini et al., 1972a & b)

$$\gamma = 75.668 - 0.1396t - (0.2885 \times 10^{-3})t^2 \qquad (2.17)$$

The entropy corresponding to equation (2.13) is

$$-d\gamma/dT = k_1 k_0 (1 - T/T_c)^{k_1-1}/T_c \qquad (2.18)$$

TABLE 2.1. Calculated γ [from Equation (2.16)] and Measured Values of Different n-Alkanes at Various Temperatures[a]

n-Alkane	T (°C)	$\gamma_{meas.}$	$\gamma_{cal.}$
C-5	0	18.23	18.25
	50	12.91	12.8
C-6	0	20.45	20.40
	60	14.31	14.3
C-7	30	19.16	19.17
	80	14.31	14.26
C-9	0	24.76	24.70
	50	19.97	20.05
	100	15.41	15.4
C-14	‘10	27.47	27.4
	100	19.66	19.60
C-16	50	24.90	24.90
C-18	30	27.50	27.50
	100	21.58	21.60

[a] From Birdi *et al.*, to be published.

and the corresponding enthalpy, h_s is

$$h_s = \gamma - T(d\gamma/dT)$$
$$= k_0(1 - T/T_c)^{k_1-1}[1 + (k_1 - 1)T/T_c] \tag{2.19}$$

The reason heat is absorbed on expansion of a surface is that molecules must be transferred from the interior against the inward attractive force, to form the new surface. In this process, the motion of the molecules is retarded by this inward attraction, so that the temperature of the surface layers is lower than that of the interior, unless heat is supplied from outside.

Further, extrapolation of γ to zero surface tension in the data given in Fig. 2.2 gave values of T_c that were 10%–25% lower than the measured values (Birdi *et al.*, to be published). However, when T_c was estimated for $\gamma = -5$ mN/m, then the estimated T_c values were approximately 5% of measured values. This observation is useful in showing that the γ vs. T plots can provide much useful information as regards the interfacial forces of liquids, even near the critical temperature, T_c. This means that T_c can be easily estimated by merely measuring γ vs. T in some convenient temperature range.

The surface tension, γ, of any liquid would be related to the pressure, p, as follows (Lewis and Randall, 1923; Defay *et al.*, 1966):

$$(d\gamma/dP)_{A,T} = (dV^\sigma/dA)_{P,T} \tag{2.20}$$

Since the quantity on the right-hand side would be positive, the effect of an increase in pressure should be to give an increase in γ. Preliminary analyses indicate that the term $d\gamma/dP$ is dependent on the alkane chain length (Birdi, Sanchez and Fredenslund, to be published).

The following relationship relating γ to density was given in early work (Ramsay and Shields, 1938):

$$\gamma[M/\rho]^{2/3} = k(T_c - T - 6) \qquad (2.21)$$

where M is the molecular weight, ρ is the density, and M/ρ is the molar volume. The quantity $\gamma(M/\rho)^{2/3}$ is called the "molecular surface energy."

2.2.2. HEAT OF SURFACE FORMATION AND EVAPORATION

Several attempts have been made to relate the γ of a liquid to the latent heat of evaporation. It was argued a century ago (Stefan, 1886) that when a molecule is brought to the surface of a liquid from the interior, the work done overcoming the attractive forces near the surface should be related to the work expended when it escapes into the much less dense vapor phase. It was suggested that the first quantity should be approximately half the second. According to the Laplace theory of capillaries, the attractive forces acts only over a small distance equal to the radius of the sphere (see Fig. 1.1), and in the interior the molecule is attracted equally in all directions

TABLE 2.2. Ratios of the Enthalpy
of Surface Formation,
h_s (ergs/molecule), to the Enthalpy
of Vaporization (h_{vap})

Molecules (liquid)	h_s/h_{vap}
Hg	0.64
N_2	0.51
O_2	0.5
CCl_4	0.45
C_6H_6	0.44
Diethyl ether	0.42
$Cl-C_6H_5$	0.42
Methyl formate	0.40
Ethyl acetate	0.4
Acetic acid	0.34
H_2O	0.28
C_2H_5OH	0.19
CH_3OH	0.16

and experiences no resultant force. On the surface, it experiences a force due to the liquid in the hemisphere, and half the total molecular attraction is overcome in bringing it there from the interior.

Accordingly (Stefan, 1886), the energy necessary to bring a molecule to the surface of a liquid should be half the energy necessary to bring it entirely into the gas phase. This corresponds—with the most densely packed top (surface) monomolecular layer half filled and the next layer completely filled—to a very dilute gas phase (the distance between gas molecules is approximately 1000 times greater than in liquids or solids).

The ratio of the enthalpy of surface formation to the enthalpy of vaporization, $h_s : h_{vap}$, for various substances is given in Table 2.2. Substances with nearly spherical molecules have ratios near $\frac{1}{2}$, while substances with a polar group on one end give a much smaller ratio. This difference indicates that the latter molecules are oriented with the nonpolar end toward the gas phase and the polar end toward the liquid. In other words, molecules with dipoles would be expected to be oriented perpendicularly at gas/liquid interfaces.

2.2.3. WETTING AND SPREADING

It is desirable at this stage to consider the properties of liquid–solid interfaces. The interfacial forces that are present between a liquid and a solid can be estimated by studying the shape of a drop of liquid placed on any smooth solid surface (Fig. 2.3). The balance of forces indicated in this figure was analyzed very extensively in the last century by Young, who related the different forces at the solid–liquid boundary and the contact angle, θ, as follows (Young, 1855; Chattoraj and Birdi, 1984):

$$\gamma_{lv} \cos \theta = \gamma_{sv} - \gamma_{sl} \qquad (2.22)$$

where γ is the interfacial tension at the various boundaries between the solid, s, liquid, l, and vapor, v, phases, respectively. To understand the

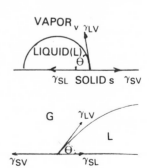

FIGURE 2.3. Interfacial forces at gas–liquid–solid inter-faces (Young's model). From Chattoraj and Birdi (1984).

degree of wetting between the liquid, l, and the solid, s, it is convenient to rewrite equation (2.22) as follows:

$$\cos \theta = (\gamma_{sv} - \gamma_{sl})/\gamma_{lv} \tag{2.23}$$

which would then allow one to understand the variation of θ. The latter is important because complete wetting occurs when there is no finite contact angle, and thus $\gamma_{lv} \leqslant \gamma_{sv} - \gamma_{sl}$. However, when $\gamma_{lv} > \gamma_{sv} - \gamma_{sl}$, then $\cos \theta <$ 1, and a finite contact angle is present. The latter is the case when water, for instance, is placed on a hydrophobic solid, such as Teflon, polyethylene, or paraffin. The addition of surfactants to water, of course, reduces γ_{lv}; therefore, θ will decrease on the introduction of such surface–active substances. A complete discussion of wetting of solids is beyond the scope of this book, and the reader is therefore encouraged to look up other standard textbooks on surface chemistry (Adamson, 1982; Chattoraj and Birdi, 1984). However, in this text, we are interested in the spreading behavior of a drop of one liquid placed on the surface of another liquid, especially when the two liquids are immiscible. Harkins (1952) analyzed the spreading phenomenon by introducing a quantity, the spreading coefficient, S, defined as

$$S_{b/a} = \gamma_a - (\gamma_b + \gamma_{ab}) \tag{2.24}$$

where $S_{b/a}$ is the spreading coefficient for liquid b on liquid a, γ_a and γ_b are the respective surface tensions, and γ_{ab} is the interfacial tension between the two liquids. If the value of $S_{b/a}$ is positive, spreading will take place spontaneously, while if it is negative, liquid b will rest as a lens on liquid a.

However, the value of γ_{ab} needs to be considered as the equilibrium value, and therefore if one considers the system at nonequilibrium, then the spreading coefficients would be different. For instance, the instantaneous spreading of benzene is observed to give a value for $S_{b/a}$ of 8.9 dynes/cm, and therefore benzene spreads on water. On the other hand, as the water becomes saturated with time, the value of $\gamma_{(water)}$ decreases and benzene drops tend to form lenses. The lower saturated hydrocarbons such as hexane and hexene also have positive initial spreading coefficients and spread to give thicker films. Longer-chain alkanes, on the other hand, do not spread on water, e.g., the $S_{b/a}$ for C-16 (hexadecane) on water is -1.3 dynes/cm at 25°C.

It is also obvious that since impurities can have very drastic effects on the interfacial tensions in equation (2.24), the value of $S_{b/a}$ would be expected to vary accordingly (see Table 2.3).

The spreading of a solid, e.g., cetyl alcohol, on the surface of water is discussed in more detail in Chapter 4. Generally, however, the detachment of molecules of the amphiphile into the surface film occurs only at the

TABLE 2.3. Calculation of Spreading Coefficients, S, for Air–Water Interfaces (20°C)[a]

Oil	$\gamma_{w/a}$	$-$	$\gamma_{o/a}$	$-$	$\gamma_{o/w}$	$=$	S	Conclusion
n-Hexadecane	72.8	–	30.0	–	52.1	=	−9.3	Will not spread
n-Octane	72.8	–	21.8	–	50.8	=	+0.2	Will just spread
n-Octanol	72.8	–	27.5	–	8.5	=	+36.8	Will spread

[a] Subscripts: (a) air; (w) water; (o) oil.

periphery of the crystal in contact with the air–water surface. In this system, the diffusion of amphiphile through the bulk water phase is expected to be negligible, because the energy barrier now includes not only the formation of a hole in the solid, but also the immersion of the hydrocarbon chain in the water. It is also obvious that this diffusion through the bulk liquid phase is a rather slow process.

Furthermore, the value of S would be very sensitive to such impurities in the spreading of one liquid on another.

2.2.4. INTERFACIAL TENSION OF LIQUID₁–LIQUID₂

When water comes in contact with an oil (e.g., alkanes), the interface between the two liquids has a contractile tendency. For *n*-butanol–water,

TABLE 2.4. Interfacial Tension (IFT) between Water and Various Organic Liquids

Organic liquid	T (°C)	IFT (mN/m)
n-Hexane	20	51.0
n-Octane	20	50.8
CS_2	20	48.0
CCl_4	20	45.1
$Br-C_6H_5$	25	38.1
C_6H_6	20	35.0
$NO_2-C_6H_5$	20	26.0
Ethyl ether	20	10.7
n-Decanol	25	10
n-Octanol	20	8.5
n-Hexanol	25	6.8
Aniline	20	5.85
n-Pentanol	25	4.4
Ethyl acetate	30	2.9
Isobutanol	20	2.1
n-Butanol	20	1.6

the interfacial tension is 1.8 mN/m, and such low figures are characteristic of oils that contain polar groups. This shows that molecules of n-butanol, with $\gamma = 24$ mN/m, must concentrate at the interface, where the repulsion between the packed and oriented molecules offsets somewhat the usual contractile tendency of an interface. Interfacial packing occurs because the hydroxyl heads of the n-butanol molecules can escape from the oil into the water, while the alkyl part remains in the oil, thus resulting in a state of low standard free energy. As can be seen from the data on interfacial tension in Table 2.4, there seems to some correlation between mutual insolubility and interfacial tension.

2.2.5. ANTONOW'S RELATIONSHIP

It was suggested by Antonow (1907, 1932) that the interfacial tension of two mutually saturated liquids is equal to the difference between their surface tensions, the latter being measured when each liquid has become saturated with the other. For the water (w) phase saturated with oil (o) phase, the surface tension, $\gamma_{w(o)}$, is related to the interfacial tension, $\gamma_{o/w}$, as

$$\gamma_{o/w} = \gamma_{w(o)} - \gamma_{o(w)} \qquad (2.25)$$

Some typical data are given in Table 2.5.

TABLE 2.5. Antonow's Rule and Interfacial Tension Data (mN/m)

Oil phase	$\gamma_{w(o)}$	$\gamma_{o(w)}$	$\gamma_{o/w}$	$\gamma_{w(o)} - \gamma_{o(w)}$
Benzene	62	28	34	34
Chloroform	51.7	27.4	23	24.3
Ether	26.8	17.4	8.1	9.4
Toluene	63.7	28.0	35.7	35.7
n-Propylbenzene	68.0	28.5	39.1	39.5
n-Butylbenzene	69.1	28.7	40.6	40.4
Nitrobenzene	67.7	42.8	25.1	24.9
i-Pentanol	27.6	24.6	4.7	3.0
n-Heptanol	29.0	26.9	7.7	2.1
CS_2	71.9	52.3	40.5	19.6
Methylene iodide	72.2	50.5	45.9	21.7

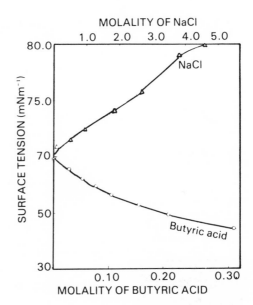

FIGURE 2.4. Plot of surface tension, γ, vs. concentration of NaCl or n-butyric acid. From Chattoraj and Birdi (1984).

2.3. GIBBS ADSORPTION THEORY

The surface tension, γ, of water changes significantly with the addition of organic or inorganic solutes, at constant temperature and pressure. Some typical plots are given in Fig. 2.4.

The γ values of aqueous solutions generally increase with increasing electrolyte concentrations. The γ values of aqueous solutions containing organic solutes invariably decrease. To analyze these data, one has to use the well-known Gibbs adsorption equation (Chattoraj and Birdi, 1984). A liquid column containing i components is shown in Fig. 2.5, according to

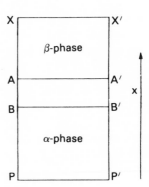

FIGURE 2.5. Liquid column in a "real" system (schematic). From Chattoraj and Birdi (1984).

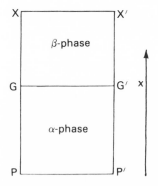

FIGURE 2.6. Liquid column in an ideal system (schematic). From Chattoraj and Birdi (1984).

the Gibbs treatment of two bulk phases, i.e., α and β, separated by the interfacial region $AA'BB'$.

Gibbs considered that this interfacial region is inhomogeneous and difficult to define, and he therefore also considered a more simplified case in which the interfacial region is assumed to be a mathematical plane GG' (Fig. 2.6).

In the actual system (Fig. 2.5), the bulk compositions of the ith component in the α and β phases are c_i and c_i', respectively. However, in the idealized system (Fig. 2.6), the chemical compositions of the α and β phases are imagined to remain unchanged right up to the dividing surface, so that their concentrations in the two imaginary phases are also c_i and c_i', respectively.

If n_i and n_i' denote the total moles of the ith component in the two phases of the idealized system, then the Gibbs surface excess n_i^x of the ith component can be defined as

$$n_i^x = n_i^t - n_i - n_i' \qquad (2.26)$$

where n_i^t is the total moles of the ith component in the real system.

In an exactly similar manner, one can define the respective surface excess internal energy, E^σ, and entropy, S^σ, by the following mathematical relationships (Chattoraj and Birdi, 1984):

$$E^\sigma = E^t - E^\alpha - E^\beta$$
$$S^\sigma = S^t - S^\alpha - S^\beta \qquad (2.27)$$

Here, E^t and S^t are the total energy and entropy, respectively, of the system as a whole for the actual liquid system in Fig. 2.5. The energy and entropy terms for the α and β phases are denoted by the respective superscripts.

The real and idealized systems are open so that the following equation can be written:

$$dE' = T\,dS' - (p\,dV + p'\,dV' - \gamma\,dA) + \mu_1\,dn_1' + \mu_2\,dn_2'$$
$$+ \cdots + \mu_i\,dn_i' \tag{2.28}$$

where V_α and V_β are the actual volumes of each bulk phase and p_α and p_β are the respective pressures. Since the volume of the interfacial region was considered to be negligible, $V' = V_\alpha + V_\beta$. Further, if the surface is almost planar, then $p_\alpha = p_\beta$, and $(p_\alpha\,dV_\alpha + p_\beta\,dV_\beta) = p\,dV'$.

The changes in the internal energy for the idealized phases α and β may similarly be expressed as follows:

$$dE_\alpha = T\,dS_\alpha - p\,dV_\alpha + \mu_1\,dn_1 + \cdots + \mu_{i,\alpha}\,dn_{i,\alpha} \tag{2.29}$$

and

$$dE_\beta = T\,dS_\beta - p\,dV_\beta + \mu_1\,dn_1 + \cdots + \mu_{i,\beta}\,dn_{i,\beta} \tag{2.30}$$

In the real system, the contribution due to the change of the surface energy, dA, is included as additional work. Such a contribution is absent in the idealized system containing only two bulk phases without the existence of any physical interface. By subtracting equations (2.29) and (2.30) from (2.28), the following relationship is obtained:

$$d(E' - E^\alpha - E^\beta) = Td(S' - S^\alpha - S^\beta) + \gamma\,dA + \mu_1\,d(n' - n_1^\alpha - n_1^\beta)$$
$$+ \cdots + \mu_i\,d(n_i' - n_i^\alpha - n_i^\beta) \tag{2.31}$$

or

$$dE^x = T\,dS^x + \gamma\,dA + \mu_1\,dn_1^x + \cdots + \mu_i\,dn_i^x \tag{2.32}$$

This equation, on integration at constant T, γ, and μ_i, etc., gives

$$E^x = TS^x + \gamma A + \mu_1 n_1^x + \cdots + \mu_i n_i^x \tag{2.33}$$

This relationship may be differentiated in general to give

$$dE^x = T\,dS^x + \gamma\,dA + \sum_i (\mu_i\,dn_i^x) + \sum_i n_i^x\,d\mu_i + A\,d\gamma + S^x\,dT \tag{2.34}$$

Combining equations (2.32) and (2.34), gives

$$-A\,d\gamma = S^x\,dT + \sum_i n_i^x\,d\mu_i \tag{2.35}$$

Let $S^{\sigma,x}$ and Γ_i^x denote the surface excess entropy and moles of the ith component per surface area, respectively. This gives

$$S^{\sigma,x} = S^x/A$$
$$\Gamma_i^x = n_i^x/A \tag{2.36}$$

and

$$-d\gamma = S^{\sigma,x}\, dT + \Gamma_1^x\, d\mu_1 + \Gamma_2^x\, d\mu_2 + \cdots + \Gamma_i^x\, d\mu_i \tag{2.37}$$

This equation is similar to the Gibbs–Duhem equations for the bulk liquid system.

To make this relationship more meaningful, Gibbs further pointed out that the position of the plane may be shifted parallel to GG' along the x direction and fixed in a particular location when n_1^t becomes equal to $(n_1^\alpha + n_1^\beta)$. Under this condition, n_1^x (or Γ_1^1 by convention) becomes zero. The relationship in equation (2.37) can be rewritten as

$$-d\gamma = S_1^\sigma\, dT + \Gamma_2^1\, d\mu_2 + \cdots + \Gamma_i^1\, d\mu_i \tag{2.38}$$
$$= S_1^\sigma\, dT + \sum \Gamma_i^1\, d\mu_i \tag{2.39}$$

At constant T and p, for a two-component system [say water (1) + alcohol (2)], we thus obtain the classic Gibbs adsorption equation as

$$\Gamma_2 = -(d\gamma/d\mu_2) \tag{2.40}$$

The chemical potential μ_2 is related to the activity of alcohol by the equation

$$\mu_2 = \mu_2^0 + RT\ln(a_2) \tag{2.41}$$

If the activity coefficient can be assumed to be equal to unity, then

$$\mu_2 = \mu_2^0 + RT\ln(C_2) \tag{2.42}$$

where C_2 is the bulk concentration of solute 2.

2.4. SURFACE TENSION, INTERFACIAL TENSION, AND SURFACE PRESSURE

Surface tension, γ, is the differential change of free energy with change of surface area at constant temperature, pressure, and composition. If the bulk phases on both sides of the surface are liquid or solid, then this tension is called the interfacial tension. If the surface area of a liquid is increased, molecules will move from the interior to the surface so that the surface density is unchanged.

The surface pressure of a monolayer is the lowering of surface tension due to the presence of a monomolecular film. This film arises from the orientation of the amphiphile molecules at the air–water or oil–water interface, where the polar group would be oriented toward and the nonpolar part (hydrocarbon) away from the aqueous phase. This orientation produces a system with minimum free energy.

2.5. ADSORPTION OF IONS AT LIQUID SURFACES

Liquid interfaces can be expected to be very sensitive to the effects of ions, since the dipoles of water molecules are asymmetrical. Detailed descriptions of these phenomena can be found in the standard texts.

In this section, we shall briefly discuss those phenomena that are directly applicable to the study of insoluble monolayers at liquid–gas or liquid–liquid interfaces.

Let us consider the interface between a polar liquid, water, and another liquid phase, an alkane. There will be a tendency for molecules near the liquid surface to have a specific orientation, as we have already seen. Since water molecules behave as dipoles, the orientation of water molecules at interfaces will produce an asymmetrical electrical field near the surface. Accordingly, any change in the system that produces a change in the orientation at the interface will produce a change in the nature of this field. Such a change is exactly what one observes when a monolayer is spread at an interface. These changes in the electrical field can thus be measured by suitable methods, as described below.

The electrical fields present at interfaces are not as clearly understood as are, for example, those at charged metal surfaces (Bockris *et al.*, 1980). As is known from classic electrostatics, the potential of a point in space is defined as the work required to bring a unit charge from infinity to that point. The potential difference between these two points thus gives the field. According to this concept, there would thus be a so-called "Galvani potential difference" across any phase boundary. Unfortunately, it is not easy to determine these potential differences experimentally.

However, another potential, the so-called "Volta potential" of a phase, is very useful. The Volta potential is defined as the potential of a point immediately outside the phase, i.e., the work required to bring a unit charge from infinity just up to, but not into, the phase.

2.6. SURFACE VISCOSITY OF LIQUID SURFACES

A monomolecular film is resistant to shear stress in the plane of the surface, as also is the case for the bulk phase: A liquid is retarded in its flow by viscous forces. The viscosity of the monolayer may indeed be measured in two dimensions by flow through a canal on a surface or by its drag on a ring in the surface, corresponding to the "Ostwald" and "Couette" instruments for the study of bulk viscosities. The surface viscosity, η_s is defined by the relationship

(Tangential force per centimeter of surface) = η_s × (rate of strain)

and is thus expressed in units of (mt^{-1}) (surface poise), whereas bulk viscosity, η, is in units of poise $(ml^{-1}t^{-1})$. The relationship between these two is

$$\eta = \eta_s/d \tag{2.43}$$

where d is the thickness of the "surface phase," approximately 10^{-7} cm ($=10^{-9}$ m) for many films. That the magnitude of η_s is of the order of 0.001-1 surface poise implies that over the thickness of the monolayer, assuming the thickness to be uniform, the surface viscosity is about 10^{4}-10^{7} poises. This viscosity has been compared to that of butter. The η_s is given in surface poises (g/sec or kg/sec).

It is easily realized that if a monolayer is moving along the surface under the influence of a gradient of surface pressure, it will carry some of the underlying water with it. In ther words, there is no slippage between the monolayer molecules and the adjacent water molecules. The thickness of such regions has been reported to be of the order of 0.003 cm. It has also been asserted that the thickness would be expected to increase as the magnitude of η_s increases.

Monolayers of long-chain alcohols exhibit η_s's approximately twenty times greater than those of the corresponding fatty acids (Fig. 2.7). If the

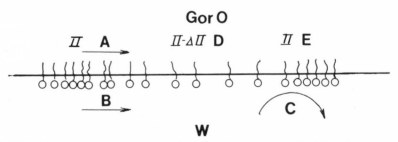

FIGURE 2.7. Drag by a moving monolayer on the underlying liquid, and the inverse phenomena (see the text for details). Redrawn from Davies and Rideal (1963) with changes.

monolayer is flowing along the surface under the influence of π (at A), it carries with it some underlying water (at B). This transport is a consequence of no slippage between the monolayer and the bulk liquid adjacent to it. For a monolayer of oleic acid moving at 1-5 cm/sec, the direct measurement gives the thickness of the entrained water as approximately 0.003 cm. If the bulk viscosity increases, then the thickness of the aqueous layer also increases in direct proportion. Conversely, if liquid is moving (at C), it carries the surface film molecules along with it, giving rise to compression at E, and back-spreading pressure tends to move from E to D.

It is also obvious that many such films will exhibit complex viscoelastic behavior, the same as one finds in bulk phases. The flow behavior then can be treated in terms of viscous and elastic components. This situation is analogous to combinations of springs and shock absorbers. The elastic modulus, G^*, is defined as follows (Tschoegl and Alexander, 1960; Tschoegl, 1962):

$$G_s^* = G^* + jw^* \eta_s \qquad (2.44)$$

where $j = \sqrt{-1}$ and w is the frequency of deformation.

Further, the equilibrium elasticity of a monolayer film is related to the *compressibility* of the monolayer (analogous to the bulk compressibility) by

$$C_s = -(1/A)(dA/d\pi) \qquad (2.45)$$

where A is the area per molecule of the film. The *surface compressional modulus*, K_s $(= 1/C_s)$, is the reciprocal of C_s. Since there is no change in surface tension with a change in the area of a pure liquid surface (i.e., $dA/d\pi = $ infinity), the elasticity is thus zero.

The *interfacial dilational viscosity*, k_s, is defined as

$$\gamma = k_s(1/A)(dA/dt) \qquad (2.46)$$

where k_s is the fractional change in area per unit time per unit surface tension difference. From this relationship C_s and E_s can be derived.

2.7. DIVERSE SURFACE PROPERTIES OF LIQUID SURFACES

2.7.1. OPTICAL AND RELATED METHODS

OPTICAL ABSORPTION

The optical reflection from a monomolecular film must be measured from the interface with a very small amount of material present. Therefore, in these methods, repreated interaction of the light beam with one or more identical films is generally used. The simplest way to observe a light beam that has passed through several identical monolayers is to transfer portions of the layer to suitable transfer end plates, which are then stacked and examined in a conventional spectrophotometer. This method was used with monolayers of chlorophyll deposited on glass slides by the Langmuir-Blodgett method (see Chapter 9) (Jacobs *et al.*, 1954).

Ferrodoxin and chlorophyll monolayers were investigated by measuring the spectra (550-750 nm) of these films at the interface (Brody, 1971).

To examine monolayers on liquid surfaces *in situ*, multiple interactions by mounting mirrors above and below the surface have been used (Kaufer and Schibe, 1955; Tweet, 1963).

ELLIPSOMETRY

The structure of liquid surfaces with monomolecular films can be studied by measuring the light reflected from the surface. The range of thickness that one generally considers to be measured varies from 100 to 1000 Å. However, in monolayers in which the molecules are oriented and the thickness involved is 5-10 Å, the methods have not been easily pursued. In a differential method in which two beams of light from the same incandescent lamp were directed to two similar troughs, reflected light from each was detected by a separate photocell; the photocells were arranged in a sensitive bridge circuit so that very small differences could be determined. With nonabsorbing fatty acid monolayers, the observed changes in reflectance were 0.3%-3% ($\Delta r/r$) (Markel and Vanderslice, 1960). Various other reflectance methods have been described by other investigators (Rothen, 1945; Bouhet, 1931; Tromstad and Fichum, 1934). The ellipsometers currently in general use have much better sensitivity than earlier models (Cuypers *et al.*, 1978).

ELECTRON MICROSCOPY AND ELECTRON DIFFRACTION

The first application of the electron microscope was in the study of monomolecular films, and it was shown that monolayers could be investigated after the films were deposited on glass slides and shadowed with evaporated metallic films. The techniques have been refined and extended considerably by later investigators (Epstein, 1950; Ries and Kimble, 1955; Ries and Walker, 1961).

To examine a sample in the electron microscope, it must be supported in some way; a thin evaporated carbon or plastic film is used. A thin layer of organic matter on such films gives insufficient contrast for its detection.

To achieve contrast, heavy atoms may be incorporated into the sample, either as a stain or by coating the sample with them to give a silhouette of the specimen. The staining of biological specimens, for example, is typically carried out by treating them with OsO_4, $KMnO_4$, or uranyl salt solutions. More detailed information about the organization of monomolecular film layers has been obtained by the shadowing method. In this procedure, the monolayer is deposited on a suitable solid support, and a film of Cr or Pt metal is evaporated in a vacuum chamber. It is these shadows that reveal the monolayer structure.

Whereas the technique of examining monolayer structures by these methods seems to be quite straightforward for the experienced investigator, the intrepretation of these structures requires extreme caution (Sheppard *et al.* 1964, 1965; Ries and Kimble, 1955).

The main criticism one may raise is that the monolayers originally present on the surface of water are deposited on a solid metal surface. This state is obviously not the same as that on the liquid (water) surface; therefore, on a molecular level, there may or may not be interactions that would give rise to different results.

2.7.2. GAS TRANSPORT AND EVAPORATION

Measurement of the evaporation of water through monolayer films was found to be of considerable interest in the study of methods for controlling evaporation from large bodies of water (see Chapter 4). In the original procedure, the box containing the desiccant is placed over the water surface, and the amount of water sorbed is determined by simply removing the box and weighing it (Archer and La Mer, 1954, 1955; Langmuir and Schaefer, 1943). The results are generally expressed in terms of specific evaporation resistance, r (Chapter 4). The methods for calculating r from the water uptake values, together with the assumptions involved, are described in detail in the aforementioned references.

A more advanced arrangement was described that made use of recording film balances, in which the temperature difference between two cooled sheet metal probes was measured. If both probes are over clean water and the rate of moisture condensation is the same, then there is no difference in the temperature; however, if a monolayer is present, the retardation of evaporation gives rise to a temperature difference (Heller, 1954).

Techniques of measuring the diffusion of gas through monolayers at the liquid interface have also been investigated (Blank and Roughton, 1960; Blank, 1962). In these methods, a differential manometer system was used to measure the adsorption of gases, such as CO_2 and O_2, into aqueous solutions with and without the presence of monolayers. A Geiger–Müller counter with a suitable sorbent and a radioactive trace gas were used to measure the reduction of evolution of H_2S and CO_2 from the surface of the solution when a monolayer was present (Hawke and Alexander, 1960, 1962; Hawke and Parts, 1964).

3

EXPERIMENTAL METHODS AND PROCEDURES IN MONOLAYERS

3.1. FORMATION OF MONOLAYERS

When a very small amount of a virtually insoluble and nonvolatile organic substance is carefully placed on the surface of water, which has a relatively high surface tension, either of the following results may be observed: (1) The substance may remain as a compact drop (or as a solid mass), leaving the rest of the liquid surface clean, or (2) it may spread out over the entire available surface of the water. The formation of a stable monolayer by any substance is determined by the interactive forces between that substance and the subphase, water.

In other words, a stable monolayer is formed when the work of adhesion between the substance and the water is greater than the work of cohesion of the substance itself. Under these conditions, the substance spreads over the entire available surface of the water and forms a monomolecular film. Similar energetic considerations would apply for other interfaces, e.g., oil–water, mercury–water, or solid–water.

The study of the structural properties of these ordered monomolecular films has been found to provide much useful information about other complicated systems, because one can obtain from these studies information about the molecular arrangement, as regards the size, shape, and molecular interaction with the subphase. These interactions are of a direct nature.

Further, the thermodynamics of these spread films in two dimensions have been extensively described by various investigators (Rideal, 1930; Marcelin, 1931; Freundlich, 1930; Heymann, 1931; Adam, 1941; Harkins, 1952; Davies and Rideal, 1963; Gaines, 1966; Defay *et al.*, 1966; Adamson, 1982; Chattoraj and Birdi, 1984).

Analogous to the various states found in a three-dimensional system, e.g., gas–liquid–solid, similar kinds of two-dimensional states have been

27

observed. The principal forces that determine whether one of these exists are related to the magnitude of the energy of interaction with the subphase, which is an attraction force perpendicular to the surface, while the principal forces that determine the state of the film are related to the magnitude and distribution of the adhesive forces, which act laterally between the molecules.

When the perpendicular attraction between the film molecules and the water (subphase) molecules is weak, the stability of the two-dimensional film is very weak, and it breaks down under even very small lateral compression (e.g., Brownian motion), and in most cases no film structures at all will be formed.

On the other hand, if there are some interactions between the polar groups of the film-forming molecules and the water molecules, through hydrogen bonds, and the lateral van der Waals forces are weak, then the film molecules move about independently on the surface, partaking in the translational movement of the substrate molecules. Such a film is called a "gaseous film," in two dimensions. Conversely, if the van der Waals forces are very strong, then the film molecules cohere in large islands of film, in which the thermal translational movements of the molecules are restrained. The film molecules thus escape into the gas phase as a continuous phase, this state being analogous to the vapor pressure of a solid or liquid substance in three dimensions.

In a film in which the magnitudes of the van der Waals forces are very large, the film is not able to move about freely, and thus resembles a "solid" in a two-dimensional state. However, if the van der Waals forces between the molecules are weaker, the result is a "liquidlike" film, which behaves as a two-dimensional liquid film. It is thus obvious that the role of the interaction of the subphase, i.e., water, with the film-forming substance is very important. The effect of temperature will therefore become very significant, since it affects the hydrogen bonds.

The presence of a molecular film on the surface of a liquid has a pronounced effect on the physical properties of the surface (interfacial) region of the system, i.e., liquid–air or $liquid_1$–$liquid_2$.

The various physical properties that have been the subjects of investigations are described in this chapter.

3.1.1. LANGMUIR FILM BALANCE

Almost all the methods in use at present are modifications of the original Langmuir method, which was based on directly measuring the outward force exerted on a floating barrier that divides the film-covered surface from a clean surface. The most essential points to which one must pay particular attention for maximum accuracy are as follows:

1. The barrier, preferably of Teflon, used to confine the film must provide perfect leak-proof alignment. This criterion must be maintained under any surface pressure, since leakage of the film is more likely at higher π.
2. The barrier should be stable and maintain position when stopped.
3. The magnitude of π can be measured by two different procedures: (a) direct force measurement using a transducer or Langmuir's method or (b) the more modern procedure, where the magnitude of $\pi = \gamma_{\text{water}} - \gamma_{\text{film}}$. Since γ can be conveniently measured by using a sensitive electrobalance, the accuracy of π can be very high (± 0.001 mN/m = dyne/cm).

To measure a very low π, i.e., of the order of 0.001 mN/m (dyne/cm), an apparatus of the surface micromanometer type has been used (Guastala, 1938, 1942; Saraga, 1957; Kalousek, 1949; Tvaroha, 1954; Chen, 1976). The π is measured by recording the displacement of a fine silk thread, lightly covered with petroleum jelly, that separates the film-covered surface from the clean surface in the trough (Fig. 3.1). A somewhat different method was used in more recent studies (Jalal, 1978).

In the original design of the Langmuir method, a "boom" of waxed mica floating on the surface was used to compress the film. However, in the recent versions, the trough and the barrier have been made of Teflon. Leakage of the film around the barrier has also been reduced to a negligible

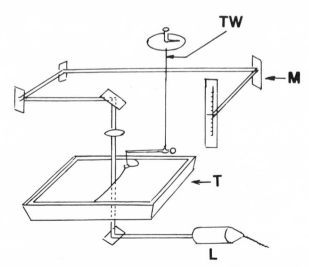

FIGURE 3.1. Surface micromanometer balance. (L) Lamp; (T) trough; (M) mirror; (TW) torsion wire. Redrawn from Davies and Rideal (1963) with changes.

amount by the smooth Teflon surfaces; in fact, there is practically no leakage past a Teflon barrier placed on a smooth Teflon (or Teflon-coated) trough (as used in the author's laboratory) (see Fig. 3.2).

A few newer versions based on Langmuir's original design have been described (Trurnit and Lauer, 1959; Mann and Hansen, 1963; Malcolm and Davies, 1965; Anderson and Evett, 1952).

Several general descriptions of automatic balances mostly based on a Wilhelmy plate have been reported (Mauer, 1954; Elworthy and Mysels, 1966; Mendenhall and Mendenhall, 1963; Hawke and Alexander, 1960; Gini *et al.*, 1972; Golian *et al.*, 1978).

In some cases, the apparatus used has been designed to include various simultaneous measurements, e.g., surface pressure, surface potential, and constant pressure (Golian *et al.*, 1978).

A monolayer trough with two movable barriers and several compartments has been described (Fromherz, 1971). In this apparatus, the spread monolayer can be transferred into another compartment containing a subphase of different composition. Ordinary compressed monolayers can be investigated under different compression rates by electronic control. The monolayer reactions can be investigated. The double-barrier trough can be applied as a single-beam spectrophotometer by reflecting a light beam through the water (subphase) surface by using three mirrors. A dye monolayer enclosed between two barriers can be moved into the light beam. The surface potential, ΔV, can be monitored simultaneously with π.

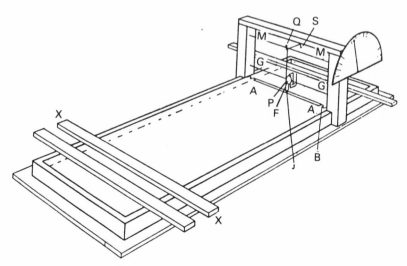

FIGURE 3.2. A typical Langmuir balance. (GG) Torsion wire; (XX) sweep; (AA) float; (F) mirror; (S) calibration arm; (MM) main torsion wire; (B) gold foil barrier; (J) wire for mirror; (PQ) rigid framework soldered to MM. Redrawn from Adam (1968) with changes.

Monolayers can be transferred onto solid supports using the depositing lift and the constant-π mode of the trough. The decrease of film area can be recorded and, with the use of an x-y recorder, compared directly to the area of the covered solid support (for making Langmuir–Blodgett films on solids, see Chapter 9).

In the author's laboratory, the subphase composition was changed by using a pump to circulate the subphase, with the possibility of adding different substances. If the circulation speed is slow, there is no effect on the π measurements.

COMPRESSION BARRIER

A brief description of the construction of the barrier and the trough is useful. The trough can be made either of Teflon or of Teflon-coated metal. The latter is preferable for large troughs, i.e., approximately 70×15 cm (as used by the author). The barriers can be made of solid Teflon (or of Teflon reinforced with stainless steel rods inside). In Langmuir's original instrument, leakage of the film past the ends of the barriers was prevented by leaving a narrow gap, into which jets of air were directed. These jets are not easily controlled, however, and later instruments have not used this method, instead using a gold or platinum strip or a light waxed thread for the barrier. The leakage is best determined by measuring a monolayer of a typical lipid, e.g., cholesterol. If there is film leakage, the π vs. A (area/molecule) isotherm will not agree with the published data. Furthermore, the isotherm will show an abrupt drop in π as the leakage becomes appreciable at some high π; also, the π vs. A isotherms will not be reproducible. A high degree of reproducibility actually confirms the absence of leakages.

The barrier drive system must be as vibrationless as possible. It can be similar to that used by many potentiometric chart recorders and can be powered by a miniature dc gear motor. A pulley on the motor output shaft operates a drive cord, and a ten-turn potentiometer is applied to indicate the barrier position and speed.

3.1.2. EXPERIMENTAL METHODS OF SPREADING FILMS AT THE OIL–WATER INTERFACE

The surface pressure, π, as measured at the oil–water interface is equal to

$$\pi_{ow} = \gamma_{ow} - \gamma_{film} \qquad (3.1)$$

where γ_{ow} is the interfacial tension of the clean interface and γ_{film} is the interfacial tension of the film interface. The experimental procedure needed to perform investigations at the oil-water interface requires special methods, as compared to that for investigating the air-water interface.

Procedures used to change the area at the oil-water interface have been described by different investigators, with more or less success (Davies and Rideal, 1963; Pethica *et al.*, 1981).

However, one can easily adapt the usual π vs. C_s method to measurements on the oil-water interface (Birdi, 1982). This adaptation consists of measuring the pressure on the injection of small amounts of the lipid molecule (dissolved in a suitable solvent: lipids in chloroform or hexane; proteins in water).

3.1.3. MEASUREMENT OF MONOLAYER PROPERTIES

SURFACE PRESSURE

The change in surface tension of a liquid surface allows one to measure the surface pressure, π. This being the most fundamental physical quantity, extensive data on such systems have been reported. In the earlier instruments, π was measured by a simple bell-crank balance in which weights were hung in a pan. Later, torsion wires were used. The surface tension can be measured by various methods, as described in detail in most surface chemistry textbooks (Gaines, 1966; Adam, 1941; Adamson, 1982; Chattoraj and Birdi, 1984).

Because of the specific procedure needed in monolayer studies, as described below, the most commonly used method is based on the so-called "Wilhelmy plate method."

WILHELMY PLATE METHOD

The early basic method (Wilhelmy, 1863) is based on measuring the pull on a thin plate (usually platinum; glass, quartz, mica, and paper have also been used) suspended in the liquid. As shown in Fig. 3.3, the forces acting on the plate when placed in the liquid surface are: (gravitational force) + (surface tension) + (buoyant effect). The net downward force for a rectangular plate of length L_p, width W_p, thickness T_p, and plate density D_p, when immersed to depth H_1 in a liquid of density D_1 is

$$F = (D_p)g(l_p w_p t_p) + 2\gamma(T_p + W_p)(\cos \theta) - D_1 g T_p W_p h_1 \qquad (3.2)$$

where γ is the surface tension of the liquid, θ is the contact angle of the liquid with the solid plate, and g is the gravitational force constant.

The relationship given here neglects the effect from the air phase, but this effect can be included if instead of air the other phase is another liquid

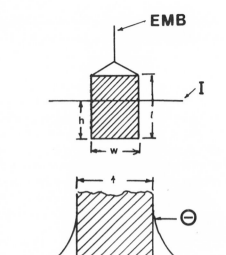

FIGURE 3.3. Wilhelmy plate (platinum or glass) at an air–water or oil–water interface (I) connected to an electromicrobalance (EMB). (h) Height; (l) length; (w) width; (t) thickness; (θ) contact angle.

(oil). If one maintains a completely wetted plate (i.e., $\theta = 0$, $\cos \theta = 1$) and constant H_1, then one obtains

$$\pi = \gamma_{\text{water}} - \gamma_{\text{film}} \tag{3.1-a}$$

$$= -\Delta F / [2(T_p + W_p)] \tag{3.3}$$

$$= -\Delta F / (2 W_p) \qquad \text{if } W_p \gg T_p$$

which suggests that the sensitivity can be increased by using very thin plates. Use has been made in the author's laboratory of platinum plates of varying thicknesses: 0.1–0.02 mm.

Some refined analyses have been proposed (Paddy, 1951) in which plate end corrections were included. However, in most cases, where the accuracy of γ is of the order of ± 0.001 mN/m, the use of a thin platinum plate suffices without any end corrections.

This method has been found to be comparable with earlier methods in which apparatuses other than a Wilhelmy plate and electronic balances were used for measuring π with a sensitivity of ± 0.001 mN/m (Puddington, 1946; Alexander, 1949; Anderson and Evett, 1952; Tvaroha, 1954; Allan and Alexander, 1954; Trurnit and Lauer, 1959; Mann and Hansen, 1963; Gaines, 1966; Vroman et al., 1968; Pagano and Gershfeld, 1972; Albrecht and Sackmann, 1980).

Furthermore, modern electrobalances allow very little change in the movement of the plate depth during measurements, which gives enhanced sensitivity. The major advantages of using a Wilhelmy plate are that the plate is easily cleaned and easily maintained.

The effect of the vessel wall on the determination of γ using a flat plate has been analyzed (Furlong and Hartland, 1979). However, this effect is negligible when the plate length is many orders smaller than the trough.

In a recent study, low π vs. A isotherms were obtained by using a Wilhelmy plate method (Matsumoto *et al.*, 1977). The π vs. A plots of alcohols and fatty acids are given in Figs. 3.4 and 3.5. The π of n-tetradecanol ($C_{14}OH$) and pentadecanoic acid increased with decreasing A to approximately 2000 $Å^2$, and was constant thereafter at 0.145 and 0.143 mN/m, respectively. On the other hand, n-pentadecanol ($C_{15}OH$) and palmitic acid ($C_{15}OOH$) showed almost no change in π over the same molecule/area region.

In this study, a Teflon trough was filled with water. Measurements were made with a glass plate 0.005 cm thick with a periphery of 9.535 or 9.995 cm connected to a microbalance. The plate was raised and lowered until a constant reading was obtained. In the author's laboratory (Birdi and Nikolov, 1979), a platinum plate was used. It was observed that there was no need to raise or lower the plate, since stable values were obtained,

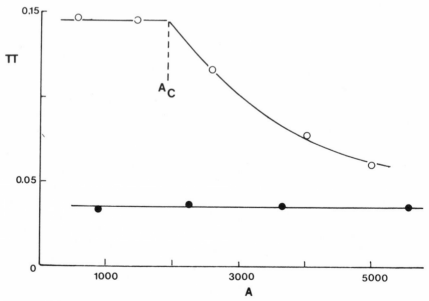

FIGURE 3.4. π vs. A isotherms of $C_{14}OH$ (○) and $C_{15}OH$ (●) on subphase water. Redrawn from Matsumoto *et al.* (1977) with changes.

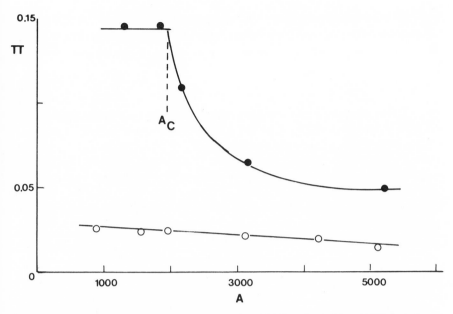

FIGURE 3.5. π vs. A isotherms of n-pentadecanoic acid (●) and palmitic acid (○) (subphase pH 5.8). Redrawn from Matsumoto *et al.* (1977) with changes.

indicating complete wetting. However, it is very important that the platinum plate should not be exposed to the air for more than a minute or so after it has been cleaned with a gas flame. A solution of lipid in chloroform was applied to the surface and after 10 min the film was compressed. The sensitivity of this apparatus was approximately ±1 mdyne/cm (0.001 mN/m).

An important problem in these studies is the determination of the solvent effects (if any) on the measured π. This problem was investigated

TABLE 3.1. Effect of Spreading Solvents
on the Measured π^a

Solvent	Time (min) needed for π to return to 0 (± 0.001 mN/m)
Methanol	30
Benzene	5
Petroleum ether	1
Chloroform[b]	5

[a] From Pagano and Gershfeld (1972).
[b] From Birdi (unpublished).

for different solvents (Pagano and Gershfeld, 1972), and the data were as shown in Table 3.1. Since π can be measured with very high sensitivity, it can be concluded that there are no solvent effects on the data.

RING METHOD

The surface tension determines the forces required to detach a metal ring from the surface of a liquid (Fig. 3.6). The vertical pull, f, is equal to the surface tension force as

$$f = 4\pi r\gamma \tag{3.4}$$

where r is the radius of the ring. It is necessary to introduce a correction factor, β, as given by

$$\gamma = \beta f/(4\pi r) \tag{3.5}$$

where β depends on r and on the density of the liquid. The values of β are generally available in different textbooks (Adamson, 1982).

DROP WEIGHT METHOD

The weight (or volume) of each liquid drop that detaches itself from the tip of a vertical tube (see Fig. 3.7) is determined by the γ of the liquid. Assuming that the drops are formed extremely slowly, they will detach completely from the tip when the gravitational force just reaches the restraining force of surface tension:

$$Mg = Vdg = 2\pi a\gamma \tag{3.6}$$

and

$$\gamma = Mg/(2\pi a) = (Vdg)/(2\pi a) \tag{3.7}$$

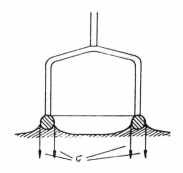

FIGURE 3.6. Ring method.

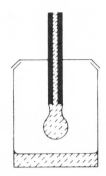

FIGURE 3.7. Drop weight method. From Chattoraj and Birdi (1984).

where g is the gravitational acceleration, M and V are the mass and volume of each drop, d is the density of the liquid, and a is the radius of the tip of the tube. The magnitude of Φ is needed for various corrections:

$$\gamma = (\Phi \, Vdg)/(2\pi a) \qquad (3.8)$$

3.1.4. EFFECT OF SOLVENTS ON MONOLAYER FILMS

Langmuir's method of spreading monolayers at the air–liquid interface of solutions of surfactants in volatile solvents has received a great deal of attention in recent years. For example, a study was carried out in which the effect of a number of highly purified solvents on the γ of a water surface was measured as a function of time (Jaffe, 1954). However, in these investigations, the wax with which the trough was coated was suspected to have dissolved (Robbins and La Mer, 1960).

In another study (Saraga, 1949), the effect of solvent on myristic acid monolayers was investigated. The solvents (water, 2-propanol, benzene) were found to have no effect on the monolayers of Aerosol-OT (Brady, 1949).

In a later investigation (Barnes et al., 1968), it was quite conclusively shown that the solvent (^{14}C-labeled decane) disappeared completely within 10–15 min. This result confirms the observations carried out by others with pentadecane as solvent (Gaines, 1961).

Monolayers of octadecyltrimethylammonium bromide were investigated using different solvents by Mingins et al. (1969).

3.1.5. RELATIVE METHODS

In addition to the aforedescribed methods, there is another procedure, in which a "piston oil" is used to compress the film being studied. These piston oils are substances that exert a fixed spreading pressure rather

TABLE 3.2. Surface Pressure (π) of Various Substances
Used as Piston Oils

Material	π (at $\approx$25°C) (mN/m)
Oleic acid	30
Ethyl myristate	20
Triolein	15
Heavy mineral oil + ethyl laurate	
40% 60%	17
60% 40%	14
70% 30%	11
85% 15%	7

rapidly—e.g., the case of applying oleic acid on the surface of water, which yields $\pi = 30$ mN/m.

The substances that have been used for this purpose are listed in Table 3.2.

3.2. STATES OF MONOLAYERS

In this section we shall discuss the various states of monomolecular films as measured from the surface pressure, π, vs. area, A, isotherms, in the case of simple amphiphile molecules. The π-A isotherms of biopolymers will be described separately, since they have been found to be of a different nature. But before presenting this analysis, it is necessary to consider some parameters of the two-dimensional states that are of interest. We need to start by considering the physical forces acting between the alkyl–alkyl groups (parts) of amphiphiles, as well as the interactions between the polar head groups. In the process whereby two such amphiphile molecules are brought closer together during the measurement of π vs. A, the interaction forces would undergo certain changes, which would be related to the packing of the molecules in the two-dimensional plane at the interface in contact with water (subphase).

This change in packing is thus conceptually analogous to three-dimensional P vs. V isotherms, as is well known from classic physical chemistry. We know that as pressure, P, is increased on a gas in a container, when $T < T_{cr}$, the molecules approach closer together and transition to a liquid phase takes place. Furthermore, compression of the liquid state results in the formation of a solid phase.

In the case of alkanes, the distance between the molecules in the solid–state phase is approximately 5 Å, while in the liquid state it is 5-6 Å.

The distances between molecules in the gas phase, in general, are approximately 1000 times larger than in the liquid state (volume of 1 mole liquid water = 18 cm^3; volume of 1 mole gas = 22.4 liters).

It is found that the isotherms of two-dimensional films also resemble three-dimensional P vs. V isotherms and that one can use the same classic molecular description as regards the qualitative analyses of the various states. However, it is also obvious that the correspondence between the two-dimensional and the three-dimensional structures will not be complete, since there are very subtle differences in these two systems, as described later.

In the three-dimensional structural buildup, the molecules are in contact with near neighbors, as well as with molecules that may be 5-10 molecular dimensions apart (as determined by X-ray diffraction). This is apparent from the fact that in liquids there is a long-range order up to 5-10 molecular dimensions.

On the other hand, in two-dimensional films, the state is much different. The amphiphile molecules are oriented at the interface such that the polar groups are pointed toward the water (subphase), while the alkyl groups are oriented away from the subphase. This orientation gives the minimum surface energy. The structure is stabilized through lateral alkyl group–alkyl group, polar group–subphase, and polar group–polar group interactions.

The main differences between these two-dimensional and the three-dimensional structure are shown in Table 3.3.

The most convincing results were those obtained with normal fatty alcohols and acids. Their monomolecular films were stable and exhibited very high surface pressures. A steep rise in π is observed around 20.5 Å^2, regardless of the number of carbon atoms in the chain. The volume of a $-CH_2-$ group is 29.4 Å^3. This volume gives the length of each $-CH_2-$ group perpendicular to the surface, or the vertical height of each group, as approximately 1.42 Å. This value compares very satisfactorily with the value of 1.5 Å obtained by X-ray diffraction.

Since high pressures lead to transitions from gas to liquid to solid phases in three-dimensional systems, a similar state of affairs would be expected in two-dimensional film compression π vs. A isotherms (see Fig. 3.8), as described below.

TABLE 3.3. Intermolecular Interactions

System	Alkyl–alkyl	Polar–polar
Two-dimensional	Interactions only in one plane	Interaction with subphase hydration of polar groups
Three-dimensional	Interactions in three-dimensional space in multiple layers; no subphase present	

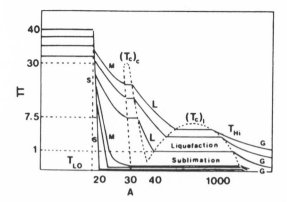

FIGURE 3.8. Schematic representation of states of two-dimensional films. π (mN/m) vs. A (area/molecule) isotherms of a spread lipid monolayer at the air–water interface at different temperatures. (T_{Lo}) Low; (T_{Hi}) high; (T_c) critical transition temperature; (G) gas; (L) liquid; (S) solid; (M) mesomorphic. From Chattoraj and Birdi (1984).

3.2.1. GASEOUS FILMS

The simplest type of amphiphile film or polymer film is a "gaseous" state. This film consists of molecules that are sufficiently far apart that lateral adhesion (van der Waals) forces are negligible. However, there is sufficient interaction between the polar group and the subphase that the film-forming molecules cannot be easily lost into the gas phase and amphiphiles are almost insoluble in water (subphase).

When the area available for each molecule is many times larger than the molecular dimension, the gaseous-type film (state 1) is present. As the available area per molecule is reduced, the other states, e.g., liquid-expanded (L_{ex}), liquid-condensed (L_{co}), and finally solidlike (S or S_{co}) states are present.

The molecules will have an average kinetic energy of $\frac{1}{2} kT$ for each degree of freedom, where k is the Boltzmann constant ($=1.372 \times 10^{-16}$ ergs/T) and T is the temperature. The surface pressure measured would thus be equal to the translational kinetic energy, derived from collisions between the amphiphiles and the float, for the two degrees of freedom in the two dimensions. It can thus be seen that the ideal gas film obeys the relationship

$$\pi A = kT \quad \text{("ideal film")} \quad (3.9)$$

This relationship is analogous to the three-dimensional gas law (i.e., $PV = kT$). At 25°C, the magnitude of kT is 411×10^{-16} ergs. If π is in mN/m and A in Å^2, then the magnitude of kT is 411. In other words, if one has a system with $A = 400$ Å^2, then $\pi = 1$ mN/m for the ideal gas film.

In general, ideal gas behavior is observed only when the distances between the amphiphiles are very great and the value of π is thus very small, i.e., less than 0.1 mN/m.

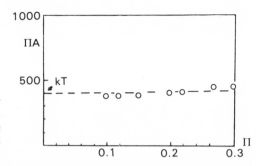

FIGURE 3.9. πA vs. π plot for valinomycin (mol. wt. 1·100) monolayer (25°C) at the air–water interface (π, mN/m; A, $10^{-20}\,m^2 = \mathring{A}^2$). From Chattoraj and Birdi (1984).

The latter observation requires an instrument with very high sensitivity, approximately ±0.001 mN/m. The π vs. A isotherms of n-tetradecanol, pentadecanol, pentadecylic acid, and palmitic acid in the low-π region were given in Figs. 3.4 and 3.5 (Matsumoto et al., 1977). Similar isotherms were reported by other workers using different methods (Jalal, 1978; Gershfeld and Tajima, 1979).

In Fig. 3.9, the plot of πA vs. π for a monolayer of valinomycin [a dodecacylic peptide (see Chapter 8)] shows that the relationship as given in equation (3.9) is valid. In this equation, it is assumed that the amphiphiles are present as monomers. However, if any association takes place, then the measured values of πA would be less than $kT < 411$, as has also been found (see Chapter 8).

3.2.2. LIQUID (EXPANDED AND CONDENSED) FILMS

With respect to simple amphiphiles (e.g., fatty acids, fatty alcohols, lecithins), there have been observed in several cases transition phenomena between the gaseous and the coherent states of films that show a very striking resemblance to the condensation of vapors to liquids in three-dimensional systems. Liquid films show various states in the case of some amphiphiles, as was shown in Fig. 3.8. In general, there are two distinguishable types of liquid films.

LIQUID EXPANDED FILMS (L_{ex})

The first state is called the liquid expanded (L_{ex}) (Adam, 1941; Harkins, 1952; Adamson, 1982; Chattoraj and Birdi, 1984). If one extrapolates the π vs. A isotherm to zero π, the value of A obtained is much larger than that obtained for close-packed films. This difference shows that the distance between the molecules is much greater than in solid films, as discussed later. These films exhibit very characteristic elasticity, which is described further below.

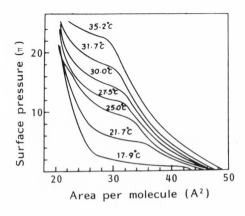

FIGURE 3.10. π vs. A isotherms of myristic acid as a function of temperature. From Chattoraj and Birdi (1984).

LIQUID CONDENSED FILMS (L_{co})

As the area per molecule (or the distance between molecules) is further decreased, there is observed a transition to a so-called "liquid condensed" (L_{co}) state. This state has also been called a "solid expanded" film (Adam, 1941), which is discussed in greater detail later.

The π vs. A isotherms of myristic acid (an amphiphile with a single alkyl chain) as a function of temperature are given in Fig. 3.10. The π vs. A isotherms for two-chain alkyl groups, such as lecithins, also show a similar behavior (Fig. 3.11), as discussed later.

3.2.3. SOLID FILMS

The π vs. A isotherms below the transition temperatures show the liquid-to-solid-phase transition. These solid films have also been called

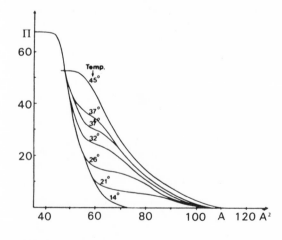

FIGURE 3.11. π vs. A isotherms of dipalmitoyl lecithin as a function of temperature. From Chattoraj and Birdi (1984).

TABLE 3.4. Different States of Monolayers of Lipids[a]

	Gaseous		Expanded		Condensed
Adam	G	Ve	Le	Tr	Cond Close-packed heads Chains
Harkins	G		L_1	Int	L_2 LS S CS
Dervichian	G		L	Exp Mes	Mes S

[a] Abbreviations: (G) = gaseous; (Ve) vapor expanded = L_1 = L; (Tr) transition = (Int) intermediate = (Exp Mes) expanded mesomorphic; (Cond) condensed; (L_2) liquid condensed; (Mes) mesomorphic; (LS) super liquid; (S) solid; (CS) condensed solid.

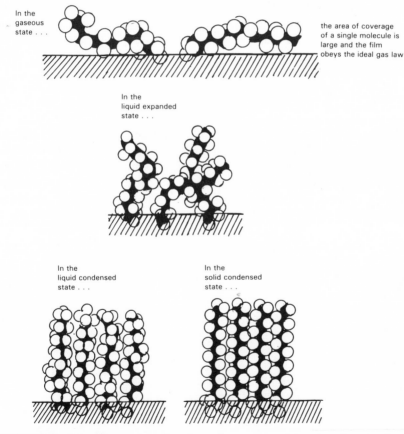

FIGURE 3.12. Schematic representation of monolayer states and the orientation of molecules at the air–water interface. From Cadenhead (1971).

"condensed films" (Adam, 1941). These films are observed in systems such as those in which the molecules adhere very strongly to each other through the van der Waals forces; the π vs. A isotherm generally shows no change in π at high A, while a sudden increase in π is observed at a rather low A value, as was shown in Fig. 3.8. In the case of straight-chain molecules, such as stearyl alcohol, the sudden increase in π is found to take place at $A = 20$-22 Å^2, at room temperature (which is much lower than the phase-transition temperature, to be described later).

From these descriptions, it can be seen that the films may under given experimental conditions show three first-order transition states (see Fig. 3.8 and Table 3.4): (1) transition from the gaseous film to the liquid–expanded (L_{ex}); (2) transition from the L_{ex} to the liquid-condensed (L_{co}); and (3) from the L_{ex} or L_{co} to the solid state, if the temperature is below the transition temperature.

The temperature above which no expanded state is observed has been found to be related to the melting point of the lipid monolayer. The orientation of the molecules in these states is depicted schematically in Fig. 3.12.

3.2.4. COLLAPSE STATES

Measurements of π vs. A isotherms, when compressed, generally exhibit a sharp break in the isotherms, which has been connected to the collapse of the monolayer under the given experimental conditions. The collapse states of the mechanism are depicted schematically in Fig. 3.13.

In general, the collapse pressure, π_{col}, is the highest surface pressure to which a monolayer can be compressed without a detectable movement of the molecules in the film to form a new phase (see Fig. 3.13).

In other words, this pressure will be related to the nature of the substance and the interaction between the subphase and the polar part of

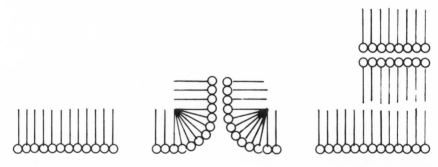

FIGURE 3.13. States of collapse after overcompression of condensed monolayers.

the lipid or the polymer molecule. However, there exists some misinterpretation of the collapse phenomena as found in the older literature (Gaines, 1966).

As described in the next section, the monolayer of a lipid can be formed by different spreading methods, and comparison of the data on collapse pressure, π_{col}, is therefore discussed later. The thermodynamics of the π_{col} analyses are considered in Chapter 4. Monolayer collapse will also be shown to provide much information in the case of protein monolayers (Chapters 5 and 8).

3.3. DIFFERENT MONOLAYER SPREADING METHODS

In the current literature, many procedures other than the Langmuir film balance method are used to form monolayers. These different methods are described below.

π_{eq} *vs. A Isotherms.* These isotherms are obtained by the Langmuir film balance, as described above. These isotherms will therefore be expected to resemble three-dimensional P vs. V isotherms.

π *vs. Surface Concentration,* C_s, *Isotherms.* These isotherms are obtained by applying an increasing amount of the film-forming substance (such as a solution or in liquid or solid form) to a surface of constant area (generally 100 cm^2). These π vs. C_s isotherms are thus different from those obtained as π vs. A, since in the former the surface concentration, C_s, is varied.

π_{eq} (*Equilibrium Surface Pressure*). The value of the equilibrium surface pressure, π_{eq}, is obtained when a bulk solid or liquid lipid is brought into equilibrium with the monolayer film at the surface of a liquid (or the oil–water interface). The magnitude of π_{eq} will be shown to be the same as the π_{col} only when the bulk phase is in liquid form.

Accordingly, the lipid-phase transition data as obtained from these different spreading methods will be expected to be different under varying experimental conditions. These different data are analyzed in detail later in this text.

3.4. TRANSITIONS IN INSOLUBLE MONOLAYERS

The thermodynamic phase transitions in one-component systems can be classified into n different orders if the nth derivatives of the molar free energy, G_n, with respect to the intensive variables T and P show a discontinuity at the transition point, and the $G^{-(n-1)}$ derivatives are still continous, but show a bend (Ehrenfest, 1933). This classification cannot be used, however, in the case of so-called "λ transitions" (Pippard, 1957; Berry *et al.*, 1980). The first derivatives of $\bar{G}$ are generally experimentally known, such that for systems with any number of phases (Schwarz, 1982)

$$(d\bar{G}/dT)_{P,x_i} = -\bar{S} \qquad (3.10)$$

$$(d\bar{G}/dP)_{T,x_i} = \bar{V} \qquad (3.11)$$

$$(d\bar{G}/dx_1)_{T,P,x_i \neq x_1,x_2} = \mu_1 - \mu_2 \qquad (3.12)$$

These relationships were derived for systems with several components, x_i, where $\bar{S}$ is the molar entropy, $\bar{V}$ is the molar volume, and μ_i are the chemical potentials.

It has been shown that the thermodynamic classification of phase transitions, as described above (Ehrenfest, 1933), can be extended to insoluble monolayers (Dervichian and Joly, 1939). The successive derivatives of the surface chemical potential, μ_s, with respect to surface pressure, π, determine the order of transition (Chattoraj and Birdi, 1984):

$$(d\mu_s/d\pi)_{T,P} = A \qquad (3.13)$$

and

$$(d^2\mu_s/d\pi^2)_{T,P} = (dA/d\pi) \qquad (3.14)$$

Hence, a first-order transition is characterized by a discontinuity in A.

In support of this argument, one can note that the transitions have been reported for lipid films at the mercury–nitrogen interface (Aellison, 1962), for tridecylic–myristic acid films (Boyd, 1958; Harkins and Nutting, 1939), and for synthetic lecithin films at the air–water interface (Phillips and Chapman, 1968).

In each instance, $\pi_{tr} \geq \pi_{eq}$, indicating that for these systems, the monolayer in the transition region is metastable with respect to the bulk liquid phase. The implication of these studies is that the transition arises because of the slow rate of the transformation from the liquid state to the bulk liquid.

3.5. SURFACE POTENTIAL MEASUREMENTS

The surface potential at any interface, air–water or liquid–liquid, changes when a film-forming molecule is oriented with its polar end toward the liquid phase and the alkyl part is situated away from it. This orientation would thus give rise to a change in the dipoles of water near the interface, μ_1. The other dipole to be considered would be the polar group of the amphiphile, μ_2. The alkyl part of the molecule will contribute with its dipole, μ_3. If one applies the Helmholtz equation for an array of n dipoles per unit area (cm^2), and assumes that these dipoles are additive, one obtains the following expression for the surface potential at any interface, ΔV (Davies and Rideal, 1963):

$$\Delta V = 4\pi n(\mu_1 + \mu_2 + \mu_3) \qquad (3.15)$$

Since it is not easy to determine the magnitude of μ_1 and μ_2, one generally combines these two terms. The orientations of the different dipoles are shown in Fig. 3.14. The expression for ΔV can be written as follows if one combines all three dipoles, μ_{total}:

$$\Delta V = 4\pi n\mu_{total} \qquad (3.16)$$

It is thus clear that ΔV measurements would provide much sensitive information regarding the changes in any of the three different dipoles in monolayer studies. These equations were derived in analogy to the parallel plate condenser theory. If the amphiphile is charged with potential ψ_0, then one can write (Davies and Rideal, 1963; Gaines, 1966; Adamson, 1980; Chattoraj and Birdi, 1984)

$$\Delta V = 4\pi n\mu_{total} + \psi_0$$

$$= 4\pi n\mu_{ov} \qquad (3.17)$$

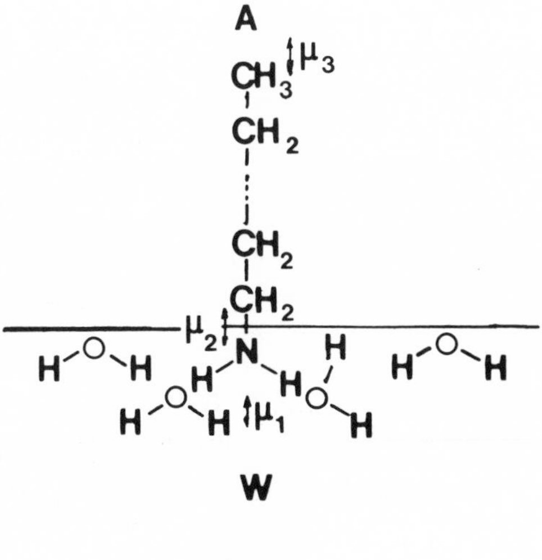

FIGURE 3.14. Orientation of dipoles of molecules, e.g., subphase water (μ_1), polar (μ_2), and alkyl (μ_3) part of the film-forming molecule. Redrawn from Davies and Rideal (1963) with changes.

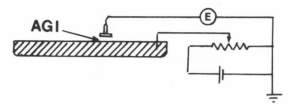

FIGURE 3.15. Apparatus for measuring surface potential, ΔV. (AGI) Air gap ionized α-emitter; (E) voltmeter.

where dipole μ_{ov} denotes the overall dipole. Since the Gibbs excess, Γ, would be equal to n, and for a condenser with separation d, one obtains

$$\Delta V = 4\pi\sigma d / D \tag{3.18}$$

where the charge density is σ and D is the dielectric constant (assumed to be equal to 1). It can be seen that $\sigma d = ned = n\mu_{ov}$, or that equations (3.15)-(3.18) are equivalent.

The magnitude of ΔV can be measured by placing an electrode coated with an α-emitter, such as polonium or americium (≈ 1 mC) (Fig. 3.15). The resulting air ionization (≈ 1000 times) makes the gap between the probe and the liquid surface (≈ 5 mm) sufficiently conducting that the potential difference can be measured by means of a high-impedance ($>10^{16}\,\Omega$) voltmeter (or an ordinary pH meter). The reference electrode placed in the bulk phase can be an Ag–AgCl or Pt electrode (Gaines, 1966; Adamson, 1980; Chattoraj and Birdi, 1984).

Typical plots of π vs. A, ΔV vs. A, and μ_{ov} vs. A are shown in Fig. 3.16.

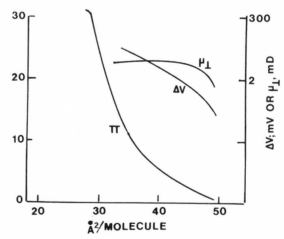

FIGURE 3.16. Variation of π, ΔV, and μ_1 ($\equiv\mu_{ov}$) with Å^2/molecule (A) of a lipid monolayer.

A more extensive description of the measurement of ΔV of polypeptides is given below.

3.6. SURFACE VISCOSITY OF MONOLAYERS

Surface viscosity, η_s, in qualitative terms is very easily demonstrated if one considers what happens when one blows on the surface of a liquid. If one blows on a clean surface of a liquid on which some particles are floating about, the particles will be easily blown to the edge of the container and will remain there. However, if the surface is not clean, as the surface of tea or coffee or beer, the surface film present will not easily be able to move the particles, which means that the surface contains a rigid film.

A great variety of methods for measuring these characteristics arising from the interfacial or surface viscosities, η_s, have been described in the literature (Davies and Rideal, 1963; Gaines, 1966; Jarvis, 1962; Joly, 1964; Goodrich, 1973). In some cases, the method has been a kind of adaptation of the three-dimensional bulk viscosity techniques, since it has been shown that viscosity measurements pose a big problem in obtaining reliable data. The data reported in the literature are found in some cases to be highly disputable. For instance, in a recent study (Abraham and Ketterson, 1985), the viscosity of DPPC (dipalmitoylphosphatidylcholine) was reported to be in disagreement with literature data by a few orders of magnitude.

However, an attempt will be made here to present and compare the literature data in order to arrive at a more reliable analysis than one finds in the current literature.

CANAL VISCOSIMETER

In this method, the monolayer is allowed to flow through a slit or a so-called "canal" in the liquid surface, from a region of high surface pressure, π_{hi}, to one of lower pressure, π_{lo}. Obviously, the surface viscosity would be related to or can be obtained from the rate of film flow by an equation that has been found to be analogous to the Poiselle formula for the flow of a liquid through a capillary tube (Meyers and Harkins, 1937; Harkins and Kirkwood, 1938).

The surface viscosity, η_s, is given as follows:

$$\eta_s = (\pi_{hi} - \pi_{lo})a^3/(12Ql) - (a\eta_0/\pi) \tag{3.19}$$

where a and l are the width and length of the canal, Q is the area of monolayer flowing through the canal per second, and η_0 is the bulk viscosity of the subphase, i.e., water. The surface viscosity is obtained in surface

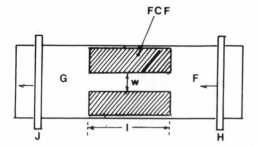

FIGURE 3.17. Canal method for measuring η_s. (FCF) Floating canal frame; (F) higher π; (G) lower π; (w) width; (l) length. Redrawn from Gaines (1966) with changes.

poises (sp) when cgs units are used. The arrangement used for the canal method is shown in Fig. 3.17. This method with narrow canals allows surface viscosity to be measured below 10^{-4} sp.

DAMPING METHODS

In this method, the damping of a body moving on the surface is related to the surface viscosity, η_s. In the simplest case, successful use has been made of the so-called "Brookfield viscosimeter" (as used for measuring bulk fluid viscosities), by placing the moving cylinder or the disk on the surface as described by various investigators (Blank, 1970). The surface viscosities of beer were investigated by this method (Birdi *et al.*, to be published).

The other type of analogous instrument is the torsion pendulum, which is the fixed suspension type with a stationary trough (Fig. 3.18). The

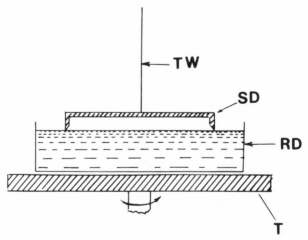

FIGURE 3.18. Torsion method for measuring η_s. (T) Turntable; (RD) rotating dish of water; (SD) stationary disk; (TW) torsion wire. Redrawn from Davies and Rideal (1963) with changes.

pendulum is set in motion by a momentary deflection, which is usually produced by twisting the torsion wire through some definite angle. The pendulum then oscillates freely (Tschoegl, 1962). The relative surface viscosity, η_s, is given as follows:

$$\eta_{s,re} = (l_0 p)/(l p_0) \qquad (3.20)$$

where l and l_0 are logarithmic decrements in the presence and absence of the film, respectively, and p and p_0 are the periods of oscillation in the presence and absence of the film, respectively.

LASER-LIGHT SCATTERING MEASUREMENT

The light scattering of the capillary wave motion of aqueous surfaces covered with monomolecular surface films has been measured. In this method, the sum of a dilational (i.e., compressional) and a shear component is determined (Hard and Neuman, 1981; Langevin, 1981).

OTHER METHODS FOR MEASURING SURFACE VISCOSITY

Other methods that have been reported are as follows:

- Circular knife-edge surface viscosimeter (Mannheimer and Burton, 1970).
- Rotating wall knife-edge viscosimeter (Goodrich et al., 1975).
- Deep-channel viscosimeter (Wasan et al., 1971).
- Surface shear viscosity from decay of surface motions at a rotated liquid–gas or liquid–liquid interface (Krieg et al., 1981; Birdi, unpublished; Hassager and Westborg, 1987).
- Oscillatory method: The monolayer was compressed by withdrawing subphase (water) from the bottom of a Teflon container (truncated cone) (Abraham et al., 1982, 1983a,b). However, the π vs. A data were found to be in discord with the conventional Langmuir film balance method (Birdi, 1987). This disagreement indicates that the conical container might lead to artifacts due to possible adsorption of the monolayer on the walls of the container.

3.7. RETARDATION OF SURFACE EVAPORATION RATES BY MONOLAYERS

It has been known for some time that the presence of a monolayer of an amphiphile on the surface of water retards the rate of evaporation (Rideal, 1925). It is not easy, however, to observe this phenomenon, since the natural evaporation of water is already enormously retarded by the resistance caused by the slow diffusion of the evaporated molecules away from the surface. Evaporation rates must therefore be investigated with the water under vacuum (Rideal, 1925). It has been found that certain films retard the rate of evaporation up to 50%.

The evaporation rates can be ascribed to contributions from the various parts of the interface, e.g., water itself (negligible) + film (major) + vapor above.

3.8. CALMING OF WAVES BY SPREAD MONOLAYERS

Spread monolayers have been known since ancient times to damp out the small ripples that are constantly being formed by the action of wind, and to lead to dangerous breaking of large waves, by their cumulative disturbance of the surface. It was suggested nearly two centuries ago by Benjamin Franklin that the film "lubricates" the water surface (Franklin, 1976), so that the wind cannot grip it as well as it does a clean surface. This, of course, cannot be the actual mechanism in view of the compact character of surface films (Lucassen and Hansen, 1966).

3.9. MONOLAYERS, BILAYERS, AND VESICLES

The monolayer is in equilibrium with other structures, and the monolayer may therefore be accepted as the basic structure for the other complicated systems.

The study of the main component of biological membranes, lecithin, has been useful as a model of membrane systems, e.g., monolayers, bilayers, and vesicles. These three model systems comprise a very attractive experimental approach, as documented by the vast literature devoted to this subject. Furthermore, although monolayers, bilayers, and vesicles do not exhibit exactly the same physical properties as biological membranes, these three systems do exhibit, collectively, very subtle physical properties that provide a very exact molecular picture of the interactions. For instance, bilipid membranes (BLMs), as described below, exhibit some physical properties that are similar to those of biological membranes, such as water permeability, interfacial tension, electrical resistance, and capacitance. In contrast to biological membranes, BLMs appear to be virtually impermeable to ions and to other water-soluble compounds, while the permeability can be increased after interaction with such substances as hormones, drugs, and divalent metal ions.

In the early stages of research, most of the work was carried out on air–water or oil–water interfaces. These model systems can be expected to be limited, since they cannot provide information on transport phenomena. The formation of BLMs and vesicles is discussed briefly below.

BILIPID MEMBRANES

BLM formation can be compared to the formation of black soap films. The early BLMs were formed by applying the membrane-forming solution [generally brain lipids dissolved in an organic solvent (tetradecane)] with a brush over a circular hole ($0.5–5$ mm^2) in a Teflon or polyethylene septum separating two identical solution chambers (Fig. 3.19).

In some recent studies, it has now been verified that by using a lecithin solution in an organic solvent, one can form stable BLMs (Hanai *et al.*, 1964). After the application of the lecithin solution, the film thins to a black lipid; the thickness varies from 62 to 77 Å. The electrical (DC) resistance has been reported to range from 10^6 to 10^9 Ω/cm^2 (Ohki *et al.*, 1973).

More recently (Montal and Muller, 1972), stable BLMs were formed from lipid monolayers. This procedure also eliminates the necessity for

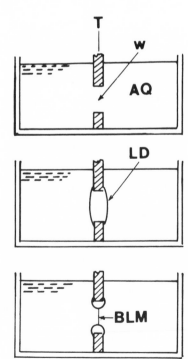

FIGURE 3.19. Schematic representation of the BLM-formation setup. (w) Water; (AQ) aqueous phase; (LD) drop of lipid solution in organic solvent; (BLM) bilipid membrane.

organic solvents, although they can be used. Bilayers of a few square micrometers in area were formed from lipid-protein vesicles (Schuerholz and Schindler, 1983). Two monolayers were first formed from vesicles and then brought into local contact by use of a micropipette to form a bilayer.

However, this observation indicates that the lipid monlayer is in equilibrium with the BLM structure. The attractive van der Waals forces are the main reason for the stability of the BLM, which is analogous to monolayer formation.

PHOSPHOLIPID VESICLES (LIPOSOMES)

Phospholipids are known to form a variety of liquid crystalline structures when they interact with varying amounts of water. These structures are in some ways analogous to the soaps formed from long-chain aliphatic and other amphiphilic molecules (with exceptions) (Luzzati, 1968). Phospholipids form lamellar structures when the water content is greater than 50%; this phenomenon is also called "smectic mesophase." These structures are stacked lamellae of molecules in a bilayer configuration, with water being situated between the polar ends. The tendency of phospholipid molecules to form such hydrated lamellar structures is related to the transition temperature (i.e., melting of the hydrocarbon chains).

The vesicles can be made by simple mechanical shaking of a dry phospholipid sample with aquaeous phase in a flask. By this procedure (Blume, 1979), one obtains a heterogeneous mixture of particles of varying radii. Cholesterol by itself does not form vesicles, but can be incorporated into lecithin vesicles up to a ratio of 1:1.

On exposure to ultrasonic radiation, the coarse vesicles break up into smaller bilayer vesicles (200–500 Å). Egg lecithin produces aggregates of molecular weight 1.5–2.1×10^6, which gives 1.9–2.7×10^3 molecules/particle. However, large unilammellar and oligolamellar vesicles were formed when an aqueous buffer was introduced into a mixture of phospholipid and organic solvent and the solvent was subsequently removed by evaporation under reduced pressure (Szoka and Papahadjopoulos, 1978).

Recent studies have further shown that vesicles are in equilibrium with monolayers (which are in equilibrium with BLMs) (Schuerholz and Schidler, 1983).

In recent years, the quasi-elastic light-scattering (QLS) method has been used to investigate the size and radius of vesicles (Schutenberger et al., 1986). These studies were carried out on mixtures of bile salt and lecithin vesicles. The critical packing radius, R_c, of such structures is given as (Israelachvili et al., 1977):

$$R_c = l_c \{3 + [3(4V/(a_0 l_c)) - 1]^{0.5}\}/\{6[1 - V/(a_0 l_c)]\} \qquad (3.21)$$

where l_c is the maximum length of the hydrocarbon region of the lipid molecule, V is the volume of the hydrocarbon region, and a is the so-called "surface area per molecule." By using the following values for the molecule DML (dimyristoyl lecithin):

$$a_0 = 62.2 \text{ Å}^2 \qquad \text{(from X-ray diffraction)}$$

$$V = 808 \text{ Å}^3$$

$$l_c = 15.3 \text{ Å} \qquad \text{for two C-14 hydrocarbon chains}$$

one gets a value of $R_c = 98$ Å. This value is in close agreement with the QLS data.

4

LIPID MONOLAYERS AT LIQUID INTERFACES

4.1. STRUCTURE OF LIPIDS

In living systems, the structures of the cells and tissues are based on large molecules—proteins, polysaccharides, and complex lipids—while the organization of the systems appears to be a function of the nucleic acid complexes. Most lipid molecules are not as complex or as large as proteins. A large variety of compounds exist, many of them differing only in the composition of the long-chain fatty acid or aldehyde moieties. The word "lipid" (from the Greek *lipos*, "fat") covers what seems to be an ever-expanding group of compounds, the classification of which has been a subject of some controversy. One of the most typical classifications of lipids is that of Deuel (1951):

1. Simple lipids:
   a. Neutral fats (glycerol esters of largely long-chain fatty acids).
   b. Waxes (solid esters of long-chain monohydric alcohols).

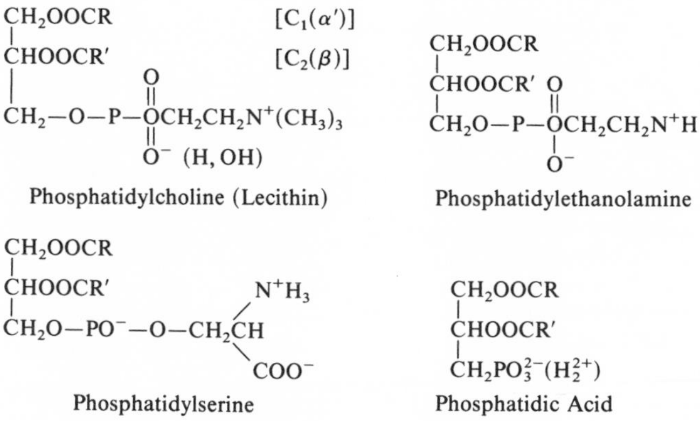

CH_2OOCR $[C_1(\alpha')]$
$CHOOCR'$ $[C_2(\beta)]$ CH_2OOCR
$\qquad\qquad\qquad O$ $CHOOCR'\quad O$
$CH_2-O-P-OCH_2CH_2N^+(CH_3)_3$ $CH_2O-P-OCH_2CH_2N^+H$
$\qquad\qquad O^-\ (H, OH)$ $\qquad\quad O^-$

Phosphatidylcholine (Lecithin)   Phosphatidylethanolamine

CH_2OOCR
$CHOOCR'$ N^+H_3 CH_2OOCR
$CH_2O-PO^--O-CH_2CH$ $CHOOCR'$
$\qquad\qquad\qquad\qquad COO^-$ $CH_2PO_3^{2-}(H_2^{2+})$

Phosphatidylserine   Phosphatidic Acid

FIGURE 4.1. Lipid structures.

57

2. Compound lipids (esters of fatty acids with alcohols, which contain additional groups):
   a. Phospholipids (lipids containing a phosphate residue) (see Fig. 4.1).
   b. Cerebrosides and gangliosides (lipids containing a carbohydrate residue).
   c. Sulfatides (lipids containing a sulfate residue).
3. Derived lipids: This group contains products derived from groups 1 and 2 that still possess lipidlike characteristics. They include fatty acids, long-chain alcohols, sterols, hydrocarbons (carotenoids, vitamins D, E, and K), and sphingosine.

4.1.1. MONOLAYERS OF LONG-CHAIN LIPID MOLECULES (SINGLE CHAIN)

For obvious reasons, the lipid monolayers that have been investigated in greater detail are the long-chain amphiphiles derived from paraffins. As mentioned earlier, it is necessary that the hydrophile–lipophile balance (HLB) of the amphiphile be of the right value to be able to form a stable monolayer. It is also noteworthy that most of the naturally occurring lipids are composed of linear and even-numbered chains. Some typical such lipids are listed in Table 4.1.

TABLE 4.1. Various Single-Alkyl-Chain Fatty Alcohols and Fatty Acids Used for Monolayer Studies[a]

C-n	Alcohol	Trivial name	m.p. (°C)	Trivial name	m.p. (°C)
12	1-Dodecanol	Lauryl	23	Lauric acid	44
13	1-Tridecanol	—	31	—	40
14	1-Tetradecanol	Myristyl	38	Myristic acid	58
15	1-Pentadecanol	—	44	—	51
16	1-Hexadecanol	Cetyl	49	Palmitic acid	63
18	1-Octadecanol	—	58	Stearic acid	70
				Oleic acid	14
				Linoleic acid	−5
				Linolenic acid	−11
19	1-Nonadecanol	—	62	—	66
20	1-Eicosanol	—	71	Arachidic acid	75
22	1-Docosanol	—	—	Behenic acid	80
24	1-Tetracosanol	—	—	Lignoceric acid	84
26	1-Hexacosanol	—	80	Cerotic acid	88
30	1-Triacontanol	—	86	Melissic acid	92

[a] (C-n) Number of carbon atoms; (m.p.) melting point.

Furthermore, single-chain lipids have been found to exhibit monolayer properties that are different from those of the two-alkyl-chain lipids (Birdi, unpublished). These differences are discussed separately because of their importance in both industrial and biological systems.

Some examples are given below, to make the reader aware of the differences that are found even among such simple lipids as the straight long-chain alcohols and acids (in undissociated form). The π vs. A isotherms of various fatty alcohols and acids have been investigated. However, these data are presented in the various sections, whenever analyses have been given.

Let us consider the effect of temperature on monolayers, as studied by the π_{eq} method.

π_{eq} vs. T: The π_{eq} vs. temperature, T, isotherms of fatty acids (Fig. 4.2) and fatty alcohols (Fig. 4.3) are very much alike in general. However, the temperature at which a break is observed is not as distinct near the bulk melting point for fatty alcohols (see Fig. 4.3) as it is for fatty acids (see Fig. 4.2). This observation has not been carefully analyzed in the current literature. The differences among isotherms between the π_{eq} vs. T for fatty acids, fatty alcohols, and two-chain lecithins are described later in this chapter.

The monolayers of fatty alcohols are stable when the temperature is below the melting point.

It is essential to note that given the strong dependence of π_{eq} on temperature, both the melting point and the temperature must be considered before comparing these systems. In other words, it is not easy to obtain any comparison when considering the π_{eq} data at 25°C for different fatty alcohols (Table 4.1).

Unsaturated Fatty Acids. Unsaturated fatty acids are widely distributed in biological tissues as a main constituent of biomembranes. In most cases,

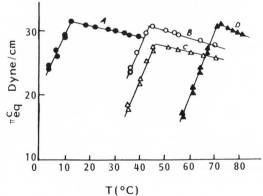

FIGURE 4.2. π_{eq} vs. T isotherms for fatty acids. (A) Oleic acid; (B) elaidic acid; (C) 9-stearolic acid; (D) stearic acid. From Chattoraj and Birdi (1984).

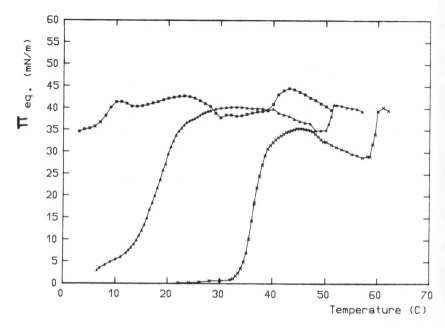

FIGURE 4.3. π_{eq} vs. T isotherms for various long-chain alcohols on water subphase. (■) n-Tetradecanol; (▲) n-hexadecanol; (×) n-octadecanol. From Birdi *et al.* (to be published).

they are linked at position 2 of phospholipids and generate a large variety of characteristics of membranes; i.e., they promote fluidity and control the phase transition behavior depending on the environmental conditions. From the technological viewpoint, these compounds as well as their derivatives have attracted great attention because of their unique activities or functions as medical and industrial materials.

The data in Table 4.2 suggest that π_{eq} decreases by approximately 3 mN/m per $-CH_2-$ group. This is a safe conclusion, since the plots of π_{eq} vs. T are almost 1.0 mN/m per degree (Figs. 4.2 and 4.3). Accordingly, the π_{eq} value of any n-alcohol would then be determined by the temperature, the chain length, and the melting point.

The π_{eq} vs. T curves for all fatty alcohols are linear below their bulk melting point (Fig. 4.3). At the bulk melting point, there is a transition, after which the magnitude of π_{eq} decreases. In some studies (Boyd and Schubert, 1957; Trapeznikov, 1941; Birdi *et al.*, to be published), peculiar π_{eq} vs. T (below the melting point) isotherms have been observed, even though the π vs. A isotherms of these fatty alcohols give normal curves, i.e., the same as for fatty acids (as described later).

Subphase pH over the range of 2–10 has been found to have no effect on the properties of fatty alcohol monolayers. Addition of electrolytes has

TABLE 4.2. π_{eq} Values of Various *n*-Fatty Alcohols at 25°C[a]

Alcohol	π_{eq} (mN/m)	T (m.p., °C)
Dodecanol	—	—
Tetradecanol	46	38
Pentadecanol	43	44
Hexadecanol	40	49
Heptadecanol	41	—
Octadecanol	35	58
Nonadecanol	—	—
Eicosanol	33	71
Docosanol	28	—

[a] From Deo *et al.* (1962).

also been found to have no effect on these monolayers (Birdi, unpublished). However, addition of ethanol to the subphase does have some effect on these plots (Fig. 4.4).

Fatty Acid Monolayers (*Un-ionized*). The presence of a carboxyl group, instead of an alcohol, gives rise to much different π vs. A isotherms in un-ionized fatty acid monolayers (when spread on acidic subphases, i.e., pH < 2). At high pH (> 6), ionization takes place, thus altering the π vs. A isotherms. The un-ionized monolayers are quite insensitive to pH (< 2). The magnitudes of π_{eq} are found to be much lower than those for the corresponding fatty alcohols at the same temperature. This difference is attributable to the higher bulk melting points for fatty acids than for alcohols (Table 4.2).

Surface Potential (ΔV). The data on surface potential, ΔV, have not been extensively studied (Schulman and Hughes, 1932; Harkins and Fischer, 1933; Harkins and Copeland, 1942). In some of these studies, breaks in ΔV vs. T were reported at the bulk melting point.

Monolayers of Cis/Trans. Both the introduction of *cis* double bonds into the hydrocarbon chain and chain-branching introduce considerable film expansion. The π vs. A data for C-18 acid, *cis*, and *trans* forms of 13-docosenoic acid (erucic and brassadic) have been reported (Marsten and Rideal, 1938; Lagaly *et al.*, 1976). The differences between *cis*- and *trans*-retinal monolayers (Fig. 4.5) also show the same correlation to the configuration (Yasuaki and Toshizo, 1967).

Furthermore, in most cases, the magnitude of ΔV (≈ 350 mV) for a lipid monolayer increases from low surface concentration to high surface concentration (≈ 450 mV). At the collapse point, the magnitudes of ΔV were found to depend on the number of carbon atoms (≈ 15 mV$/-CH_2-$) in the case of fatty acids and alcohols (Birdi, unpublished).

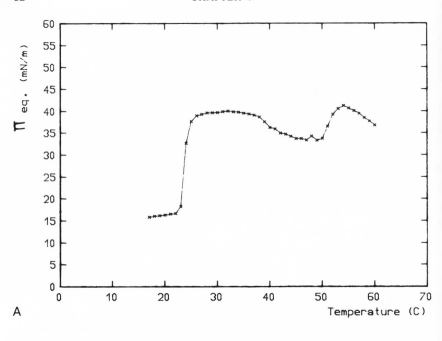

A

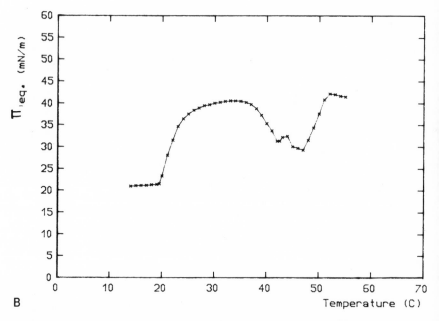

B

FIGURE 4.4. π_{eq} vs. T plots of n-hexadecanol ($C_{16}OH$) on a subphase with ethanol concentrations of 5% (A) and 10% (B). From Birdi *et al.* (to be published).

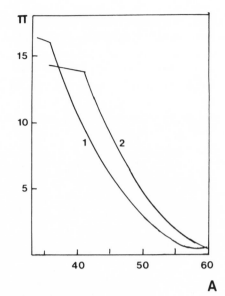

FIGURE 4.5. π vs. A isotherms of all-*trans* retinal (1) and 11-*cis* retinal (2). π_{col} and A_0 (limiting area) are, respectively, 16.1 mN/m and 35.7 Å^2 (1) and 13.8 mN/m and 41 Å^2 (2). Redrawn from Yasuaki and Toshizo (1967) with changes.

4.2. THERMODYNAMICS OF MONOLAYERS

4.2.1. STATISTICAL MECHANICAL DERIVATION OF SURFACE PRESSURE (π)

Statistical mechanical theoretical derivations have been given by various investigators (Ree and Hoover, 1964; Hill, 1962, 1963; Defay *et al.*, 1966). The surface free energy, G^σ, was given elsewhere as:

$$G^\sigma = -S^\sigma\, dT - \pi\, dA + \sum_i \mu_i\, dN_i \qquad (4.1)$$

From the Bragg–Williams approximation, the surface energy is given as (Fowler and Guggenheim, 1949)

$$G^\sigma = RT\{-N_A \ln q(T) + (\underline{w}_{AA}/kT)(N_A/N_S)^2$$
$$+ N_A \ln(N_A/N_S) + (N_S - N_A) \ln[(N_S - N_A)/N_S]\} \qquad (4.2)$$

and

$$\mu_A = kT\{\ln[N_A/(N_S - N_A)] - \ln q(T) + [(2\underline{w}_{AA}N_A)/(kTN_S)]\} \qquad (4.3)$$

since $\mu_A = (dG/dN_A)_{T,A}$ where N_S is the total number of sites on the surface, N_A is the number of sites occupied by the A molecules, and

$(N_S - N_A)$ is the number of unoccupied sites; the interaction energy between pairs of nearest neighbors of A molecules is denoted by w and $w_{AA} = (z/2)w$, where z is the coordination number and the factor of $\frac{1}{2}$ is to take account of symmetry. The partition function is $q(T)$ for the internal degrees of freedom of an adsorbed molecule. We can rewrite equation (4.3) as follows:

$$\pi = -(1/A_0)(dG/dN_S)_{T,N}, \qquad A = A_0 N_S \qquad (4.4)$$

or

$$\pi = -(kT/A_0)[(w_{AA}N_A)^2/(kTN_B)^2 + \ln(N_S - N_A)/N_S] \qquad (4.5)$$

By writing $X_i = N_i/N_S$, one obtains

$$\mu_A = kT\{\ln[X_A/(1 - X_A)] - \ln q(T) + 2w_{AA}X_A/kT\} \qquad (4.6)$$

and

$$\pi = -kT/A_0[-(w_{AA}X_A^2/kT) + \ln(1 - X_A)] \qquad (4.7)$$

Since both μ_A and π are functions of X_A, one obtains

$$(d\mu_A/dX_A)_{T,q(T)} = kT\{1/X_A + [1/(1 - X_A)] + 2w_{AA}/kT\} \qquad (4.8)$$

and

$$(d\pi/dX_A)_{T,q(T)} = kT/A_0[(2w_{AA}/kT) + 1/X_A(1 - X_A)]X_A \qquad (4.9)$$

From these equations, one obtains

$$d\pi/d\mu_A = X_A/A_0$$

$$= \Gamma_A \qquad (4.10)$$

where Γ_A is the surface excess of component A. Gibbs adsorption can then be derived as follows: At equilibrium between surface and bulk, one has

$$\mu_A^{\text{surface}} = \mu_A^{\text{bulk}} \qquad (4.11)$$

$$= \mu_A^0 + kT \ln(f_{\text{bulk},A}C_{\text{bulk},A}) \qquad (4.12)$$

where $f_{\text{bulk},A}$ and $C_{\text{bulk},A}$ are the activity coefficient and concentration of component A, respectively. The following can be derived:

$$d\mu_A^{\text{surface}} = RT\, d \ln f_{\text{bulk},A} + RT\, d \ln C_{\text{bulk},A} \qquad (4.13)$$

At low concentrations of $C_{bulk,A}$, one can neglect the term $d \ln f_{bulk,A}$ and

$$\Gamma_A = (1/kT)(d\pi/d \ln C_A)_{T,A}, \qquad C_A = C_{bulk,A} \qquad (4.14)$$

which is the Gibbs equation of surface excess [see equation (2.40)]. To estimate the value of $\underline{w}_{AA}$, one can derive

$$\ln C_A = -\ln K + \ln[\Gamma_A/(1/A_0) - \Gamma_A] + 2\underline{w}_{AA}A_0\Gamma_A/kT \qquad (4.15)$$

where $K = \exp[(\mu A_0/kT)f_A q(T)]$ and

$$\pi = k(T/A_0)[(\underline{w}_{AA}\Gamma_A^2 A_0^2/kT) - \ln(1 - A_0\Gamma_A)] \qquad (4.16)$$

If the magnitude of $\underline{w}_{AA}$ is very small and negligible, then equation (4.16) becomes

$$\pi A_0 = -kT \ln(1 - X_A) \qquad (4.17)$$

$$= kTX_A \qquad \text{when } X_A \ll 1 \qquad (4.18)$$

$$= kTN_A/N_S \qquad (4.19)$$

$$\pi A = N_A kT \qquad (4.20)$$

which is the ideal gas equation for two-dimensional films. Further, by combining equations (4.15) and (4.16), we obtain

$$\underline{w}_{AA}A_0/kT = \ln C_A/2\Gamma_A + \ln K/2\Gamma_A - (1/2)\Gamma_A$$
$$\times [\Gamma_A A_0/(1 - \Gamma_A A_0)] \qquad (4.21)$$

$$= \pi/(kT\Gamma_A^2) + [1/(\Gamma_A^2 A_0)] \ln(1 - A_0\Gamma_A) \qquad (4.22)$$

Since experimentally one knows C_A, Γ_A, π, and A_0, one obtains

$$\ln C_A = -\ln K + \ln[\Gamma_A A_0/(1 - \Gamma_A A_0)] + (2\pi/\Gamma_A kT)$$
$$+ [2 \ln(1 - A_0\Gamma_A)/\Gamma_A A_0] \qquad (4.23)$$

4.2.2. VAN DER WAALS FORCES IN MONOLAYERS

The surface pressure, π, at the air–water interface is given as

$$\pi_{aw} = \pi_{kin} + \pi_{coh} + \pi_{el} \qquad (4.24)$$

where total π_{aw} is given as a sum of different forces, e.g., kinetic, π_{kin}, cohesive, π_{coh}, and electrostatic, π_{el}, interactions.

At the oil-water interface, the measured π_{ow} would be given as

$$\pi_{ow} = \pi_{kin} + \pi_{el} \qquad (4.25)$$

The term π_{coh} would be negligible or absent, since the alkyl chains present in the oil phase would be devoid of van der Waals interactions. In other words, the magnitude of π_{coh} could thus be estimated from the difference

$$\pi_{coh} = \pi_{aw} - \pi_{ow} \qquad (4.26)$$

The magnitude of π_{coh} can also be estimated, if π_{el} is negligible (as would be the case for neutral lipids, such as fatty alcohols), at the air-water interface, since

$$\pi_{coh} = \pi_{aw} - \pi_{kin} \qquad (4.27)$$

and

$$\pi_{kin} = kT/A \qquad \text{(ideal film)}$$

The different equations of state have been based on both empirical and theoretical models. The various equations of state for neutral lipids (or biopolymers), where both π_{coh} and π_{el} are negligible, as found in the literature are summarized in the following equations:
Ideal Films:

$$\pi A = kT \qquad (4.28)$$

Various Nonideal Films:

$$\pi(A - A_0) = kT \qquad (4.29)$$

where A_0 is the co-area, i.e., the area actually occupied by the molecules in the surface. This equation is analogous to the gas equation, where molecular volume is used instead. Thus, the magnitude of A_0 will be the area per molecule in the compact state. From linear plots of πA vs. π, one can estimate A_0, as described later. A few typical values of A_0 as determined from monolayer films are given in Table 4.3. The π vs. A data thus analyzed would provide information on A_0, i.e., the close-packed solid (or liquid) state. However, if both π_{coh} and π_{el} are present, then the procedure to analyze the films is somewhat more elaborate. In those cases in which π_{el} can be varied by subphase pH, one can estimate π_{el} by measuring the π of very dilute films (i.e., A is very large). This procedure is delineated later in this chapter. The estimation of π_{coh} is described separately.

TABLE 4.3. Magnitudes of A_0 for Different Film-Forming
Molecules on the Surface of Water

Compounds	A_0 (Å²)[a]
Straight-chain acid (on water)	20.5
Straight-chain acid (on dilute HCl)	25.1
Esters of saturated acids	22.0
n-Fatty alcohols	21.6
Long-chain p-phenols	24.0
Cholesterol	40.8
Lecithins	≈50
Dilauryl oxalate	90
Didecyl adipate	120
Dioctyl sebacate	150
Dibutyl thapsate	190
Proteins	≈1 m²/mg

[a] 1 Å = 10^{-10} m.

Furthermore, because of diminished cohesion among the alkyl chains of the monolayer molecules at the oil–water interface, the π vs. A isotherms have been reported to be more expanded than the isotherms at the air–water interface (Davies, 1951a; Phillips and Rideal, 1955; Brooks and Pethica, 1964; Mingins *et al.*, 1975; Birdi, 1972; Chattoraj and Birdi, 1984).

In a recent study (Burke *et al.*, 1973), the surface area, A, was used for the microdetermination of the naturally occurring lipids:

$$A = q \times \mu\text{moles of lipids} \qquad (4.30)$$

where q is a proportionality constant that depends on the structure of the compound (at constant $\pi = 17\,\text{mN/m}$).

4.2.3. ELECTRODYNAMIC INTERACTIONS OR THE VAN DER WAALS FORCES IN LIPID MONOLAYERS

As indicated by equation (4.26), one would expect the alkyl chain–alkyl chain interaction to be given by the π_{coh} term. A similar kind of situation arises when one considers the packing of micellar structures, as formed by the association of surfactant molecules (Tanford, 1980; Birdi, 1976b; Ben-Naim, 1980; Birdi, 1988). However, the situation in monolayer systems is somewhat more simplified than in the case of micellar systems: In the former, the lipid molecules are situated at a plane surface; in the latter, there are highly curved surfaces. Furthermore, the possibility of measuring the π of monolayers directly allows one to estimate the magnitude of molecular interactions, while such measurement is obviously not possible in systems such as micelles or vesicles or membranes.

Let us consider first the state of saturated alkyl chains in lipids, such as phospholipid molecules. The alkyl chains (two per lipid molecule) will be able to interact through the dispersion forces arising from the induced dipoles. These forces are known to arise from the following phenomena:

1. A fluctuation in the electron cloud of a given molecule creates a short-lived dipole moment.
2. This dipole moment polarizes a region of a neighboring molecule, creating a dipole moment opposite in orientation to the first.
3. These induced dipoles are thus able to attract each other.

In real systems, the situation is somewhat different and more complex than described above, since the fluctuations at a given moment point are dependent on the spontaneous fluctuations taking place at all other points in the system. In other words, one needs to know the correlation between fluctuations that could give the magnitude of dispersion forces. The van der Waals forces are particularly significant, since they are universally present, just like gravitational forces. In this respect, therefore, van der Waals forces are quite different from covalent, ionic, or dipolar interactions.

The van der Waals equation for real gases is given as

$$(P + \underline{a}n^2/V^2)(V - n\underline{b}) = nRT \tag{4.31}$$

where P is the pressure of the gas, V is the volume, and $\underline{a}$ and $\underline{b}$ are constants depending on the gas. The term $n\underline{b}$ is the excluded volume correction. The term $\underline{a}$ accounts for the attractive forces the gas molecules exert on one another. This was realized when it was considered that even nonpolar gases would condense to liquids or solids when the temperature was lowered. To compensate for this attractive force, van der Waals added the term $\underline{a}n^2/V^2$ to the P. The equations of state for the two-dimensional films have been derived on the basis of an analogous model.

The well-known relationship that gives the interaction energy due to the dispersion forces is as follows (Salem, 1960, 1962a,b; Zwanzig, 1963; Chu, 1967; Israelachvili and Tabor, 1976):

$$W_{\text{dis}} \propto \alpha_i\alpha_j/r_{ij}^6 \tag{4.32}$$

where α_i is the polarizability of the ith molecule and r_{ij} is the distance between the ith and jth molecules. The quantity polarizability gives the magnitude of the induced dipole arising from the applied external field. If the polarizability is small, then the dispersive interaction must also be small, since it depends on field-induced distortion of electron clouds. In general, there are two basic methods of calculating the magnitude of the dispersive interactions. The first method is based on the equation derived for determining the dispersive interaction between two particles in vacuum. It involves summation of interacting pairs of subunits making up the system (Hamaker, 1937; London, 1930). It is known that this calculation is not rigorous.

The other method for calculating the dispersive force interaction is based on the analyses given by Lifshitz and co-workers (Lifshitz, 1956; Dzyaloshinski *et al.*, 1961). According to this theory, the interacting bodies are considered as continuous media. The interaction is considered to take place through a fluctuating electromagnetic field present in the interior of material bodies and extending beyond their boundaries. This theory is considered to be rigorous, but its extension to nonpolar geometrics has been found to be difficult. To obtain some estimate of the magnitudes of dispersion energies in bilayer systems, the London–Hamaker summation technique was applied (Salem, 1960). Consider two parallel, saturated linear hydrocarbon chains, such that L is the chain length, λ is the length of each subunit in the alkyl chain, N is the number of subunits in the chain, and D is the distance between chains $(D > \lambda)$. It is further assumed that the individual subunits have an interaction energy

$$w_i = \mathcal{E}/d^6 \qquad (4.33)$$

where d is the distance between each subunit and $\mathcal{E}$ is the Hamaker constant. The interaction between all subunits can be summed to give

$$W_{\text{dis}} = [\mathcal{E}L/(4\lambda^2 D^5)]3 \tan^{-1}(L/D) + (L/D)/[1 + (L^2 + D^2)] \qquad (4.34)$$

The subunit $_H{-}C{-}C{-}^H$ used here has the value of $\lambda = 1.26$ Å. The interaction between two subunits, w_{bond}, is obtained by calculating the following bond–bond interactions:

$$CC{-}CC \qquad \text{(one interaction)}$$

$$CH{-}CH \qquad \text{(four interactions)}$$

$$CC{-}CC \qquad \text{(four interactions)}$$

From these relationships, one obtains

$$w_{\text{bond}} = (-e^2 \bar{\alpha}\bar{\alpha}'/d^6)\left\{ \bar{\alpha} \Big/ \left\langle \left(\sum_i r_{ii}^2 \right) \right\rangle + \bar{\alpha}' \Big/ \left\langle \left(\sum_i r_i'^2 \right) \right\rangle \right\}^{-1} \qquad (4.35)$$

where the primes indicate the different bonds, the term $\langle (\sum_i r_i^2) \rangle$ is the expectation value of the operator $\langle(\sum_i r_{ii}'^2)\rangle$ in the ground state, and r_i and r_i' refer to electronic coordination of the bonds. This expression is an approximation of London's formula for the dipole-dipole contribution to the dispersion energy of two molecules. Using experimental values for the mean bond polarizabilities and calculating $\langle(\sum_i r_i^2)\rangle$ for the C—H and C—C bond wave functions, it was found (Salem, 1960) that

$$w_i = -1.34 \times 10^3/d^6 \quad (\text{kcal/mole}) \quad (4.36)$$

where d is in angstroms. Hence, for two linear alkyl chains, we have

$$W_{\text{dis}} = 3\pi\!\!\!\not\!E N/8\lambda D^5 \quad (4.37)$$

$$= -(N \times 1.24 \times 10^3)/D^5 \quad (4.38)$$

$$= -1.24 \times 10^3 n_C/(2\sqrt{A}/\sqrt{3})^5 \quad (4.39)$$

where the magnitude of $\not\!E$ is -1.34×10^3 kcal/mole. The relationship in equation (4.39) is obtained on the assumption that hexagonal molecular close packing in such monolayers is present [an assumption generally accepted by various investigators in such structures (Salem, 1962a, b; Ohki, 1976; Mingins et al., 1975; Birdi, 1976b; Chattoraj and Birdi, 1984)], which gives from the geometric relationships $A = \sqrt{3}D^2/2$. Other kinds of geometric packing of molecules can be easily obtained. Unfortunately, the relationship given by equation (4.39) has not been exhaustively applied in the current literature (Salem, 1952; Ohki and Ohki, 1976; Birdi and Sørensen, 1979; Chattoraj and Birdi, 1984) to estimate the magnitude of the cohesive forces, π_{coh}, in monolayers. The magnitude of W_{dis} per n_C has been estimated to be approximately 0.5 kcal/mole n_C (Table 4.4). It is of interest to observe that the van der Waals forces in isostearic acid monolayers are much smaller than in stearic acid monolayers. The reason is that the close-packed

TABLE 4.4. Calculated Values [Equation (4.38)] of van der Waals (vdW) Forces between Lipid Molecules in a Monolayer[a]

Lipid	Formula	D (Å)	vdW (kJ/mole)
Stearic acid	$CH_3(CH_2)_{16}COOH$	4.8	-35
Hexatriacontanoic acid	$CH_3(CH_2)_{34}COOH$	4.8	-70
Isostearic acid	$CH_3\!-\!CH\!-\!(CH_2)_{14}COOH$ $\quad\quad\;\; |$ $\quad\quad\;\;CH_3$	6.0	-12

[a] From Salem (1962b).

area/molecule is 31 Å^2 for isostearic acid, while it is 20 Å^2 for stearic acid, as measured from monolayer studies. Analogous to this cause, one finds that differences between *cis* and *trans* double bonds arise from the fact that the magnitude of the van der Waals forces is larger in *trans* than in *cis*. The *trans* forms closed-packed films with lower values of A than the *cis* (see Fig. 4.5).

It is of interest to consider that the assumed existence of an abrupt boundary between two phases with a corresponding change in dielectric constant is probably not realized in practice (Buff *et al.*, 1972). It is more likely that the magnitude of the dielectric constant varies smoothly near the interfacial region (Chapter 8). The magnitude of van der Waals forces would thus vary according to this concept.

In liquid-expanded films, L_{ex}, it is safe to assume that at moderate π values, the van der Waals forces are the most prominent in neutral lipid films. It would also be assumed that the lipid molecules pack as a hexagonal structure, although other packing arrangements can be considered in a more rigorous analysis. Furthermore, it can also be assumed as a first approximation that the term arising from the cohesive force contribution, π_{coh}, is a function of the distance between the molecules, as follows (Birdi, unpublished; Chattoraj and Birdi, 1984):

$$\pi_{coh} = C_1 + C_2/A^{2.5} \qquad (4.40)$$

where C_1 is a constant and C_2 is proportional to W_{dis} [equation (4.39)]. However, the main approach in this analysis is that the only major parameter to which the cohesive forces are related is the distance between molecules, which in turn can be accurately estimated from area/molecule ($= A$) from π vs. A isotherms.

The analysis of various single- and double-chain lipid films is given in Table 4.5. It is remarkable that in all cases, the magnitude of π_{coh} varies

TABLE 4.5. Equation of State [Equation (4.40)] for Lipid Monolayers at the Air–Water Interface (for Liquid-Expanded Films)[a]

Lipid	T (°C)	C_1	C_2	Correlation coefficient
Myristic acid	17	0.42	−10,048	0.99
	25	1.96	−143,000	0.989
	35	9.31	−150,784	0.999
Dimyristoyl lecithin	17	−12.9	644,055	0.989
Dipalmitoyl lecithin	25	−2.83	394,294	0.905
	35	12.25	−419,685	0.996

[a] From Birdi (to be published); Chattoraj and Birdi (1984).

linearly with $1/A^{2.5}$, with correlation coefficients of approximately 0.9. In other words, the theoretical basis for the van der Waals forces is very satisfactory at this stage. The equation of state for neutral amphiphile films then can be written as (Birdi and Sanchez, to be published)

$$[\pi - (C_1 + C_2/A^{2.5})](A - A_0) = kT \qquad (4.41)$$

The molecular interpretation of constants C_1 and C_2 remains to be analyzed.

In a recent study, the direct van der Waals attractive adhesion forces between lipid bilayers in aqueous solutions were reported. Bilayers of dipalmitoylphosphatidylcholine (DPPC), dipalmitoylphosphatidylethanolamine (DPPE), and dipalmitoylphosphatidylglycerol (DGDG) were investigated (Marra, 1986a,b). The dependence of the long-range van der Waals interaction on the bilayer separation was found to be in agreement with the Hamaker equation.

4.3. PHOSPHOLIPID POLYMORPHISM

In earlier studies during the mid- and late 1960s, the polymorphic nature of phospholipids, i.e., their lyotropic and thermotropic phase properties, were investigated by X-ray diffraction techniques (Luzzati, 1968; Small, 1967; Reiss-Husson, 1967). Phase diagrams constructed as a function of temperature (thermotropism) and water content (lyotropism) clearly indicated that the dominant phase to be considered as equivalent to the biological membrane, i.e., the bilayer phase, was the smectic phase. It was also found that this class of lipids, i.e., two-alkyl-chain amphiphiles, exhibited other phases, which could also be expected in membrane structures. Perhaps this result could be an indication for the dependence of life on lipid polymorphism (Chapter 8). This observed transition was, of course, the phase transition from the solid or gel state (solid in three dimensions) to the fluid or liquid crystalline state [fluid in the plane of the lipid bilayer, but having some solidlike character in the plane normal to the bilayer (Larsson, 1969, 1973, 1976)].

This transition as studied by X-ray diffraction shows a change in distance between alkyl chains (carbon–carbon) from 4.2 to 4.6 Å; the membrane goes from the solid to the liquid phase. Assuming hexagonal packing [see equation (4.39)], this transition gives an area/molecule change of 30.55 Å^2 ($= 0.3055 \text{ nm}^2$) to 37 Å^2 (one can assume for phospholipids that the alkyl chains have area values of the same magnitude).

Highly detailed calorimetric studies have also been carried out for these structural melting states. Other techniques that have also been used are infrared and Raman spectroscopy (Bulkin and Krishnamachari, 1970;

Lippert and Peticolas, 1971) and fluorescence polarization with the use of probes (Lusccan and Fancon, 1971; Sackman and Trauble, 1972; Petersen and Birdi, 1983). CNDO/2 investigations on the conformation of the α-chain of lecithin indicated a strong preference for a *gauche-gauche* arrangement about the phosphodiester group (Flaim *et al.*, 1981). Three levels of water-binding energies were estimated. The highest binding of water was to the unesterified phosphate oxygens. By investigating complete incorporation of nine water molecules into a lipid structure, a plausible lecithin-water geometry was deduced for a liquid crystalline system.

In a recent study, data on transitions and molecular packing in highly purified dipalmitoyl lecithin (DPL)-water phases were described (Albon, 1983). The purity criterion is important when it is considered that the complex phase diagrams of these systems are very sensitive to very minute impurities. A new procedure was used to purify (DPL). This study used a sophisticated method of solvent crystallization that gave a large supply of single crystals of purity greater than 99.94%. The single crystals were studied by X-ray crystallography.

The thermal analysis of DPL-water systems is given in Table 4.6. As already observed, the effect of water on DPL phases is striking and complex. The adsorbed water in DPL can be removed only by heating to 200°C. Thus, anhydrous DPL can be obtained only by prolonged exposure to vacuum.

TABLE 4.6. Data on Thermal Analysis of DPL-Water Phases[a]

Phase no.	H_2O (%)	Preparation method[b]	Transition (°C)[b]	Powder (d Å)	Comments[b]
2a	10	Crystal solv. Excess water	—	3.914 4.367	— —
6	5	Heat phase 2 to 63.2°C	66.5	4.21–57.46 4.07–4.93	Stable
7	5	Heat phase 6 to 66.5°C	67.5	—	Stable
8	2–5	Heat phase 7 to 67.5°C (cr.)	175	4.5–54	Peak at 104°C
9	2.5	Heat phase 8 to 175°C	200 dec.	—	—
10	>15	Add water to phases 2–5	35	4.2–64	Hexagonal cell
11	>15	Heat phase 10 to 35°C	41.5	4.27–66 140	Gel
12	>25	Heat phase 11 to 41.5°C; add water	—	4.6 (broad) 54	Disord. Liq. cry.

[a] From Albon (1983).
[b] (solv.) solvent; (cr.) critical; (dec.) decompose; (disord.) disordered chains; (liq. cry.) liquid crystalline.

When excess water is present, as in phases 10, 11, and 12 (Table 4.6), the polar groups of different bilayers do not interact appreciably, and in each layer, the charged groups will pack in contact within a planar array. The forces between polar groups across each bilayer will tend to compress the bilayer. This compression will give rise to the reduction in bilayer thickness above the main transition, in phase 12, when the forces between the chains are weak.

4.4. LIPID MONOLAYERS AT THE OIL–WATER INTERFACE

Investigations of monolayer spread at the oil–water interface are of much importance for understanding emulsion formation (and stability) where an oil–water interface is present. As mentioned above van der Waals forces would be expected to be absent (or negligible) at the oil–water interface. Saturated diacylphosphatidylethanolamine gives a clear plateau region on its π vs. A isotherms at the oil–water interface, analogous to the isotherms at the air–water interface (Yue *et al.*, 1976; Taylor *et al.*, 1973; Bell *et al.*, 1978; Phillips and Chapman, 1968; Hayashi *et al.*, 1972, 1975, 1980).

The π vs. A curves at the n-hexadecane–water interface of phosphatidylserine (PS) were compared with those at the air–water interface (Seimiya and Ohki, 1972; Ohki and Ohki, 1976). The magnitude of $\pi_{ow} = 40$ mN/m corresponded to $A = 70$ Å^2, while the value of π_{aw} was 20 mN/m at $A = 70$ Å^2. It is thus seen that the PS monolayer at the oil–water interface is much more expanded than that of the air–water interface, owing to the absence of π_{coh}. Similar observations have been reported for other lecithins by other investigators (Taylor *et al.*, 1973). The area/molecule of 70 Å^2 is approximately the same area/molecule estimated from the X-ray diffraction studies of liquid crystalline phospholipid bilayers (Luzzati and Husson, 1962; Reiss and Husson, 1967).

Surface pressure, π, and surface viscosities, η_s, of DPPE (dipalmitoyl-phosphatidylethanolamine) and its polar group analogues were investigated at the hexane–water interface as a function of bulk pH (Hayashi *et al.*, 1980). The π vs. A isotherms of Na-octadecyl sulfate at the air–water and oil (n-heptane)–water interfaces (0.01 M NaCl) (Goddard, 1975) were reported (Mingins *et al.*, 1975). π_{ow}'s were measured by the dipping plate method, which uses a carbon-black-coated platinum plate and a microelectronic balance. The monolayers were compressed by using the oil–water trough method (Brooks *et al.*, 1964). The magnitude of π_{ow} was read to ±0.001 mN/m. The $\pi_{ow}A$ vs. π_{ow} plots were found to give $\pi A = kT$ as $\pi \to 0$. The data for dipalmitoyl lecithin (DPL) spread at the heptane–water

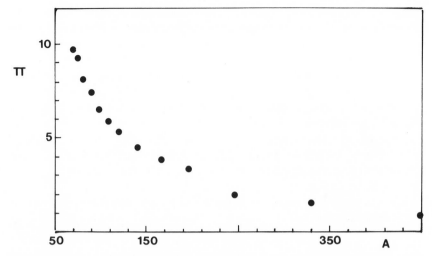

FIGURE 4.6. π vs. A isotherms of a C_s monolayer of DPL spread at the heptane–water interface (25°C). From Birdi *et al.* (to be published).

interface (Birdi, 1988) as shown in Fig. 4.6 and also gave πA vs. A plots where $\pi_{ow} A = kT$ as $\pi_{ow} \to 0$. The π_{ow} was measured by using a platinum plate Wilhelmy method. It was found that highly reproducible results were obtained when the plate was placed at the oil–water interface in such a manner that it barely touched the aqueous phase.

π_{ow} vs. A plots of monolayers of distearyl lecithin (DSL) at the interfaces between 0.1 M NaCl and n-pentane, n-hexane, cyclohexane, n-heptane, i-octane, n-nonane, and n-undecane have been investigated

TABLE 4.7. Magnitudes of A_{tr} and π_{tr} at the Phase Transition in DSPL Monolayers at Various Hydrocarbon–Aqueous NaCl (0.1 M) Interfaces[a]

	Temperature			
	3°C		20°C	
Solvent	A_{tr} (Å^2)	π_{tr} (mN/m)	A_{tr} (Å^2)	π_{tr} (mN/m)
n-Pentane	116	11.2	86	23.7
n-Hexane	110	13.4	81	25.8
Cyclohexane	—	—	83	30.4
n-Heptane	106	13.5	82	26.5
i-Octane	119	10.6	84	23.3
n-Nonane	114	12.4	80	24.9
n-Undecane	128	9.3	85	21.6[b]

[a] From Jackson and Yui (1975).   [b] At 40.9°C.

(Jackson and Yui, 1975). It can be clearly seen that the area and π at which the phase transition begins depend on the hydrocarbon component of the oil–water system. The area, A_{tr}, and π_{tr} at which the phase transition of these and other solvents occur are given in Table 4.7. It can be seen that around 20°C, the phase transition varies over approximately 6 Å^2, which is within the experimental error.

The statistical mechanical model plotted in Fig. 4.7 was described for phospholipid monolayers at the oil–water interface (which display second-order phase transitions between the L_{co} and L_{ex} phases) (Bell et al., 1978). In this model, each head group is localized on N available sites in a plane lying parallel to the interface. The expression for π was derived as

$$\pi A_0 / kT = (z/2)\{\ln[(u + \underline{e}(1 + s^{z-1}))/u]\}$$
$$+ (z/2 - 1)\ln(1 - \rho) \tag{4.42}$$

where z ($= 4$–6) is the coordination number, ρ ($= A_0/A$) is the fraction of sites occupied, $\underline{e}$ is the Boltzmann factor, and the parameter u is dependent on ρ and $\underline{e}$.

This model fits the experimental data only qualitatively. Other investigators (Scott, 1975; Marcelja, 1974) have attributed to the presence of impurities this lack of such agreements between the first-order transitions predicted by their mean-field theories and the second-order transitions observed experimentally. However, the dependence of experimental data on chain length of lipids rules out the impurity criticism (Bell et al., 1978; Birdi, unpublished). Furthermore, π measurements using various spreading

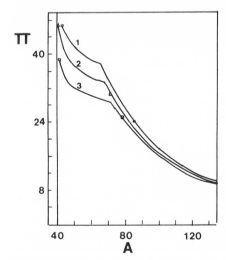

FIGURE 4.7. π vs. A isotherms for di-C_{16}-lecithin monolayers at the n-heptane-aqueous 0.01 M NaCl interface at various temperatures: (1) 10°C; (2) 15°C; (3) 20°C. From Bell et al. (1978).

procedures have given very reproducible results [even where the sensitivity was very high, i.e., ±0.001 mN/m (Birdi, unpublished)].

4.5. CHARGED LIPID MONOLAYERS

In the case of an aqueous solution consisting of fatty acid or sodium dodecyl sulfate, R-Na, and NaCl, for example, the Gibbs equation [equation (2.40)] may be written as (Chattoraj and Birdi, 1984)

$$-d\gamma = \Gamma_{RNa}d\mu_{RNa} + \Gamma_{NaCl}d\mu_{NaCl} \qquad (4.43)$$

Further

$$\mu_{RNa} = \mu_R + \mu_{Na} \qquad (4.44)$$

$$\mu_{NaCl} = \mu_{Na} + \mu_{Cl} \qquad (4.45)$$

It can be easily seen that the following will be valid:

$$\Gamma_{NaCl} = \Gamma_{Cl} \qquad (4.46)$$

and

$$\Gamma_{RNa} = \Gamma_R \qquad (4.47)$$

It can also be seen that the following equation will be valid for this solution:

$$-d\gamma = \Gamma_R d\mu_R + \Gamma_{Na}d\mu_{Na} + \Gamma_{Cl}d\mu_{Cl} \qquad (4.48)$$

This is the form of the Gibbs equation for an aqueous solution containing three different ionic species. The more general form for solutions containing i ionic species would be

$$-d\gamma = \sum \Gamma_i d\mu_i \qquad (4.49)$$

In the case of a charged film, the interface will acquire surface charge. This charge may be positive or negative, depending on the cationic or anionic nature of the lipid or polymer ions. The surface potential, ψ, would thus also have a positive or negative charge. The interfacial phase must be electroneutral, which can be possible only if the inorganic counterions are also preferentially adsorbed in the interfacial phase.

The surface phase can be described by the Helmholtz double-layer theory (Fig. 4.8). If a negatively charged lipid molecule, $R-Na^+$, is adsorbed at the interface AA', the interface will be negatively charged (air–water or oil–water). According to the Helmholtz model for a double layer, Na^+ on

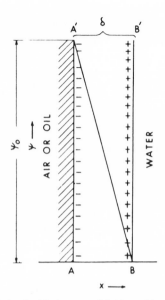

FIGURE 4.8. Helmholtz double-layer model. From Chattoraj and Birdi (1984).

the interfacial phase will be arranged in a plane BB' toward the aqueous phase. The distance between the two planes, AA' and BB', is given by δ. The charge densities, σ (charge per unit surface area), are equal in magnitude, but with opposite signs in the two planes. The (negative) charge density of the plane AA' is related to the surface potential (negative), ψ_0, at the Helmholtz charged plane:

$$\psi_0 = 4\sigma\pi\delta/D \tag{4.50}$$

where D is the dielectric constant of the medium (aqueous). According to the Helmholtz double-layer model, the potential ψ decreases sharply from its maximum value, ψ_0, to zero as δ becomes zero. The variation of ψ is depicted in Fig. 4.8. The Helmholtz model was found not to be able to give a satisfactory analysis of measured data.

　　　Later, another theory of the diffuse double layer was proposed, known as the Guoy–Chapman theory (Chattoraj and Birdi, 1984). The interfacial region for a system with charged lipid, $R\text{-}Na^+$, with NaCl is shown in Fig. 4.9. As in the case of the Helmholtz model, the plane AA' will be negative due to the adsorbed R^- species. Because of this potential, the Na^+ and Cl^- ions will be distributed nonuniformly owing to the electrostatic forces. The concentrations of the ions near the surface can be given by the Boltzmann distribution, at some distance x from the plane AA', as

$$\dot{c}^s_{Na^+} = \dot{c}_{Na^+}\, \underline{e}^{-(\varepsilon\psi/kT)} \tag{4.51}$$

$$\dot{c}^s_{Cl^-} = \dot{c}_{Cl^-}\, \underline{e}^{+(\varepsilon\psi/kT)} \tag{4.52}$$

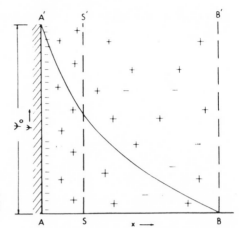

FIGURE 4.9. Gouy-Chapman double-layer model. From Chattoraj and Birdi (1984).

where $\dot{c}_{Na^+}$ and $\dot{c}_{Cl^-}$ are the number of sodium and chloride ions per milliliter, respectively, in the bulk phase. The magnitude of ψ varies with x as shown in Fig. 4.9, from its maximum value, ψ_0, at plane AA'. From equations (4.51) and (4.52), we thus find that the quantities $\dot{c}_{Na^+}^s$ and $\dot{c}_{Cl^-}^s$ will decrease and increase, respectively, as the distance x increases from the interface, until their values become equal to $\dot{c}_{Na^+}$ and $\dot{c}_{Cl^-}$, where ψ is zero. The variation of $\dot{c}_{Na^+}^s$ and $\dot{c}_{Cl^-}^s$ with x is given in Fig. 4.10. The extended region of x between AA' and BB' in Fig. 4.10 may be termed the diffuse or the Gouy-Chapman double layer.

The volume density of charge (per milliliter) at a position within $AA'BB'$ may be defined as equal to

$$\varepsilon(\dot{c}_+^s - \dot{c}_-^s) \qquad (4.53)$$

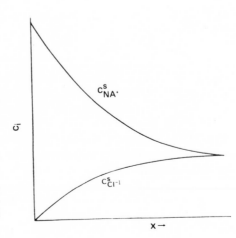

FIGURE 4.10. Variation of the concentrations of ions with distance from the interface (AA'). From Chattoraj and Birdi (1984).

which can be expressed by the Poisson relationship:

$$d^2\psi/d^2x = -(4\pi)/D \qquad (4.54)$$

In this derivation, it is assumed that the interface is flat, such that it is sufficient to consider only changes in ψ in the x direction normal to the surface plane. The following relationship can be derived from equation (4.54):

$$d^2\psi/dx^2 = d(d\psi/dx)/dx \qquad (4.55)$$

$$= -(4\pi N\varepsilon/1000D)c(\underline{e}^{-\varepsilon/kT}\,\underline{e}^{+\varepsilon\psi/kT}) \qquad (4.56)$$

where c is the bulk concentration of the electrolyte. In a circle of unit surface area on the charged plane A-A', the negative charges acquired by the adsorbed organic ions (amphiphiles) within this area represent the surface charge density, σ:

$$\sigma = -\int_0^\infty \rho\,dx \qquad (4.57)$$

$$= (D/4\pi)(d\psi/dx) \qquad (4.58)$$

At $x = 0$, the magnitude of ψ reaches ψ_0:

$$\sigma = (2DRTc/1000\pi)^{0.5}[\sinh(\varepsilon\psi_0/2kT)] \qquad (4.59)$$

The average thickness of the double layer, $1/k$, i.e., the Debye-Hückel length, is given as (Chattoraj and Birdi, 1984)

$$1/k = (1000DRT/8\pi N^2\varepsilon^2c)^{0.5} \qquad (4.60)$$

At 25°C, for univalent electrolytes, one obtains

$$k = 3.282 \times 10^7\sqrt{c} \qquad (\text{cm}^{-1}) \qquad (4.61)$$

For small values of ψ, one obtains the following relationships:

$$\sigma = (DRTk/2\pi N\varepsilon)\sinh(\varepsilon\psi_0/2kT) \qquad (4.62)$$

This relates the potential charge of a plane plate condenser to the thickness $1/k$. The expression based on the Gouy model is derived as

$$\sigma = 0.3514 \times 10^5\sqrt{c}\,\sinh(0.0194\psi_0) \qquad (4.63)$$

$$= \Gamma_R zN\varepsilon \qquad (4.64)$$

where the magnitude of Γ_R can be experimentally determined and the magnitude of ψ_0 can thus be estimated. The free energy change due to the electrostatic work involved in charging the double layer is (Chattoraj and Birdi, 1984)

$$F_e = -\int_0^{\psi_0} \sigma \, d\psi \tag{4.65}$$

By combining these equations, one can write the expression for π_{el} (Chattoraj and Birdi, 1984):

$$\pi_{el} = 6.1c^{1/2}[\cosh \sinh^{-1}(135/A_{el}c^{1/2}) - 1] \tag{4.66}$$

The quantity (kT) is approximately 4×10^{-14} erg at ordinary room temperature, and $(kT/\varepsilon) = 25$ mV.

ION-SPECIFIC EFFECTS AND ION EXCHANGE BY MONOLAYERS

The adsorption of ions to charged lipid monolayers would thus give rise to subtle effects on the various properties of the monolayers, e.g., π, ΔV, η_s (Payens, 1955; Davies, 1956; Chattoraj and Birdi, 1984). The electroneutrality requirement thus requires an ion-exchange process of adsorption.

The effects of ions on various lipid monolayers (stearic acid, arachidic acid, behenic acid) have been reported for monovalent counterions (Goddard et al., 1967). The adsorption of the ion gives rise to an expansion or swelling of the fatty acid monolayer. The degree of expansion is related to the size of the *unhydrated* cation for the series Li^+, Na^+, K^+, trimethyl-ammonium (TMA), or tetraethylammonium (TEA) when the fatty acid monolayer is of the condensed type. The π vs. A isotherms shown in Fig. 4.11 give the results of counterion adsorption (Goddard et al., 1967a,b). It is to be observed that the Li^+ salt monolayers are the least expanded, with the lowest ΔV isotherms. This could be explained by saying that because of their small size, Li^+ ions can penetrate more deeply into the monolayer and thus exhibit greater discharging. Since the degree of expansion is related to the size of the counterion, i.e., $Li^+ < Na^+ < K^+ < TMA < TEA$, the degree of penetration would in return determine the van der Waals distance of separation between the alkyl chains. The expanded films are indicative of the degree of penetration. These findings are in accord with those for other monolayer systems, i.e., counterion adsorption·to docosyl (C_{22}) TMA ions. The degree of expansion was in increasing order: $SCN^- < I^- < NO^{3-} < Br^- < Cl^- < F^-$ (Goddard et al., 1968). The double-layer theory

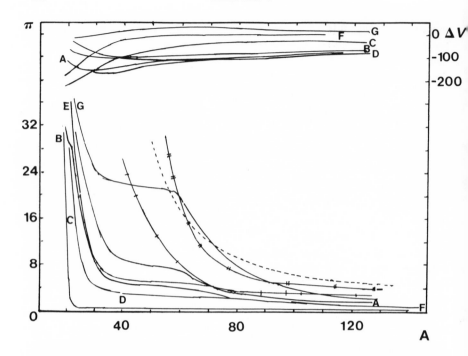

FIGURE 4.11. π vs. A and ΔV vs. A isotherms at 26°C for stearic acid monolayers on alkaline subphases (0.1 and 0.01 M): (A, D) Na^+; (B, E) K^+; (C, F) Li^+; (G) TMA; curve of the equation 4.29 where $A_0 = 40$ Å^2 (----); curve of the equation 4.66 where $A_{el} = 20$ Å^2 (——+——) or $A_{el} = 40$ Å^2 (——#——). From Goddard *et al.* (1967).

based on diffuse Davies does not explain these data satisfactorily (Chattoraj and Birdi, 1984).

π and ΔV vs. A isotherms of nonadecylbenzenesulfonate monolayers were investigated on a variety of substrates (Dreher and Wilson, 1970). In the ionized form, the benzene ring did not exhibit close packing in the monolayer. The collapse of the film was observed at approximately the same area per molecule as that of the alkyl sulfonate. Monolayers were expanded in the order $K^+ < NH_4^+ < Na^+ < Li^+$, and $Sr^{2+} < Ca^{2+} < Mg^{2+}$.

PHOSPHOLIPID MONOLAYERS

The effect of pH on lipid [phosphatidylethanolamine (PE), phosphatidylisopropanolamine (PIP), and phosphatidyl-*n*-propanolamine (PNP)] monolayers was investigated in the presence of UO_2^{2+} and Th^{4+} (Hayashi *et al.*, 1972). Above pH 9.5, the quaternized amine groups are transformed to amine groups. The value of the limiting areas is 55–65 Å^2

in the presence of UO_2^{2+}, while it is 48–90 Å^2 in the absence of cations. However, the films behaved differently for Th^{4+} as compared to UO_2^{2+}. The mechanisms have been described in detail elsewhere (Saraga, 1975).

An aging effect on phosphatidylinositol (PI) monolayers on subphases of pH 3–5 has been reported (Quinn and Dawson, 1972). The monolayers were stable at pH 8. The adsorption of Ca^{2+} gives rise to an increase in π, while ΔV decreases. The aging effect could be explained as due to the hydrolysis or protonation of the phosphate group of PI, which does not occur at pH 8.

In a recent study (Abraham and Ketterson, 1985), the time-dependent response to a shearing stress of a monolayer of DPL adsorbed on the surface of water was measured as a function of both pH and the surface concentration, C_s. In these investigations, a new method was used for obtaining π vs. $C_{s,ak}$ isotherms ("ak" denotes the new method). The procedure used was different from the conventional Langmuir film balance. A conical container was used to change the surface area by raising or lowering the liquid level. The π vs. $C_{s,ak}$ isotherms showed no feature that could be attributed to a phase transition (from L_{ex} to L_{co}), as compared to π vs. A isotherms from the Langmuir balance. π vs. A isotherms using the Langmuir balance were later reported (Birdi, 1987) as a function of pH (4.6, 6.0, and 7.5), to compare these data with the data of π vs. $C_{s,ak}$. The π vs. A isotherms showed negligible pH dependence (Fig. 4.12), while the π vs. $C_{s,ak}$ isotherms had shown very significant pH dependence. The differences between π vs. A and π vs. $C_{s,ak}$ can be explained as possibly attributable to the up or down movement of the liquid surface in the conical container in the latter procedure, which could easily give rise to adsorption of the DPL monolayer on the glass walls [as is known to happen in Langmuir–Blodgett film formation (Chapter 9)].

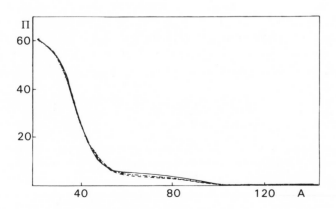

FIGURE 4.12. π vs. A isotherms of DPL on subphases of different pH. From Birdi (1987).

4.6. PHASE RULE (IN TWO DIMENSIONS) IN MONOLAYERS

To describe multicomponent systems, it is always necessary to apply the Gibbs phase-rule principle in general for either three- or two-dimensional systems. The Gibbs phase rule has been found useful in predicting equilibrium-phase relationships for systems in which the following are variable:

1. Concentrations of each component in each phase (or the composition).
2. Other intensive properties of the system (e.g., temperature and pressure).

In this derivation, it is clear that in addition to the chemical potential, μ, of every component, the temperature and pressure and any other intensive factors, such as gravitational or electrical potential, may be included, another degree of freedom being allowed for each such intensive property included. The interfacial tension or energy per unit square area (generally cm^2 or m^2) would be considered as such a property. To simplify the derivations, it is useful to consider all systems at constant temperature and external pressure (Chattoraj and Birdi, 1984).

The phase rule can be written as

$$F = C - P \tag{4.67}$$

where F is the number of degrees of freedom, C is the number of components, and P is the number of phases. Since the external pressure is assumed to be constant, only plane surfaces are considered in this case.

If each surface is considered as a separate phase, for P phases, there would be $(C - 1)P^s$ new variables, since *a priori* all components but one may be varied independently in each surface phase. There will also be P^s terms for surface energies of each surface phase, giving a total of $(C - 1)P^s + P^s = CP^s$ additional variables. Furthermore, there would be equilibrium constraints,

$$\mu_{i,s} = \mu_{i,b} \tag{4.68a}$$

where the subscripts "s" and "b" indicate the surface and bulk phases, respectively. Similarly, if there is more than one phase in a surface, one will also have

$$\mu_i^{si} = \mu_j^{sj} \tag{4.68b}$$

If one or more of the components, designated C^s, is confined to a fluid interface, and not equilibrated with the rest, this interface becomes equivalent to a phase separated by a membrane permeable to all components except C^s. The interface then possesses an independently variable energy term due to the two-dimensional osmotic pressure, π_{osm}.

Furthermore, there can be isolated "surface phases" that are quite distinct from one another and are completely delimited by mechanical constraints; other physically distinct surface phases are intimately contiguous and form a part of the same surface. The former types are called "surfaces," while the latter are called "surface phases." For example, an oil–water and an oil–air interface separated by a column of bulk phase, or two air–water interfaces separated by a mechanical barrier, would be considered separate "surfaces." On the other hand, a condensed fatty acid film and the vapor film with which it is in equilibrium would be "surface phases" in the same surface.

Without going into extensive details for the derivation of the general case, the case of a single surface, containing q surface phases, can be written as

$$F = C^s + C^b - P^b - (q - 1) \tag{4.69}$$

$$= C - P^b - (q - 1) \tag{4.70}$$

where the total number of components is $C = C^b + C^s$. Let us apply this phase rule to some typical monolayer systems:

1. In the case of a monocomponent monolayer at an air–water (A/W) or oil–water (O/W) interface, we have

A/W: $\quad C^b = 1, \quad C^s = 1, \quad P^b = 1, \quad q = 1, \quad F = 1$ (4.71)

O/W: $\quad C^b = 2, \quad C^s = 1, \quad P^b = 2, \quad q = 1, \quad F = 1$ (4.72)

2. If two surface phases are present, such as *solid* or *liquid* in equilibrium with the two-dimensional vapor film, then $q = 2$ and $F = 0$. This result implies that a flat curve would be observed in any π vs. A isotherm.

3. If there are present three phases, $q = 3$, analogous to a triple point, then $F = -1$. This result implies that this monolayer film system can exist only at a unique temperature.

The significance of the phase rule in monolayers has not been fully appreciated in the current literature. Therefore, some examples of this lack in monolayer analysis are given in the following section.

4.7. LIPID-PHASE TRANSITION IN MONOLAYERS, BILAYERS, AND MEMBRANES

It has been reported that when the pressure on vesicles is increased, the T_{tr} is shifted to higher temperatures, indicating an increase in molecular volume during the transition. It is now well established that the phase-transition characteristics, such as temperature, enthalpy, and entropy, are dependent on the alkyl group (i.e., alkyl-chain length) of the lipids; on the nature of the polar group, i.e., the degree of interaction between the polar group and the surrounding aqueous medium; and on the pH and ionic strength and the nature of the electrolytes in the medium.

It is of immediate interest that the possession of any physical property by any biological molecule found abundantly in natural species merely indicates that this particular physical characteristic is of fundamental importance in the function and behavior of that molecule. The basis for this argument is that most of the molecules found in the biological world have acquired their physical properties through selection processes during evolution (≈ 3 billion years). All the biological lipids possess hydrophile-lipophile balance (HLB) such that the lipids readily spread on the surface of water to form stable monolayers (and, with some exceptions, at oil–water interfaces as well). This observation suggests that lipid molecules possess this amphiphilic character because it is a requirement of their function in biological interfaces. Hence, these specific properties at the air–water (or oil–water) interface should be of much interest in enabling one to estimate the various interfacial forces that are responsible for the stabilization of the membrane structure, as well as for their function. Lipid monolayers have been extensively studied under various experimental conditions, and these data are analyzed in detail below.

It is now well established that lipids play a very crucial role in the function and the structural stability of biological cell membranes. It is therefore obvious that the study of physical properties of lipids in model membrane systems is vital in order to be able to describe the more complex membranes. As mentioned earlier, the various model systems that have been used are monolayers, bilayers, and vesicles (see Fig. 1.3).

It is evident that a bilayer represents the structure of biological membranes very closely, since a bilayer of lipid separates two different aqueous phases, as in biological cells. Accordingly, a vesicle also resembles the cell membrane, because of the presence of the inside and outside nature of these aqueous phases.

On the other hand, in monomolecular film systems, at the air–water interface in any case, one cannot expect the monolayer to resemble the cell membrane structure completely, for the simple reason that the system is placed at the air–water interface. In recent studies, it has been argued that lipid monolayers spread at the oil-water interface should be expected to resemble membrane behavior more closely. This view may be correct, but we must also consider that the dispersion forces, W_{dis}, are almost absent between the alkyl chains of lipids at the oil-water interface (Chattoraj and Birdi, 1984), while these forces are present both at the air-water interface and in bilayers and vesicles (and membranes). It thus seems that as regards these stabilizing forces, the air-water interface should be the more useful model. This point has been discussed elsewhere in this book.

The phase transition of lipid bilayers is an example of a two-dimensional cooperative transition. It is normally induced by raising the temperature, but can also be brought about by changing the ionic strength or pH of the surrounding medium.

4.7.1. LIPID-PHASE TRANSITION IN MONOLAYERS

In π vs. A isotherms, the transition from the L_{ex} to the L_{co} state has not been exhaustively investigated in the current literature (Birdi and Sander, 1981; Chattoraj and Birdi, 1984; Birdi, unpublished). Primarily, there is disagreement regarding whether the lipid-phase transition in the monolayer is of first or second order (see Section 3.4).

If the system is a first-order transition, then the π vs. A isotherm should exhibit a horizontal curve near the transition, π_{tr}. A second-order transition, on the other hand, would require that there be no horizontal region. All the π vs. A isotherms of lipid monolayers reported in the literature generally show no (completely) horizontal regions. It has been argued that the monolayer is a complex system as compared to pure three-dimensional systems, where horizontal curves are indeed observed. It could be argued that the reason for this difference is that not only is the monolayer π vs. A isotherm related to lipid-lipid interactions, but also the subphase determines the shape of the phase-transition region (see Section 4.3). Unfortunately, some investigators have also suggested that impurities may be the cause of the absence of horizontal curves. This argument is not valid, since most careful work reported in the literature has been carried out with lipids of high purity. It has been asserted that if impurities had significant effects, then the π vs. A isotherms could never be as highly reproducible (when different batches are used) as they are (Birdi, unpublished).

Analogous to the three-dimensional phase-transition regions, the π vs. A isotherm would be expected to show a sharp transition (ACDEB in Fig. 4.13) below the critical temperature only if an infinite number of molecules are involved in the phase transition. On the other hand, if, say, only 20, 100, or 1000 molecules are involved, then the phase-transition region would follow the curve ADB instead of ACDEB (Hill, 1962, 1963).

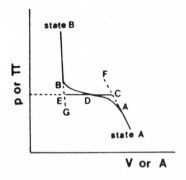

FIGURE 4.13. Schematic pressure (p) or surface pressure (π) vs. volume (V) or area (A) isotherm for a small system at a temperature at which a first-order phase transition takes place. (ACF and BEG are extrapolations of states A and B, respectively.) From Chattoraj and Birdi (1984).

It has been shown that in such transition regions with finite molecules undergoing a phase transition from state A to B (Fig. 4.13), the following equations are valid (Hill, 1962, 1963):

$$\text{Length(E--D)} = \text{length(D--C)} \tag{4.73}$$

and

$$\text{Area(ACD)} = \text{area(DEB)} \tag{4.74}$$

The π vs. A isotherms of DPL and DML do indicate that the relationships in equations (4.73) and (4.74) are valid (Birdi, unpublished). To add more information about the lipid–phase transition in monolayers, two-dimensional gas–liquid condensation was investigated (Hawkins and Benedek, 1974; Kim and Cannell, 1975). The density of the liquid film, ρ_L ($= 1/A$), and the density of the gas film, ρ_V ($= 1/A$), were estimated by fitting the π vs. A isotherm with a polynomial in or near each side of the coexistence curve and taking the intersection of these fitted curves. These analyses are given in the following equations (Kim and Cannell, 1975):

Two-dimensional compressibility

$$= (K_T)_V = 2.73 \times 10^3 (T_{tr} - T)^{-0.97} \text{ (cm/dyne)} \tag{4.75}$$

$$(K_T)_L = 1.25 \times 10^3 (T_c - T)^{-0.98} \text{ (cm/dyne)} \tag{4.76}$$

$$(K_T)_{T > T_c} = 6.73 \times 10^2 (T_c - T)^{-0.98} \text{ (cm/dyne)} \tag{4.77}$$

$$\rho_L - \rho_V = 41.9 (T_c - T)^{0.5} \text{ (molecules/10}^4 \text{ Å}^2) \tag{4.78}$$

where $\rho_c = 41.7$ molecules/10^4 Å^2, $T_c = 26.27°C$, and $\pi_c = 174$ mdynes/cm.

K is the two-dimensional compressibility. The critical exponents in these equations are in agreement with mean field theory exponents.

In an analogous manner, the $L_{ex} \rightarrow L_{co}$ isotherms of lecithin films of DML and DPL have been analyzed (Birdi and Sander, 1981; Chattoraj and Birdi, 1984):

DML Films:

$$(\rho_{ex} - \rho_{co}) = 5.91 \times 10^{-4}(T_{tr} - T)^{0.5} \text{ (molecules/Å}^2)$$

$$T_{tr}^{DML} = 22.3°C$$

(4.79)

DPL Films:

$$(\rho_{ex} - \rho_{co}) = 8.1 \times 10^{-4}(T_{tr} - T)^{0.5} \text{ (molecules/Å}^2)$$

$$T_{tr}^{DPL} = 41.2°C$$

(4.80)

The plots of $(\rho_{ex} - \rho_{co})$ vs. temperature, t, are given in Fig. 4.14A. It can be seen that the relationships in equations (4.79) and (4.80) fit the data points satisfactorily. Similar analyses have been carried out for the π vs. A isotherms of 1-monopalmitate (1-MP) and 2-monopalmitate (2-MP) (Birdi and Sanchez, to be published). The relationships given by equations (4.79) and (4.80) were also found to fit the data satisfactorily (Fig. 4.14B). The π_{eq} vs. t data are given (Fig. 4.14C) for comparison.

It can thus be seen that the limiting value of $\Delta\rho$ approaches zero as $t \rightarrow T_{tr}$ in $\pi-A$ isotherms at the same temperature at which vesicles show a transition state. This finding indicates that these monolayer structures are equilibrium isotherm systems and that the enthalpic energies are comparable to those found in vesicles and bilayers.

From the $\pi-A$ data for 1,2-di-C_{12}-PE, 1,2-di-C_{14}-PE, and 1,2-di-C_{16}-PE as a function of temperature, i.e., the plots of $\Delta\rho$ vs. T, it was found that the transition temperature, T_{tr}, varied with the number of carbon atoms, n_C, as follows:

$$T_{tr} = -48.37 + 7.10n_C$$

(4.81)

with a correlation coefficient of approximately 0.99 (Doerfler and Rettig, 1980; Birdi, unpublished). It can thus be seen that for a homologous series of lecithins the alkyl-chain-length effect is of the same order, approximately 7.5°C per -CH_2 group, as for other lecithins, e.g., phospholipids. In bilayers, this value is 8.64 [equation (4.84)].

The $\pi-A$ isotherms at the oil–water (O/W) interface at different temperatures have been reported. It is of interest to determine whether the phase-transition temperature remains unchanged at the O/W interface as compared to the air–water (A/W) interface. As mentioned earlier, the plot

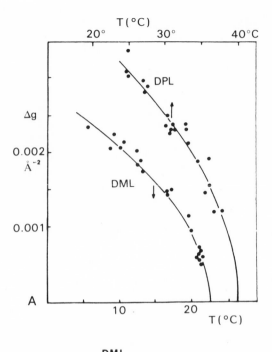

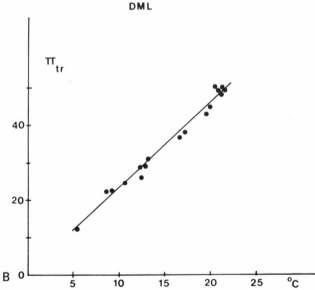

FIGURE 4.14. (A) Variation of $\Delta \rho$ ($=\rho_{ex} - \rho_{co}$) with temperature t (°C) for monolayers of DML and DPL. From Chattoraj and Birdi (1984). (B) Variation of π_{tr} with temperature (°C) for DML monolayers. From Birdi *et al.* (to be published). (C) Variation of π_{eq} with temperature for DML. From Gershfeld and Tajima (1979) (●); Chattoraj and Birdi (1984).

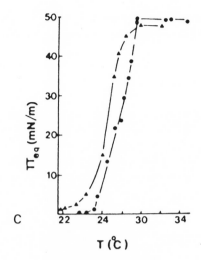

FIGURE 4.14 (*continued*).

of ΔA ($= A_{ex} - A_{co}$) vs. T gives the T_{tr} at the A/W interface. A similar analysis of ΔA for the O/W data for DSL was found to give $T_{tr} \approx 42°C$ (with a correlation coefficient of 0.95). This value is lower than the value found for A/W monolayers or vesicles ($\approx 60°C$); various investigators have considered O/W monolayers to be a more useful model than A/W monolayers for biological membranes. More investigations are indeed required at the O/W interface for such systems to determine why the magnitudes of T_{tr} for monolayers at the A/W interface and for vesicles (or membranes) are almost the same, while the values of T_{tr} at the O/W interface are much lower. The conformations of fatty acyl chains in α- and β-phosphatidylcholine (PC) and -PE derivatives have been investigated in sonicated vesicles (Pluckthun *et al.*, 1986).

4.7.2. INTERACTION OF LIPID MONOLAYERS AND VESICLES WITH SOLVENT

The mechanical properties as well as the ΔV's of different lipid liquid-expanded (L_{ex}) monolayers (e.g., ethyl myristate, ethyl pentadecanoate, myristic acid, oleic acid) on aqueous solutions of a variety of organic solutes (e.g., n-butanol, t-butanol, 1,4-butanediol, glycerol, D-glucose) were examined (MacArthur, 1977). π and ΔV vs. A data were obtained with the help of an automated Langmuir–Adam–Wilhelmy surface balance.

The data for n-butanol and t-butanol solutions were significantly distorted by relaxation processes of solute contents in subphases as low as 0.01 mole%. However, metastable equilibrium data were obtained for the other solutes on subphases containing 1.0, 1.5, and 4.0 mole%. The data for glycerol and D-glucose were limited owing to rheological effects. A quantitative analysis based on the surface excess of solute, as calculated from the Gibbs adsorption theory, and the observed expansion produced by the solute in the subphase, indicated that more surface-active solute produced greater expansion.

These monolayer investigations are worth comparing with the behavior of vesicle structures and the effect of solvent composition. It has been demonstrated that addition of water to lipid vesicles lowers the transition temperature, T_m, and simultaneously raises the ΔH_m until the water contains about 30% by weight (Chapman et al., 1967). Beyond this point, both T_m and ΔH_m remain constant. The simultaneous nature of the changes in T_m and ΔH_m on the addition of water suggests that both the gel and the liquid crystalline phase undergo the same mechanism during the melting of the hydrocarbon chains. In other words, if water plays an important role in the contribution to the changes in T_m and ΔH_m, that would suggest that water should be associated with both phases rather than just one. Hydration of the polar head groups of lipid vesicles in the gel phase probably affects the conformation of the hydrocarbon groups in that phase so as to affect ΔH_m (Chapman et al., 1967). However, such an effect could not be a major contribution to ΔH_m, since both phases should be subject to the same mechanism during the melting of the hydrocarbon groups. The head-group's hydration mechanism does not require a consideration of this effect in the gel phase or in the liquid crystalline phase.

It has been proposed that some water is present around the hydrocarbon tails of lipid vesicles in both phases (Chen, 1980, 1981; Chen et al., 1981). In the course of the phase transition, there is melting of the more structured water, which reflects the difference in the amount of the more structured water around the hydrocarbon tails between the two phases. In a recent study, the thermodynamic properties of vesicles in D_2O and H_2O were

TABLE 4.8. Values of T_m, ΔH_m, and ΔS_m for Aqueous Dispersions of DSPC, DPPC, and DMPC in Water and D_2O^a

Lipid vesicles	Solvent	T_m (°C)	ΔH_m (kcal/mole)	ΔS_m (eu)
DSPC	Water	54.65	9.09	27.7
	D_2O	55.9	9.32	28.4
DPPC	Water	41.67	7.25	23.0
	D_2O	42.08	8.42	26.7
DMPC	Water	24.22	4.85	16.3
	D_2O	24.46	5.45	18.3

^{a} From Chen (1982).

compared. The observed T_m's of distearylphosphatidylcholine (DSPC), DPPC, and dimyristoylphosphatidylcholine (DMPC) (Table 4.8) were found to be higher in D_2O solvent than in H_2O (Chen, 1982). Since looser molecular packing of the hydrocarbon tails is associated with a lower T_m, the small increase in T_m was found to suggest that the molecular packing is tighter in D_2O, which may result from stronger hydrophobic interactions in this molecule (Wen and Muccitelli, 1979; Momura and Verrall, 1981; Emerson and Holtzer, 1967).

Aqueous dispersions of lipid vesicles have a main phase-transition temperature, T_m, and a well-defined ΔH_m of transition. For the transition from the gel to the liquid crystalline phase, an equilibrium equation can be written as

$$\text{Gel phase} \rightleftharpoons \text{liquid crystalline phase}$$
$$\Delta G = G_{lc} - G_{gel} \tag{4.82}$$

where ΔG is the Gibbs free energy of the transition and "lc" and "gel" denote the liquid crystalline and the gel phase, respectively. At the transition, where $\Delta G_m = 0$, ΔS_m can be written as

$$\Delta S_m = \Delta H_m / T_m \tag{4.83}$$

assuming a first-order equilibrium transition, where $\Delta H_m = H_{lc} - H_{gel}$ and $\Delta S_m = S_{lc} - S_{gel}$.

4.8. COMPARISON OF DATA ON PHASE TRANSITION BY DIFFERENT MONOLAYER SPREADING METHODS (π vs. A; π vs. C_s; π_{eq} vs. T)

As described above, monolayers can be spread by using a solution of amphiphile in a suitable water-insoluble solvent (such as $CHCl_3$, n-C_6H_{14}, or ether) on the surface of water. The π_{eq} can be measured if the lipid (in the solid or liquid state) is placed on the surface of water. The various equilibria in this latter system have been described by different investigators (Gershfeld, 1976; Chattoraj and Birdi, 1984). The equilibrium $B_s \leftrightarrow V$ is the process of sublimation and $B_l \leftrightarrow V$ is the vaporization of the bulk lipid phase (solid or liquid). The $B_s \leftrightarrow \pi_{eq}$ or $B_l \leftrightarrow \pi_{eq}$ is the equilibrium surface pressure, π_{eq}, from the bulk phase (as a solid, "s", or a liquid, "l"). The equilibrium $\pi_v \leftrightarrow \pi_{eq}$ is the expansion or compression of the respective state in the *absence* of any bulk phase. The evaporation of monolayer lipids at π_{eq} or π_v is given as $\pi_{eq} \leftrightarrow V$ or $\pi_v \leftrightarrow V$, respectively. In the current literature, although the thermodynamic analysis of these equilibria has been extensively reported over the past few decades, there remains a fundamental misconception as regards the phase equilibria. The following equilibria, especially, have invariably been contested by various investigators:

1. Monolayer compression from a gaseous (π_v) film to a liquid film (at temperatures above the transition temperature) and to a solid film (at $< T_{tr}$), which may (in the case of fatty acids) or may not (in the case of fatty alcohols or lecithins) correspond to the bulk-phase melting point.

2. Bulk phase in equilibrium with monolayer, π_{eq}, at temperatures above or below the T_{mp} of the bulk phase (this can be realized in the case of fatty acids).

Before these descriptions as found in the literature (Fig. 4.15) can be elaborated on, it is necessary to consider some of the available data in detail.

π vs. A isotherms are obtained after compression of the lipid molecules, which are generally present as gas films (see Chapter 3). The values of the phase-transition temperature, T_c, in the lipid bilayers for different lecithins at maximum hydration are given in Table 4.9. These data on transition temperatures, T_c, can be shown to be linearly dependent on the number of carbon atoms, n_C, in the saturated acyl chain, as follows:

$$T_c = -97.8 + 8.64 n_C \qquad (r^2 = 0.9998) \qquad (4.84)$$

The data on T_m, however, are variable as reported in the literature. The ranges for DML, DPL, and DSL are 23–23.9, 41–41.8, and 55–58°C, respectively. Similarly, the ΔH_m ranges for DML, DPL, and DSL are 5.4–6.7,

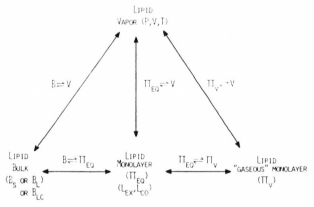

FIGURE 4.15. Various equilibria (π_{eq} or π_v) between bulk phases or monolayer phases and the vapor phase (V). From Chattoraj and Birdi (1984).

TABLE 4.9. Phase-Transition Temperatures, T_c, for 1,2-Diacyl-L-phosphatidylcholines[a]

Acyl chain	n_C	T_c (°C)
Behenoyl	20	75
Stearoyl	18	58
Palmitoyl	16	41
Myristoyl	14	23
Oleoyl	18 (−=−)	−22

[a] From Phillips et al. (1969).

8.7–9.7, and 10.6–10.7, respectively. The ΔH_m varies with n_C as follows:

$$\Delta H_m = -12.45 + 1.3 n_C \qquad (4.85)$$

and the entropy of melting, ΔS_m, as

$$\Delta S_m = \Delta H_m / T$$
$$= -20.4 + 2.95 n_C \qquad (\text{cal/mole/°K}) \qquad (4.86)$$

It can thus be seen that the entropy change per $-CH_2-$ group is 1.5 ($= 2.95/2$) cal/mole/°K/$-CH_2-$. This value in comparison to the entropy of melting of straight-chain alkanes is 2.6 cal/mole/°K/$-CH_2-$, while for nonlinear alkanes it is 1.9 cal/mole/°K/$-CH_2-$. It can thus be seen that the change in entropy in the gel–liquid-crystal phase change is less than the melting of alkanes (solid–liquid). The reason for this difference is that the polar groups in lecithins are hydrated with water molecules and the alkyl chains in lecithins therefore have a lesser degree of fluidity than do the melted chains of alkanes.

It is of interest to note that addition of small-molecule solutes to lecithin suspensions has an effect on both T_{tr} and enthalpy, ΔH_{tr} (Sturtevant, 1982). The effect of urea was negligible, while the effects of small-molecule solutes were in the order ethanol < ethyl acetate < butanol. The addition of ethanol to the subphase reduces the magnitude of π_{col}, which agrees with the calorimetry data (see Fig. 4.4).

In the case of monoglycerides in bilayers, the transitions were investigated by using light scattering (Crawford and Earnshaw, 1986).

4.8.1. EQUILIBRIUM SURFACE PRESSURE, π_{eq}, OF DIFFERENT LIPIDS

The literature values for π_{eq} are inconsistent. An explanation that has been given is that the measurements were not performed correctly. The π_{eq}

is defined as the surface pressure at equilibrium between a crystal or a droplet and its monolayer. By a simple calculation, it can be shown that a crystal less than 0.01 mg on an area of 50 cm^2 will be sufficient to reach equilibrium spreading to satisfy the assumption that A_c is negligible in relation to A_e. However, the experiments reported in recent studies indicate that the reproducibility of π_{eq} is very high (Birdi, unpublished). In earlier studies, several hundred crystals were added to the liquid surface, and the criterion for equilibrium spreading was the constancy of π for 15 min. With such a technique, values of π above π_{eq} are usually obtained, since smaller amounts of solutions carry the compression to above π_{eq}. Overall, compression is achieved by increasing the area available for the monolayer by lateral compression or by increasing the number of crystals. With allowance of sufficient time, which is much longer than 15 min, depending on the melting point, such pressure will decline to π_{eq}.

4.8.2. THERMODYNAMICS OF EQUILIBRIUM SURFACE PRESSURE (π_{eq}) OF DIFFERENT LIPID MONOLAYERS

MONOLAYER SPREADING THERMODYNAMICS FROM SPREADING BULK PHASE

In the study of phase transitions, the Clausius–Clapeyron equation is the most fundamental thermodynamic relationship. In monolayers, the heat of spreading, Q_s, of an insoluble film is related to the temperature coefficient of its π_{eq}:

$$d\pi_{eq}/dT = Q_s/T(A_{ex} - A_{co}) \tag{4.87}$$

where A_{ex} is the molar area of the spread monolayer and A_{co}, which is negligible compared to A_{ex}, is the molar area of the solid or liquid bulk phase, placed at the interface. This formula was used in earlier studies (Cary and Rideal, 1925) and has been used by later investigators (Boyd, 1958; Harkins and Nutting, 1939; Boyd and Schubert, 1957; Brooks and Alexander, 1962a).

The derivation of the Clausius–Clapeyron equation can be carried out on more explicit grounds (Alexander and Goodrich, 1964). The interfacial or spreading (s) region is a mixture of n_w^s moles of subphase, i.e., water (w), and n_i^s moles of lipid (i), separated mathematically from the adjacent bulk phases by boundaries drawn as dotted lines in Fig. 4.16. The thermodynamic relationships will be independent of the location of the boundaries as long as they enclose the entire region in which deviations from bulk-phase properties are to be expected (Guggenheim, 1951; Chattoraj and Birdi, 1984).

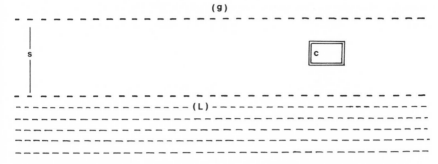

FIGURE 4.16. Mathematical separation of interfacial region (s) from the gas phase (g) and the liquid phase (L). (c) Bulk crystal phase. From Alexander and Goodrich (1964).

At equilibrium, the intensive quantities will be uniform throughout the system. If the temperature, T, is changed by dT, then the equilibrium will be reestablished, and the changes in the intensive properties of the system will be related by the Gibbs–Duhem equation:

$$S^s \, dT - V^s \, dP + A \, d\gamma_{eq} + n_w^s \, d\mu_w + n_i^s \, d\mu_i = 0 \tag{4.88}$$

$$d\mu_w = -\bar{S}_w^L \, dT + \bar{V}_w^L \, dP \tag{4.89}$$

$$d\mu_w = -\bar{S}_w^g \, dT + \bar{V}_w^g \, dP \tag{4.90}$$

$$d\mu_i = -\bar{S}_i^c \, dT + \bar{V}_i^c \, dP \tag{4.91}$$

where S^s and V^s are the total entropy and volume, respectively, of this interface of area A, γ_{eq} is the equilibrium surface tension, P is the vapor pressure of water, and μ_w and μ_i are the chemical potentials of water and lipid. The entire system is composed of vapor (g), liquid phase (L), crystal (c), and interface (s). In equations (4.89)–(4.91), the bars denote the partial molar quantities. By combining these equations and neglecting the pressure changes, one obtains

$$d\gamma_{eq}/dT = -(S^s - n_w^s \bar{S}_w^L - n_i^s \bar{S}_i^c)/A \tag{4.92}$$

which is the Clausius–Clapeyron equation relating the temperature coefficient of the surface tension, γ_{eq}, to the temperature of the film-covered surface. The enthalpy change can be written as

$$d\gamma_{eq}/dT = -(H^s - n_w^s \bar{H}_w^L - n_i^s \bar{H}_i^c)/TA \tag{4.93}$$

$$= \Delta H_m/TA \tag{4.94}$$

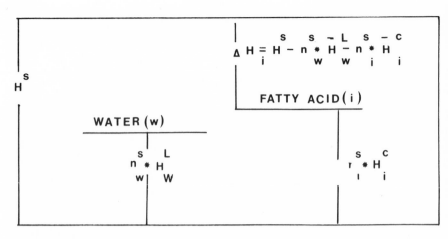

FIGURE 4.17. Monolayer formation from the elements of the bulk phases (monolayer ↔ bulk) at 25°C. From Jalal (1978).

The numerator in equation (4.94), ΔH_m, is the enthalpy of formation of a film-covered interface from its bulk constituents; i.e., it is the enthalpy change involved in bringing n_w^s moles of substrate from the bulk phase (L) and n_i^s moles of film-forming molecules from its bulk phase [crystal (c) or liquid] into the interface (s) (Fig. 4.17).

From similar considerations, the enthalpy of formation of a pure substrate interface is given as

$$d\gamma_0/dT = -(H_0^s - n_0^s \bar{H}_w^L)/TA \qquad (4.95)$$

where γ_0 and H_0^s are the surface tension and the enthalpy of a pure liquid interface, respectively. Further

$$\pi_{eq} = \gamma_0 - \gamma_{eq} \qquad (4.96)$$

Then one obtains

$$d\pi_{eq}/dT = d\gamma_0/dT - d\gamma_{eq}/dT \qquad (4.97)$$

From these equations, one can derive

$$TA(d\pi_{eq}/dT) = [(H^s - H_0^s) - (n_w^s - n_0^s)\bar{H}_w^L - n_i^s \bar{H}_i^c] \qquad (4.98)$$

This is the enthalpy of formation of a film-covered interface of area A minus the enthalpy of an equivalent area of pure substrate. Thus, the quantity

Q_s actually refers to a two-step isothermal reversible process:

1. n_w^s moles of w and n_i^s moles of i are transferred from their bulk phases, L and c, to the interface, s, and mixed.
2. n_0^s moles of w are transferred from a clean interface, s, to L.

The heat of fusion, ΔH_f, of the crystal can be derived as

$$\Delta H_f = (\bar{H}_i^l - \bar{H}_i^c) \tag{4.99}$$

$$= \{(H_l^s - H_c^s) - \bar{H}_w^L(n_{w,l}^s - n_{w,c}^s)$$

$$+ T_m A[(d\pi_{eq}^c/dT) - (d\pi_{eq}^l/dT)]\}/n_i^s \tag{4.100}$$

If the films in equilibrium with lipid crystals and liquid at the melting point, T_m, are identical, then $H_l^s = H_c^s$ and $n_w^{l,s} = n_w^{c,s}$, from which one obtains

$$\Delta H_f = T_m A_m[(d\pi_{eq}^c/dT) - (d\pi_{eq}^l/dT)] \tag{4.101}$$

where A_m is the molecular area at π_{eq} at T_m and $A = n_i^s A_m$. The magnitude of A_m can only be estimated from π vs. A isotherms.

This relationship has been derived rigorously when the lipid crystal and liquid are in equilibrium with identical films. The magnitude of ΔH_f can be determined by the independent calorimetric method, which allows one to verify the relationship in equation (4.101). Typical π_{eq} vs. T data are given in Fig. 4.3. The data in Table 4.10 show the comparison of ΔH_f from monolayer measurements and calorimetry. The magnitude of A_m in the aforementioned films was estimated from the plots of A_{tr} vs. temperature (Fig. 4.18). Accordingly, the magnitudes of π_{eq} are compared with those

TABLE 4.10. T_m, A_m, $d\pi_{eq}^c/dT$, $d\pi_{eq}^l/dT$, ΔH_f Data for Various Fatty Acid Monolayers[a]

Fatty acid	T_m (°C)	$d\pi_{eq}^c/dT$	$d\pi_{eq}^l/dT$	ΔH_f[b] (kcal/mole)	$\Delta H_{f,calorim.}$ (kcal/mole)
Myristic	55	0.71	−0.16	10.7	10.7
				(14.3)	
Palmitic	63	0.86	−0.17	12.6	12.9
				(16.6)	
Stearic	71	1.01	−0.19	14.6	14.9
Oleic	13	0.77	−0.10	9.1	9.6

[a] From Jalal (1978). $A_m = 26$ Å^2; ΔH_f [from π_{tr} (midpoint of the A_{ex}-A_{co} transition) $- T$] $= -1.80 + 1.15 n_C$; $\Delta H_{f,calorim.} = -3.23 + 1.0 n_C$.
[b] Values calculated from π_{tr} are in parentheses.

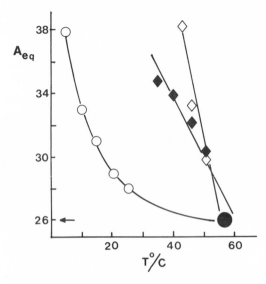

FIGURE 4.18. Plots of A_{tr} (area at the transition between the L_{ex} and L_{co} states) vs. temperature for monolayers of myristic acid ($\bigcirc$), palmitic acid ($\blacklozenge$), and stearic acid ($\diamondsuit$). Data from Jalal (1976).

of π_{tr} for different fatty acids in Fig. 4.19. It can be seen that the magnitudes of the quantities are comparable ($d\pi_{eq}^s/dT \approx d\pi_{tr}/dT$). The differences among ΔH_f values for the series of fatty acids are the same when π_{tr} vs. T data are used (Table 4.10).

The formation of film-covered interfaces from bulk-phase constituents always involves an absorption of heat. The spreading process from a crystalline lipid is always an endothermic process, but spreading from molten or liquid crystalline lipid is exothermic. Furthermore, the magnitude of H_s spreading was reported to be approximately half of ΔH_f for trilaurin, while for di-C_{12}-PE monolayers, these two quantities were almost equal

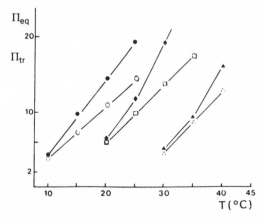

FIGURE 4.19. Variation of π_{eq} ($\bigcirc$, $\square$, $\triangle$) and π_{tr} ($\bullet$, $\blacklozenge$, $\blacktriangle$) (from π vs. A isotherms) with temperature for myristic acid ($\bigcirc$, $\bullet$), cis-3-decanoic acid ($\square$, $\blacklozenge$), and palmitic acid ($\triangle$, $\blacktriangle$). From Chattoraj and Birdi (1987).

(Phillips and Hauser, 1974). These comparative data allow one to conclude that fusion of trilaurin gives rise to a three-dimensionally disordered melt, while chain melting in the case of PE occurs in the two-dimensional bilayer. This conclusion is supported by the finding that the heat of chain melting in mesomorphic structures is smaller by a factor of approximately 2 than those of ordinary fusion processes (Phillips *et al.*, 1969). The molar entropy of spreading, S_s (Eriksson, 1972)

$$S_s = A(d\pi_{eq}/dT) \tag{4.102}$$

is related to the changes in the configurational freedom of the hydrocarbon part of the lipid. For instance, in the case of a rigid molecule such as cholesterol, the change in π_{eq} with temperature is negligible; i.e., the entropy of spreading, S_s, is zero. The entropy changes on spreading indicate that the monolayer is less ordered than the crystal (bulk phase), i.e., $S_s > 0$, but more ordered than lipid phases containing melted hydrocarbon chains (i.e., $S_s < 0$). The degree of order of liquid-expanded lipid monolayers is intermediate between those of bulk melts and anhydrous liquid crystalline states (see Fig. 4.20). The entropy of spreading, S_s, of β-trilaurin (7.5 J/mole/°C per chain C atom) is less than the entropy of fusion (10.9 J/mole/°C per chain C atom). On the other hand, the magnitude of S_s for di-C_{12}-PE (6 J/mole/°C) is greater than the chain-melting entropy (5 J/mole/°C) from β-crystal to anhydrous liquid crystal. The data on equilibrium surface

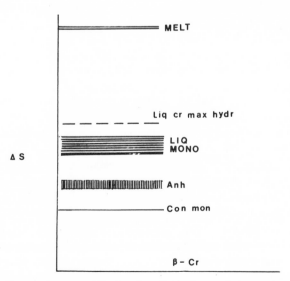

FIGURE 4.20. Entropy increase, ΔS, of various lipid phases relative to the entropy of the β-crystal form. Redrawn from Hauser (1974) with changes.

pressure, π_{eq}, of different lipids, e.g., fatty acids, fatty alcohols, dialkyl lipids, and trialkyl lipids, have not been critically analyzed in the current literature (Phillips and Hauser, 1974; Birdi, 1982 and unpublished). Since the data on π_{eq} vs. T (temperature) for various lipids are not in accord, these data are described separately for different types of lipids:

1. *Fatty Acid Monolayers.* The π_{eq} vs. T (temperature) isotherms were analyzed very extensively in a few recent studies (Jalal, 1978). These investigations clearly showed that:
   - The T_{tr} was the same as the bulk melting point, T_m (see Fig. 4.2).
   - The magnitudes of ΔH_f were the same for monolayers as found for the bulk lipid by calorimetry.
2. *Fatty Alcohol Monolayers.* Although the π_{eq} vs. T plots were the same, in general, as found for the fatty acids, the breaks in the plots were at lower temperatures than the T_m (see Fig. 4.3) (Birdi *et al.*, to be published):
   - $T_{tr} < T_m$.
   - Near the T_m, the monolayers showed strange plots and a weak break at the bulk melting point (π_{eq}/dT was dependent on n_C).
   - ΔH_f as calculated from the monolayer data was dependent on n_C (Birdi and Sanchez, to be published).
   π_{eq} vs. T plots for other *n*-alkoxy propanols (C_{16}-OC_3H_6OH, C_{18}-OC_3H_6OH, C_{20}-OC_3H_6OH, C_{22}-OC_3H_6OH) and oxy butanols (C_{18}-OC_4H_8OH) did not show any breaks (Katti and Sansare, 1970).
3. *Monoalkyl Chain Glycerides.* π_{eq} vs. T isotherms of 1-MP and 2-MP gave the following results:
   - $T_{tr} < T_m$. It was reported that the break in π_{eq} vs. T corresponded to the phase change when 5%–10% water is bound in monoglycerides (Lawrence, 1969).
   - ΔH_f was greater for 1-MP than for 2-MP.
4. *Dialkyl Chain Lipids.* The dialkyl lipids can be classified into two categories: (a) lecithins and (b) other lipids:
   a. The diglycerides (and triglycerides) show the following characteristics:
      - The π_{eq} vs. T isotherms show breaks at the fusion temperature. The β-form of trilaurin melts at 46°C, and the break in the isotherm is observed at the same temperature. The same correlation is observed for 1,3-dilaurin (β-form) with melting temperature at 57°C. The di-C_{12}-PE shows a break in π_{eq} vs. $T = 63$°C, which is identical with the chain-melting process in the bulk phase.
   b. The π_{eq} vs. T data for lecithins (DML, DPL) were as follows (Fig. 4.14C):
      - The π_{eq} vs. T isotherms were completely different from those of all the lipids mentioned under (1)–(3).

- The magnitudes of ΔH_f were extremely high (2500 kJ/mole) as compared to the calorimetry data (Table 4.10) (Chattoraj and Birdi, 1984).

π_{eq} vs. T isotherms have been reported for various lipids, e.g., lauryl (C_{12}) glycerides (1,3-dilaurin, trilaurin), PE (di-C_4-PE, di-C_{12}-PE, di-elaidoyl-PE), lecithins (DML, di-C_{12}-lecithin, dioleoyl lecithin), and cholesterol (Phillips and Hauser, 1974). In the case of glycerides and PE, the magnitude of π_{eq}/dT was found to be positive below the temperature at which monolayer transition took place. It was also observed that longer-chain lipids spread at higher temperatures. Lecithins were reported to spread only when $T > T_{cr}$, where T_{cr} is the crystal (or gel)–liquid-crystal transition temperature at maximum hydration. For example, DPL was found to give negligible π_{eq} at 22°C even after 24 hr. The cholesterol π_{eq} data were completely different from those for all the other lipids. No change was found in π_{eq} ($= 0$) with temperature from 25 to 50°C. The investigators were aware of the differences in these monolayer data.

It has been asserted that, in principle, measurements of π_{eq} as a function of temperature indicate the existence of all phase transitions, thus providing a thermodynamic basis for calculating the transition energies. Therefore, it was surprising to find that the π_{eq} vs. T plots of DML, DPL, and DSL were not in accord with this statement.

The DML data in particular were analyzed in further detail (Gershfeld and Tajima, 1979), at a higher sensitivity. The coexistence line for the equilibrium of surface film with bulk DML, either gel or liquid crystal, was estimated. T_c was estimated to be 23.5°C for DML. At temperatures below T_c, the bulk gel state is in equilibrium with a gaseous surface film. At temperatures above T_c, the liquid crystal is in equilibrium with a gaseous film that changes with increasing temperature into a liquid-expanded film. The curve for the supercooled liquid-expanded film was estimated. The phase that gives the equilibrium between liquid crystal and liquid-expanded film was described as exhibiting a rapid increase in π_{eq} with temperature. This increase has not been observed with any of the single-chain lipids (to our knowledge!), but one would expect π_{eq} to reach some finite limiting value at high temperature, i.e., greater than T_c. However, there is no way of analyzing these data as was delineated above for the single-chain lipids (except fatty acids).

The heat of fusion from crystal to melt depends on the particular polymorphic form of the liquid crystal, and the magnitude of $d\pi_{eq}/dT$ will reflect this state of equilibrium (Phillips and Hauser (1974)).

In the case of phospholipids, which are found to swell and take up water only when $T > T_{tr}$, the chain-melting process for bulk lipid at the air–water (or oil–water) interface is observed only from crystal to hydrated liquid crystal, and ΔH_f is therefore not the simple heat of chain melting.

However, it is the data on the π vs. A isotherms of fatty acids, fatty alcohols, and the dialkyl lecithins that provide the correct T_{tr}, as well as ΔH_f (Table 4.10). From the π_{tr} vs. T plots of DML (see Fig. 4.14B), one finds that $\Delta H_f \approx 10$ kcal/mole [= 7.25 kcal/mole by calorimetry (Table 4.10)]. The π vs. A isotherms thus provide the correct enthalpy estimation of the phase transition, while the π_{eq} data are not acceptable.

PHASE TRANSITION IN BRANCHED-CHAIN LIPIDS

Methyl iso- and anteiso-branched saturated fatty acids occur widely as compounds of the membrane lipids of prokaryotic microorganisms (Kaneda, 1977). In addition, branched-chain fatty acids are able to support the growth of several fatty acid-auxotrophic microorganisms in which these fatty acid classes do not naturally occur. Model membranes composed of di-isoacyl or di-anteisoacyl PCs, and natural membranes enriched in phospholipids containing branched-chain fatty acids, exhibit decreased gel-to-liquid-crystalline phase-transition temperatures in comparison to membranes containing unbranched saturated fatty acids (Silvius and McElhaney, 1980; Silvius et al., 1980). From these investigations, it has been suggested that phospholipids containing branched-chain fatty acids, and particularly methyl anteiso-branched fatty acids, act as "membrane fluidizers," in the same manner as phospholipids containing cis-unsaturated fatty acyl groups behave (McElhaney, 1976).

In Fig. 4.21, the molecular areas occupied by monolayers of DPPC, 1,2-diisoheptadecanoylphosphatidylcholine (DIHPC), and 1,2-dianteisoheptadecanoylphosphatidylcholine (DAHPC) are given as a function of temperature at a constant surface pressure ($\pi = 25$ mN/m) (Kannenberg et al., 1983). As described above, this method will not provide a complete phase-transition analysis, since the magnitude of π is kept constant. However, it can be seen that a solid-to-liquid transition takes place at 38°C for DPPC (compare with 41°C for bilayers). A similar transition is observed for DIHPC at 27°C [compare with the differential scanning calorimetry (DSC) transition at 27.6°C]. DAHPC shows a phase transition around 10°C by DSC, and there is a very weak change in slope in Fig. 4.21 around 10°C.

These data indicate that there is satisfactory agreement between the two different systems, as already described above for other straight-chain lipids. However, it is of further interest here to mention that at temperatures below the transition temperature, T_{tr}, the magnitude of A (solid) for branched-chain lipids is greater than A (solid) for straight-chain DPPC, as expected. These values for A (solid) are 51, 54, and 57 Å^2 for DPPC, DIHPC, and DAHPC, respectively. These data indicate that branched-chain fatty acyl groups are significantly more disordered in the solid state than are PC molecules containing unbranched saturated hydrocarbons.

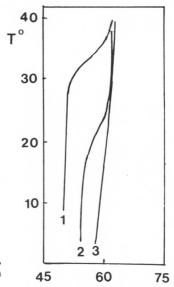

FIGURE 4.21. Variation of A vs. T for DPPC (1), DIHPC (2), and DAHPC (3). Redrawn from Kannenberg *et al.* (1983) with changes.

4.9. THEORETICAL MODELS FOR TWO-DIMENSIONAL PHASE TRANSITION

There has been much interest in the past few decades in theoretical models of fundamental physical properties as related to cooperative phenomena in lipid-phase transition in monolayers, bilayers, and membranes. In these models, the most important points that need to be considered are the contribution of entropy to the first-order endothermic melting transition and the ability to characterize the roles played by the various forces (between the lipid and the aqueous media).

One of the models (Marcelja, 1974; Jahnig, 1979) makes the principal assumption that the dominant interaction is the anisotropic part of the attractive van der Waals interaction between hydrocarbon chains in monolayers. The statistical problem was treated in a molecular-field approximation, and the statistical averages were numerically evaluated using an exact summation over all statistical configurations of a single chain in the average molecular field of the neighboring chains. The internal energy of the chain was written as

$$E(\text{conf}) = E_{\text{intra}} + E_{\text{disp}} + \pi A \qquad (4.103)$$

where E_{intra} is the intramolecular energy of a single chain and E_{disp} gives the dispersive or van der Waals anisotropic interaction. On the basis of

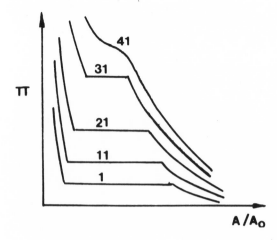

FIGURE 4.22. π vs. A/A_0 (relative area per molecule) for different temperatures. Redrawn from Marcelja (1974) with changes.

these considerations (Marcelja, 1974), π vs. A isotherms were generated (Fig. 4.22).

These isotherms show horizontal regions, a finding that is only qualitatively in agreement with the experimental data. However, this molecular-field theory cannot explain the observed rounding off of the π vs. A isotherms in the coexistence region. It has been pointed out (Jahnig, 1979) that the drawback in this theory is that it involves a large amount of computer processing and that this tends to obscure the physical meaning.

The model can be simplified in such a way that it becomes exactly solvable, while retaining the necessary essentials, e.g., steric repulsions, rotational isomerization, and attractive interactions (Nagle and Scott, 1978; Nagle, 1980). This approach leads to a first-order phase transition. However, this model is not the closest to reality, even though a sharp transition is obtained.

It has been observed that an enhanced transbilayer permeability occurs for Na^+ diffusion through DPL vesicles, near the phase-transition temperature region (i.e., 41 ± 10) (Wu and McConnell, 1973; Papahadjopolous *et al.*, 1973). This diffusion enhancement was attributed to the possible increase of the lateral compressibility in the region where fluid and solid phases coexist (Linden *et al.*, 1973), whereas other investigators (Doniach, 1978) suggested that this increase in diffusion probably results from the increase in critical fluctuations in the bilayer as the critical temperature is approached.

A molecular-dynamics calculation of the melting of a two-dimensional array of dumbbells was used to describe phase-transition phenomena (Cotterill, 1976). In this model, each dumbbell represents the projection of a lipid molecule into the bilayer, with each knob representing a hydrocarbon chain. The interaction between lipid molecules was presented by a combination of Lennard-Jones 6-12 potential and a screened Coloumb potential.

In a recent report (Caille *et al.*, 1980), a two-state Ising model was described and shown to give a good qualitative description of the main-phase transition of both mono- and bilayers and of statistical fluctuations about this transition.

In a later study, a theory was presented that explained the thermodynamically controlled striplike shapes of two-dimensional solid crystal domains as observed in phospholipid monolayers in the presence of trace amounts of cholesterol (Galler *et al.*, 1986). In this theory, dipole-dipole repulsions between lipid molecules were assumed to favor the elongation of the domains into long strips, but this elongation was supposed to be opposed by increasing interfacial free energy (line tension) associated with the parameter of the domains. An expression was derived for the dependence of line tension on the concentration of cholesterol in the monolayer, and this relationship was used to compare the predictions of the theory with experimental measurements of the width of domains as a function of monolayer compression. The theory was found to agree well at high compression, but there were found to be deviations from experimental data at low compression, where the domains were found to be short and end effects neglected in the theory became apparent.

The softening of lipid bilayer membranes in undergoing a change from the gel to the fluid phase with temperature was studied by computer simulation (Mouritsen, 1983, 1984; Mouritsen *et al.*, 1983; Mouritsen and Zuckermann, 1985; Tallon and Cotterill, 1985). These analyses also suggested the presence of nonhorizontal isotherms of lipid monolayers.

4.10. MIXED MONOLAYERS

Monolayers consisting of two or more film-forming components are of much interest in many systems (biological: lung fluid, membranes; technical: emulsions, food industry). Analogous to "mixed liquids or solids" in three-dimensional systems, these mixed films would be expected to provide much useful information as regards their orientation, packing, and interactive forces and mechanisms. However, because of certain inherent difficulties in the description of mixed monolayer data, there remains much to be resolved.

In any two-component monolayer, one can write the following expression for the average area per molecule, $A_{1,2}$, for the "ideal" mixed film consisting of two components, 1 and 2 (neglecting the substrate), at any surface pressure, $\pi_{1,2}$ (Adamson, 1982; Gaines, 1966; Chattoraj and

Birdi, 1984):

$$A_{1,2} = X_1 A_1 + X_2 A_2 \qquad (4.104)$$

where A_1 and A_2 are the areas per molecule in pure monolayers of components 1 and 2, respectively, and X_1 and X_2 are the mole fractions of these film-forming components. The excess functions can be written as for real films ($\pi_{1,2}$ and T constant):

$$A_{ex} = A_{1,2} - (X_1 A_1 + X_2 A_2) \qquad (4.105)$$

In a similar way, one can write the expressions for $\pi_{1,2}$:

$$\pi_{1,2} = X_1 \pi_1 + X_2 \pi_2 \qquad (4.106)$$

and the excess function:

$$\pi_{ex} = \pi_{1,2} - (X_1 \pi_2 + X_2 \pi_2) \qquad (4.107)$$

In these mixtures, "ideal" films for which the terms A_{ex} and π_{ex} are zero would be characterized. The expression for the excess free energy of mixing, ΔG_{ex}, has been given as (Goodrich, 1957; Gaines, 1966; Defay *et al.*, 1966; Goodrich, 1974; Chattoraj and Birdi, 1984)

$$\Delta G_{ex} = \int_0^\pi [A_{1,2} - (X_1 A_1 + X_2 A_2)] \, d\pi \qquad (4.108)$$

where the contribution from the substrate was neglected (see below). The mixed films are thus investigated by using either of these four relationships to determine the interaction between the two (or more) components. No systems with more than two components have been reported, but by keeping the ratio X_1/X_2 constant and varying X_3, one can determine the interactions [as was delineated for surface excess in three-component systems (Chattoraj and Birdi, 1984)].

In an "ideal" mixed film, a plot of $A_{1,2}$ or $\pi_{1,2}$ vs. X_1 would be a straight line. On the other hand, any deviation from a straight line would merely indicate nonideal behavior.

In a recent rigorous analysis of the thermodynamics of mixed films (Motomura, 1980), it was pointed out that earlier descriptions were not sufficiently exact (Fowkes, 1962; Gaines, 1966; Joos, 1969; Eriksson, 1976) as regards the role of the subphase.

It is known that the collapse pressure, π_{col}, of a film is characteristic of the film-forming substance as well as the degree of interaction between the lipid or polymer and the subphase molecules; thus, π_{col} is one of the most sensitive measurements obtained from monolayer studies. In the case of a two-component film, if the two components are "immiscible," then the π vs. A isotherm would exhibit the following behavior:

1. The individual components would be present separately, and as the value of π reaches a value equivalent to $\pi_{col,1}$ (if component 1 has the lower π_{col}), one will observe a collapse state.
2. The second component, i.e., the one with the higher π_{col}, would then be ejected at the corresponding surface pressure.

Since lipid–biopolymer mixed films gave results that were much different from those given by lipid films, the former are described elsewhere (see Chapters 5 and 8).

π_{eq} OF MIXED LIPID FILMS

Mixed films can also be investigated by measuring the π_{eq} (Tajima and Gershfeld, 1978; Chattoraj and Birdi, 1984; Birdi and Sanchez, to be published). The π_{eq} of DML–cholesterol (CH) and dioleoyl lecithin (DOL)–CH at 29.5°C gave interactions only at certain molar ratios. The magnitude of π_{eq} was constant and equal to the value of pure lecithin (DOL or DML) when the mole fraction of CH was 0–0.33. The phase rules (see Section 4.6) predict that two bulk lipid phases coexist, i.e., pure lecithin and lecithin–CH complex. Through rigorous thermodynamic analysis, it was established that lecithin (at 29.5°C, where both DML and DOL are above the lipid-phase transition temperatures) and CH form a stable 2:1 complex (Fig. 4.23). This finding may have some biological significance, since many membrane processes are suggested to be related to cholesterol content (see Chapter 7).

The magnitudes of π_{eq} as given in Fig. 4.23 are compared with collapse pressures of the same mixed films reported by other investigators. These data show that $\pi_{eq}^{1} = \pi_{col}$ when the experimental temperature is greater

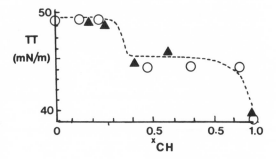

FIGURE 4.23. π_{eq} (○) and π_{col} (▲) vs. mole fraction of cholesterol (x_{CH}) in mixed films of DML–CH at 25°C. Data on π_{eq} from Tajima and Gershfeld (1978); data on π_{col} from Chattoraj and Birdi (1984).

than T_m, i.e., the phase-transition temperature. Furthermore, the π_{col} data show that the criticism that π_{col} is greater than π_{eq} is invalid under these conditions. It thus suffices to conclude that in mixed monolayers, both $(\pi^1_{eq} - T)$ and $(\pi_{col} - T)$ provide the same information when T is greater than T_{tr}. The differences observed at T less than T_{tr}, where π^s_{col} is greater than π^s_{eq}, arise from the incorrect thermodynamic considerations in the literature.

4.10.1. MISCIBILITY IN MONOLAYERS

As already mentioned in Section 4.6, it is clear that one needs to proceed as is generally done for bulk mixed phases.

As also mentioned earlier, lipid monolayers spread at the air–water interface are composed of air, water, and film-forming substances, which exist in equilibrium with bulk phases. To be able to make a consistent thermodynamic analysis of a monolayer system, the contribution of both air and water must be taken into consideration. However, it is well accepted that it is difficult to make systematic studies of these quantities (Motomura, 1980). In a recent study, approximately 80 different kinds of two-component mixed monolayer systems, composed of various types of hydrophilic and hydrophobic groups, were investigated. The effects of composition and temperature were investigated in these mixed monolayers. When the two components in the monolayer are immiscible in either the condensed state or the expanded state of a two-component mixed monolayer state, three monolayer phases exist at the transition point. According to the phase rule (see Section 4.6), the number of degrees of freedom in this case is zero under constant T and P conditions. Therefore, in such a case, the transition π_{tr} should be constant and independent of the monolayers. On the basis of this observation, it is thus useful to investigate whether the two components in the system are miscible or immiscible, both in the expanded and in the condensed state. Mixed monolayers were analyzed according to the thermodynamic treatment described below (Motomura, 1980).

Thermodynamics of Mixed Lipid Monolayers. An expression for the apparent molar thermodynamic quantity change, y^γ, associated with the phase transition from an expanded to a condensed monolayer can be written as

$$y^\gamma = x_1^{\pi,e}(\bar{y}_1^{\gamma,c}\bar{y}_1^{\gamma,e}) + x_2^{\pi,c}(\bar{y}_2^{\gamma,c} - \bar{y}_2^{\gamma,e}) \tag{4.109}$$

where

$$\bar{y}_1^\gamma = (dY/dn_1)_{T,p,\gamma,n_{j_1}}, \quad \text{where } i = 1, 2 \quad (j \neq i) \tag{4.110}$$

where Y is the thermodynamic quantity inherent in the monolayer and γ is the surface tension (Motomura, 1980).

Since the two components are immiscible in the condensed state, the apparent molar entropy changes of the first and second components are obtained by the following equations (Motomura *et al.*, 1976):

$$s_1^\gamma = [a_1^c - a^e + x_2^{\pi,e}(da^e/dx_2^{\pi,e})]_{T,p,\pi}$$
$$\times [(d\pi^{eq}/dT)_{p,x_2^{\pi,e}} - (d\gamma_0/dT)_p] \qquad (4.111)$$

$$s_2^\gamma = [a^c - a^e - x_1^{\pi,e}(da^e/dx_2^{\pi,e})_{T,p,\pi}]$$
$$\times [(d\pi^{eq}/dT)_{p,x_2^{\pi,e}} - (d\gamma_0/dT)_p] \qquad (4.112)$$

The apparent molar energy changes of the first and second components are obtained by the following equations:

$$u_1^\gamma = -(\pi^{eq} - \gamma_0)[a_1^c - a^e + x_2^{\pi,e}(da^e/dx_2^{\pi,e})_{T,p,\pi}] + Ts_1^\gamma \qquad (4.113)$$

$$u_2^\gamma = -(\pi^{eq} - \gamma_0)[a_2^c - a^e - x_1^{\pi,e}(da^e/dx_2^{\pi,e})_{T,p,\pi}] + Ts_s^\gamma \qquad (4.114)$$

In this study (Matuo *et al.*, 1981), some 80 different mixed two-component monolayers were investigated. These different monolayers gave rise to π vs. A isotherms that were classified as described below.

CIGAR-TYPE MIXED ISOTHERMS

Mixed monolayers of tetradecanoic acid ($C_{14}OOH$)-*n*-pentadecanoic acid ($C_{15}OOH$) (Siketa *et al.*, 1976; Kuramoto *et al.*, 1972) and ethyl-hexadecanoate (EH)-ethylheptyldecanoate (EHP) systems are composed of similar molecules, and they are reported to form expanded films at the experimental temperature. The hydrophilic groups are equal in these systems, and the difference in carbon number between the hydrophobic groups is only one $-CH_2-$. These two systems gave π vs. A isotherms that were designated as "cigar-type" phase diagrams (Fig. 4.24b).

This type of data is for ideal mixed systems. The monolayers are composed of molecules that form expanded films. The two components are miscible in the expanded as well as in the condensed state. In the data in Fig. 4.25, the mixed monolayers give π vs. A curves almost the same as that of the pure components, in which the π_{tr} is lower than that of the pure component, up to the transition point. However, in the case of $C_{15}OOH$-$C_{14}OOH$ (Matuo *et al.*, 1978) mixed films (Fig. 4.25), the mixed isotherms are different.

The excess apparent molar energy change is nearly equal to zero, i.e., within the experimental error, over the whole range of composition.

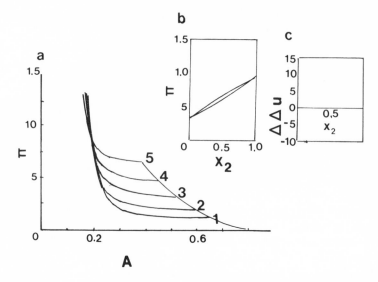

FIGURE 4.24. (a) π vs. A isotherms of EH–EHP mixed monolayers at 294.2°K. (1) $x_2^{\pi=0}$ (EHP); (2) 0.2, (3) 0.5, (4) 0.8, (5) 1 (EH). (b) Phase diagram of EH–EHP mixed monolayers at 298.2°K. (c) $\bar{\mu}^\gamma - \mu^\gamma$ vs. $x_2^{\pi,c}$ plot of EH–EHP mixed monolayers at 296.2°K. Redrawn from Matuo *et al.* 1978 with changes.

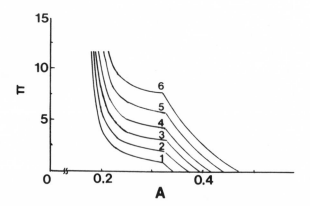

FIGURE 4.25. π vs. A isotherms of mixed monolayers of $C_{15}OOH$–$C_{14}OH$ at 25°C. (1) $x_1^{\pi=0}$ ($C_{14}OH$); (2) 0.2, (3) 0.4, (4) 0.6, (5) 0.8, (6) 1 ($C_{15}OOH$). Redrawn from Matuo *et al.* (1978) with changes.

MODIFIED CIGAR-TYPE ISOTHERMS

In Fig. 4.26, π vs. A plots of $C_{14}OOH-C_{17}OOH$ (Sekita *et al.*, 1976; Kuramoto *et al.*, 1972) are given. Mixed monolayers in which the molar fraction of $C_{14}OOH$ is greater than 0.5 give expanded states and show transition pressures, whereas in those films with molar fractions of $C_{14}OOH$ less than 0.5, the mixed films are considerably condensed and the π_{tr} disappears. The expanded phase is not present in mixed monolayers in which the molar fraction of $C_{14}OOH$ is less than 0.5. Thus, $C_{14}OOH-C_{17}OOH$ are not mixed completely in this region, and the mixed monolayer changes from an expanded to a condensed monolayer at low π. The phase diagram of these mixed systems is shown in Fig. 4.26b, and it can be seen that phases are different than in the cigar-type mixed films. Similar mixed monolayers have also been observed for mixed $C_{14}OH-C_{16}OH$ monolayers (Matuo *et al.*, 1978).

The miscibility of the two components was found to decrease gradually with an increase in the difference of chain length between the components. The apparent molar energy change, Δu^{γ}, of this system is given in Fig. 4.26c. It is found that Δu^{γ} does not change linearly with composition, and this behavior is different from that of the cigar type (Table 4.11).

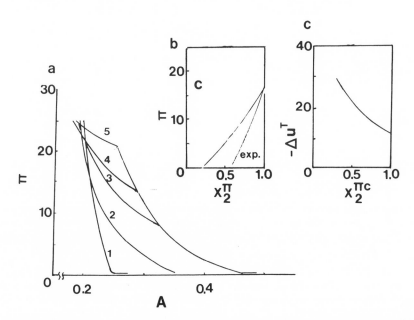

FIGURE 4.26. (a) π vs. A isotherms of mixed monolayers of $C_{14}OOH-C_{17}OOH$ at 25°C. (1) $x_2^{\pi=0}$ ($C_{17}OOH$); (2) 0.33, (3) 0.67, (4) 0.8, (5) 1 ($C_{14}OOH$). (b) Phase diagram at 21°C. (c) $\Delta\mu^{\gamma}$ vs. $x_2^{\pi,c}$. Redrawn from Sekita *et al.* (1976) with changes.

TABLE 4.11. Various Types of Phase Diagrams of Two-Component Mixed Lipid Monolayers at 298.2°K (25°C)[a]

Cigar type	Modified cigar type	Positive azeotropic	Negative azeotropic	Eutectic	Complex
(e)-(e)	(e)-(c)	(e)-(e)	(e)-(e)	(e)-(e)	(e)-(e)
C_{13}-C_{14}	C_{14}-C_{17}	C_{14}-EH	EH-PA	12HOA-C_{13}	C_{14}-BED
C_{14}-C_{15}	C_{14}-C_{18}	C_{14}-EHP	EH-HA	12HOA-C_{14}	C_{15}-PA
EH-EHP	C_{15}-C_{16}	C_{14}-PA	EH-HPA	12HOA-C_{15}	C_{15}-HA
—	C_{14}OH-C_{14}OH	C-HA	EH-HPR	—	C_{15}-DPL
C_{14}-C_{14}OH	DPL-DSL	C_{14}-HPA	EHP-HA	(e)-(c)	—
C_{15}-C_{14}OH	—	C_{14}-HPR	—	12HOA-C_{16}	(e)-(c)
HA-HPR	C_{14}-C_{16}OH	C_{15}-EH	HA-MP	12HOA-C_{18}	C_{14}-C_{19}
—	C_{15}-C_{16}OH	C_{15}-EHP	—	—	C_{14}-OH-C_{18}
C_{13}-EH	—	C_{15}-HPA	C_{14}-DPL	C_{18}-DPL	HA-C_{16}
C_{14}-MP	EH-C_{17}	C_{15}-HPR	EH-DPL	C_{14}-DSL	—
—	EH-C_{18}	—	HA-DPL	C_{15}-DSL	—
—	HA-C_{17}	EH-MP	DPL-MP	EH-DSL	—
—	HA-C_{18}	—	—	HA-DSL	—
—	—	—	C_{14}-OAC	MP-DSL	—
—	MP-C_{17}	(e)-(c)	—	—	—
—	MP-C_{19}	EH-C_{16}	—	C_{14}-CH	—
—	—	—	C_{15}OH-CH	—	—
—	—	—	C_{14}OH-CH	—	—
—	—	—	EH-CH	—	—
—	—	—	MP-CH	—	—
—	—	—	—	DPL-CH	—
—	—	—	C_{14}-CHF	—	—
—	—	—	C_{14}OH-CHF	—	—

[a] From Matuo et al. (1981). (e) Expanded; (c) condensed; (C_n) fatty acid with n carbon atoms.

POSITIVE AZEOTROPIC TYPE

This type of system is characterized by a maximum point in the phase diagram. The molar fractions of the expanded and condensed phases are equal at this point. Hence, this point may be called a "two-dimensional positive azeotropic point." π vs. A isotherms and phase diagrams of the C_{15}OOH-EH system are shown in Figs. 4.27a and b.

A similar kind of behavior has been reported for C_{15}OOH-hexadecyl proponiate (HPR) (Matuo et al., 1981) and C_{14}OOH-EHP (Motomura et al., 1979) systems. In these systems, the lipids EH, HPR, and EHP were observed to have bulky parts in their hydrophilic groups. It is thus suggested that steric hindrance of the packing of the hydrophilic groups at the interface must be the cause of this nonideal behavior. The components in these

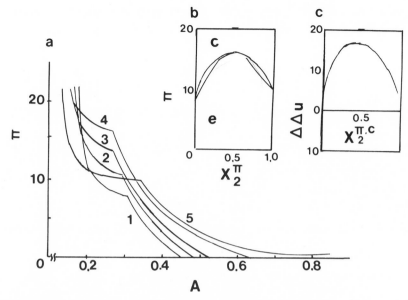

FIGURE 4.27. (a) π vs. A isotherms of mixed monolayers of $C_{15}OOH$-EH at 25°C. (1) $x_2^{\pi=0}$ ($C_{15}OOH$); (2) 0.1, (3) 0.2, (4) 0.5, (5) 1 (EH). (b) Phase diagram of $C_{15}OOH$-EH. (c) $\bar{\mu}^{\gamma} - \mu^{\gamma}$ vs. $x_2^{\pi,c}$. Redrawn from Matuo *et al.* (1978) with changes.

systems are miscible in the expanded as well as in the condensed state. The excess apparent molar energy change of the $C_{15}OOH$-EH system was found to be positive over the entire composition range, and gave a simple complex curve (Fig. 4.27c).

This analysis suggests that there may be interactions between the two compounds in the mixed monolayer that are weaker than the interaction between the pure component molecules.

NEGATIVE AZEOTROPIC TYPE

There is observed a minimum in the phase diagram, and at this point the composition of the expanded phase is equal to that of the condensed phase. Therefore, this point may be called a "two-dimensional negative azeotropic point." Systems that have been found to exhibit this behavior are EH–hexadecyl acetate (HA) (Matuo *et al.*, 1979, 1981), $C_{14}OOH$–DPL, and $C_{14}OOH$–octadecyl ammonium chloride (OAC) (Motomura *et al.*, 1980).

The π vs. A phase diagram of the EH–HA system and its apparent molar energy change are given in Fig. 4.28. Since the energy change is negative, it indicates that the mutual interactions between the components

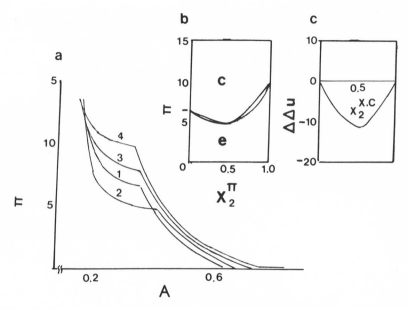

FIGURE 4.28. (a) π vs. A isotherms of mixed monolayers of EH–HA at 25°C. (1) $x_2^{\pi=0}$ (HA); (2) 0.4, (3) 0.9, (4) 1 (EH). (b) Phase diagram of EH–HA. (c) $\mu^\gamma - \bar{\mu}^\gamma$ vs. $x_2^{\pi,c}$. Redrawn from Matuo *et al.* (1978) with changes.

in the mixed monolayer are stronger than the interactions in the pure molecules themselves.

EUTECTIC TYPE

It has been shown (Motomura *et al.*, 1974, 1976) that $d\pi_{eq}/dx_2^{\pi,e} > 0$ requires that either $x_1^{\pi,c} = 1$ or $x_2^{\pi,e} > x_2^{\pi,c}$. In the former case, the two components are immiscible in the condensed state and the phase diagram is a eutectic type. In this case, the pure condensed monolayer of either component merges from an expanded two-component monolayer at a lower transition π, and another condensed monolayer is formed at a high transition π. Hence, three phases exist in the mixed monolayer at the high transition π. According to the phase rule (Motomura, 1974), the number of degrees of freedom is zero under constant temperature and pressure at the high transition π. Accordingly, there should exist two kinds of transition points in the π vs. A isotherm. Furthermore, the magnitude of π at the high transition point should be constant. In the latter case, the condensed film is composed of a homogeneous mixture of the two components, and various types of phase diagrams are considered, such as cigar type, modified cigar type, positive and negative azeotropic types, and other complicated types.

If the lipid has two hydrophilic groups, e.g., as in 12-hydroxy-octadecanoic acid (12HOA), in which one is a carboxyl group and the other is a hydroxyl group, then the mixed π vs. A isotherms are found to be different. The π vs. A isotherms of $C_{15}OOH$–12HOA were found to give two transition points, and the surface pressure of the high transition point was found to be constant. The eutectic-type diagram is therefore ascribed to the decreased miscibility of two components because of the hydroxyl group of 12HOA (Fig. 4.29). Other systems that are reported to show similar types of eutectic phase diagrams are $C_{18}OOH$–DPL, $C_{14}OOH$–DSL, and $C_{14}OOH$–CH. Most of these systems generally arise when there is a combination of expanded-state and condensed-state systems. In both DSL and CH, which are known to give condensed films, eutectic-type mixed with expanded-type films have been found. It can be seen that the energy change of $C_{15}OOH$, u_1^γ, decreases with increasing molar fraction of component 2, 12HOA, in the expanded state, $x_2^{\pi,e}$. The energy change of 12HOA, u_2^γ, increases with the increase of $x_2^{\pi,e}$. The slope of the u_1^γ vs. $x_2^{\pi,e}$ curve is larger than that of the u_2^γ vs. $x_2^{\pi,e}$ isotherm. It was concluded that the film of $C_{15}OOH$ in the mixed monolayer was more affected by the other component than by component 2 (12HOA).

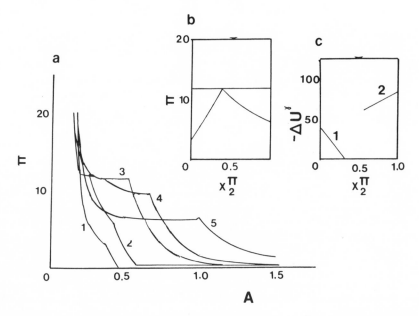

FIGURE 4.29. (a) π vs. A isotherms of $C_{15}OOH$–12HOA. (1) $x_2^{\pi=0}$, ($C_{15}OOH$); (2) 0.1, (3) 0.4, (4) 0.6, (5) 1. (b) Phase diagram of $C_{15}OOH$–12HOA. (c) $\Delta\mu^\gamma$ vs. x_2^π. (1) $C_{15}OOH$; (2) 12OHA.

COMPLICATED-TYPE MIXED MONOLAYERS

Besides the five different types of mixed-phase diagrams for two-component monolayers as described above, systems have been found that did not belong to these five different types. These systems were therefore designated as *complicated-type* phase diagrams. Systems that have been included in this type are $C_{14}OOH-C_{19}OOH$; $C_{14}OOH$–butyl-icosanedioate; $C_{15}OOH$–pentadecyl acetate (PA); $C_{15}OOH-HA$; $C_{15}OOH-DPL$; $C_{16}OOH-HA$; and $C_{18}OOH-C_{14}OOH$ (Table 4.11).

In these mixed systems, the most important difference between the molecular structures is that the hydrophilic groups are different (excepting $C_{14}OOH-C_{19}OOH$). The phase diagrams of $C_{14}OOH-C_{19}OOH$, $C_{15}OOH-$PA, and $C_{15}OOH-HA$ are given in Fig. 4.30 (Hayami *et al.*, 1980; Matuo *et al.*, 1979).

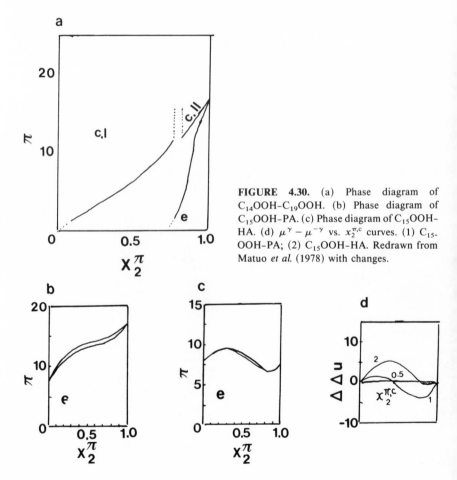

FIGURE 4.30. (a) Phase diagram of $C_{14}OOH-C_{19}OOH$. (b) Phase diagram of $C_{15}OOH-PA$. (c) Phase diagram of $C_{15}OOH-HA$. (d) $\mu^{\gamma} - \mu^{-\gamma}$ vs. $x_2^{\pi,c}$ curves. (1) $C_{15}OOH-PA$; (2) $C_{15}OOH-HA$. Redrawn from Matuo *et al.* (1978) with changes.

The excess apparent molar energy change of the $C_{15}OOH-PA$ and $C_{15}OOH-HA$ systems shows positive and negative regions [Fig. 4.30 (20)]. The three-component mixed films (1MP-2MP-DML) give a complicated type monolayer (Birdi and Sanchez, unpublished). It can thus be seen that these data show that the plots of π_{eq} vs. $x_2^{\pi,e}$ and π_{eq} vs. $x_2^{\pi,c}$ indicate that the energy change of this type of mixture depends uniquely on the composition of the monolayer.

In a recent study (Bertrand, 1982), an ideal associated gas film model was proposed for highly insoluble nonionic surface films, based on a single equilibrium constant for stepwise addition to form aggregates ranging from two molecules to infinity. This model provides a simple method for calculating the thermodynamic properties of gaseous and condensed films and is particularly useful for estimating the mixing properties of films.

4.10.2. DIVERSE MIXED FILM SYSTEMS

The surface pressure isotherms of chlorophyll a, monogalactosyl-diacylglycerol, and phytol at the air–water interface were studied (Tancrede *et al.*, 1982). The limiting areas were 115, 82, and 38 $Å^2$, respectively. The interactions in these mixed films were analyzed with the help of the additivity rule [equation (4.104)]. The data showed that chlorophyll formed ideal films with phytol, while negative deviations were observed with the diglycerol.

Monolayers of different gangliosides and related glycosphingolipids have been investigated (Maggio *et al.*, 1978a). It is known that gangliosides are present in cell membranes as substrates of neuraminidase and as receptors of hormones and toxins (Fishman and Brady, 1976).

The effect of DDT [e.g., *p,p'*-DDT; 1,1,1-trichloro-2,2-bis(*p*-chlorophenyl) ethane, and other analogues] on π vs. A isotherms obtained at the air–water interface was studied (Hilton and O'Brien, 1973). The minimum cross-sectional area of egg lecithin and DDT was studied by using photography of models. All compounds shifted the π vs. A isotherms in a direction and by an amount that was indicative of complexes with egg lecithin in a molar ratio of 1:1. However, no complex formation was reported between DDT and cholesterol; it was also concluded that the lack of difference in effects on monolayers between physiologically active and inactive analogues of DDT indicated that monolayer studies did not explain the physiological action of DDT. It is well known that DDT interferes with nerve transmission by affecting the recovery phase of the action potential; it does this by delaying the turning off of the Na^+ current (Narahashi and Haas, 1968; Hille, 1968). These findings are not conclusive until η_s has been investigated.

A number of unsaturated fatty acids and their esters are known to induce hen erythrocytes to fuse under appropriate conditions (Ahkong *et al.*, 1973). It was therefore of interest to investigate whether any specific interaction between fusogenic lipids and phospholipids could also be detected in mixed monolayers at the air–water interface (Maggio and Lucy, 1974). From mean molecular area–composition plots, it was found that fusogenic lipids—glycerylmonooleate, oleic acid, palmitoleic acid, methyloleate, and retinol—showed condensation effects (i.e., negative deviations from ideality) at 23°C and on a 0.145 M NaCl subphase with phosphatidylcholine, phosphatidylserine, or sphingomyelin.

Mixed films of DPL and estrogenic hormone (estradiol-17β) have been reported, and data have been analyzed to evaluate the biological significance of these films (Khaiat *et al.*, 1975).

SURFACE VISCOSITY OF MIXED FILMS

An extraordinarily high η_s has been observed in DPPE and other saturated phospholipids, as measured by the oscillating pendulum method (Hayashi *et al.*, 1975; Seimiya *et al.*, 1976; Nakahara *et al.*, 1978; Evans *et al.*, 1980). The high η_s values of saturated phospholipids were reduced profoundly by 1 or 2 mole% of cholesterol in mixtures. Cholesterol, which has a relatively low η_s, has a drastic effect on the η_s of saturated phospholipids in monolayers. From this finding, it was concluded that just one molecule of cholesterol can influence a large number of phospholipid molecules in the monolayer. This influence was suggested to be due to polar-group interactions, which thereby have a strong effect on the η_s.

4.11. REACTIONS IN MONOLAYERS

The fact that the molecules in a monolayer film are oriented would suggest that their reaction with other molecules (whether inside the bulk phase or in the monolayer) would be different than if the molecules were inside the bulk phase. Especially if the interface were subjected to some constraint would the reaction rates be much different. The orientation of the molecules could thus lead to a strong dependence of reaction rates on the magnitude of π. The presence of electrical charges would give additional effects.

EXPERIMENTAL PROCEDURES

The hydrolysis rate constant, k, of an alkyl ester (ethyl palmitate) on an aqueous alkali subphase (to acid and ethyl alchol) can be given as

$$-(dn_r/dt) = kn_r[OH]_s \tag{4.115}$$

where t is time, n_r is the surface concentration of reactant (molecules/cm^2) after t, and $[OH]_s$ is the concentration of hydroxyl ions in the monolayer. It is safe to assume that such reactions are pseudo-first-order, since the amount of reactant transferred from the bulk phase by the monolayer will have only an insignificant effect. If the film is neutral, then $[OH]_{bulk} = [OH]_{surface}$. By integration, one obtains

$$\int dn_r/n_r = -k[OH]_b t + \text{const} \qquad (4.116)$$

or

$$\ln n_r = -k[OH]_b t + \text{const} \qquad (4.117)$$

From the plot of $\ln n_r$ vs. t, the slope is equal to reaction velocity, k.

If surface potential, ΔV, is measured, then after time t

$$\Delta V = 4\pi n_r \mu_r + 4\pi n_p \mu_p \qquad (4.118)$$

where the subscripts "r" and "p" denote the reactant and product, respectively. From this relationship, one can derive

$$n_r = (V - V_{t=\infty})/[4\pi(\mu_r - \mu_p)] \qquad (4.119)$$

where $V_{t=\infty}$ is used as the reaction approaches completion. We thus find

$$\ln(V - V_\infty) = -k[OH]_b t + \text{const} \qquad (4.120)$$

such that from a plot of $\ln(V - V_\infty)$ vs. t, one can obtain the value of k from the slope.

The interaction of Ca^{2+} ions with mixed monolayers of stearic acid and stearyl alcohol at pH 8.8 has been reported (Shah, 1974). Hydrolysis of lipids on monolayers has been studied by various investigators. Time dependence of π and V of monolayers of alkyl esters of fatty acids spread on a substrate of 1 N NaOH has been reported (Alexander and Schulman, 1937). A similar kind of behavior was reported for monolayers of ethyl stearate on acid subphases (Muramatsu, 1959).

These phenomena were ascribed to the hydrolytic reaction leading to the corresponding fatty acid and alcohol, which were easily expelled into the subphase. The explanation was based on the tacit assumption that both the ester and the acid molecules remain in the monolayers throughout the processes. In later studies (Muramatsu and Ohno, 1971), the π and radio-activity of insoluble monolayers of methyl-[^{14}C]palmitate as a function of subphase pH were investigated. Some evaporation loss of the lipids was reported.

The remarkable sensitivity of monolayer films to the presence of metallic ions in the substrate has been reported by a number of investigators (Blodgett, 1935; Myers and Harkins, 1937; Mitchell *et al.*, 1937; Robinson, 1937; Harkins and Anderson, 1937; Langmuir and Schaefer, 1936, 1937; Trapeznikov, 1939; Wolstenholme and Schulman, 1950, 1951; Asaki and Matuura, 1951; Schulman and Dogan, 1954; Spink and Sanders, 1955).

A detailed examination of the interaction between stearic acid and copper ion with increasing pH in the substrate indicated that the first substance precipitated was a basic salt and that cupric hydroxide does not precipitate until the pH has been still further increased. In addition, the pH values at which the basic salts are formed are almost the same for a number of different salts (e.g., chloride, nitrate, sulfate, perchlorate). It was suggested that at about this pH, the copper distearate $[Cu(ST)_2]$ in the monolayer undergoes a similar reaction to form a basic stearate. In the case of a 0.002 M $CuCl_2$ solution, the probable sequence of events may be given as

Substrate: $\quad CuCl_2 \xrightarrow{pH=5.7} Cu(OH)_x Cl_{x*} \xrightarrow{pH=6.3} Cu(OH)_2$

Monolayer: $\quad Cu(ST)_2 \xrightarrow{pH=5.5} Cu(OH)ST$

The basic salt is formed by the reaction

$$Cu(ST)_2 + Cu^{2+} + 2OH^- \leftrightarrow 2Cu(OH)ST$$

The reaction accounts for the experimental observations that (1) the pH of formation of the basic salt depends on the initial copper concentration and (2) the relative proportion of copper chemically bound in the film is doubled between pH 5.3 and 6.25. The fact that the behavior of stearic acid on a solution of copper and Al^{3+} was similar suggested that the same kinds of reactions take place. The films are solid-expanded at higher pH.

OXIDATION AND DECOMPOSITION OF LIPID FILMS UNDER ULTRAVIOLET IRRADIATION AND OTHER CONDITIONS

Various monolayers have been investigated with regard to oxidation and decomposition (Giles, 1957; Felmeister and Schaubman, 1969; Rideal and Mitchell, 1937; Pilpel and Hunter, 1970).

A correlation was found between the fading and the π vs. A isotherms of various surface-active dyes investigated (Giles, 1957). Interaction between phenothiazine derivatives and films of lecithin was induced by ultraviolet (UV) irradiation (Felmeister and Schaubman, 1969). Photochemical decomposition of stearic anilide to stearic acid and aniline was investigated in monolayers by using π and ΔV (Rideal and Mitchell, 1937). Although extensive investigations have been reported on the thermal and photo-chemical decomposition of hydrocarbons and derivatives in bulk phase, determination of the mechanism of these reactions has been difficult owing to free-radical chain reactions in the process (Emanuel, 1965).

The π vs. A isotherms of stearic acid on a subphase of pH 5.4 showed some subtle changes with time of exposure to UV light (Fig. 4.31).

The transition pressure of magnitude 20 mN/m decreases to about 6 mN/m after 84 min of radiation. On analysis by thin-layer chromatography, the film material was found to indicate the presence of long-chain aldehydes. The difference in the film stability of n-octadecanol and stearic acid films was ascribed to the different ranges of UV absorption spectra. The nature of the reaction products obtained after UV irradiation of stearic acid monolayers was consistent with the oxidation of C–C bond scission, adjacent to or near the carboxyl group. The reaction kinetics did not indicate any chain reactions.

The interaction of iodine with oxidized cholesterol, DPL, and egg lecithin was investigated by the monolayer method (Szundi, 1978). Iodine

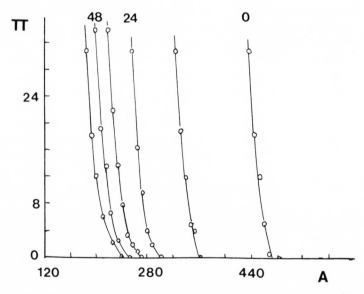

FIGURE 4.31. Changes in π vs. A isotherms of stearic acid (0.1 mg) monolayers after varying exposure to UV irradiation (min) on an aqueous subphase of pH 5.4 (22°C). Redrawn from Pilpel and Hunter (1970) with changes.

was among the first modifying agents to be applied to bilipid membranes (BLMs). It was found that the iodide ion, and especially molecular iodine, lowers the electrical resistance of BLMs by several orders of magnitude (Lauger et al., 1967; Tien and Verma, 1970; Tien, 1970). The amount of iodine incorporated was the largest for egg lecithin, and less for oxidized cholesterol, while DPL showed no measurable effect in the monolayer.

4.12. EFFECT OF MONOLAYERS ON WATER EVAPORATION RATES

The evaporation of liquids, especially water, has been of interest to scientists for many decades (Dalton and Gilberts, 1803). It has long been known that the rate of evaporation of water is considerably retarded by the presence of an amphiphile monolayer (Hedestrand, 1924; Rideal, 1925). The silent evaporation of liquids is a deceptively simple process. The rate of evaporation of water was found to be proportional to the area of the surface and to the difference in partial pressure between the liquid surface and the surroundings (Dalton, 1803). However, it was also suggested that there was an upper limit to the rate of evaporation at a given temperature (Stefan, 1873).

By the use of thin layers of molten cetyl palmitate heated from below, the cellular convection patterns, the so-called Benard cells (Benard, 1900), were observed. With nearly all liquids, an important one being water below 4°C, cooling of the surface leads to a surface layer that is denser than the layers beneath. For most liquids, the density mechanism for convection and the surface tension mechanism (Marangoni effect) reinforce each other.

At equilibrium between a liquid and its vapor, the absolute rate of evaporation must equal the absolute rate of condensation. From the kinetic theory, an expression for the rate of collision of vapor molecules with the surface can be derived. The total resistance to the transport of a molecule, r_t, is given as (Langmuir and Langmuir, 1927; La Mer, 1962; La Mer et al., 1964)

$$r_t = r_L + r_G + r_I \tag{4.121}$$

where r_L, r_G, and r_I are the resistances of the liquid, gas, and interface, respectively. In a typical study, it is necessary to measure the rate of evaporation across a clean surface (i.e., $r_I = 0$) and then measure a monolayer film.

The retardation effect is also given as a specific evaporation resistance (La Mer, 1962):

$$R = A(C_0 - C)(1/v - 1/v_0) \tag{4.122}$$

where A is the surface area, C and C_0 are the equilibrium concentrations of water vapor with and without desiccant, respectively, and v and v_0 are the evaporation rates with and without spread film, respectively.

At the molecular level, ΔR depends on the fluctuating number of holes of size sufficient to allow the passage of colliding water molecules. These theoretical treatments have been discussed by different authors (Dickinson, 1978).

The retardation of evaporation by monolayers was investigated by laser interferometry and the appropriate conventional method (O'Brien *et al.*, 1976a,b) (Fig. 4.32). In the current literature, little has been reported as regards gas transport through monolayers. In these times of much concern about pollution, it is clearly important to investigate the effects of monolayers (e.g., on seas and lakes, from spills of oil or other spills, and other pollutants with amphiphilic character) on the rate of oxygen adsorption in the underlying waters. As is obvious, if the rates are affected by the presence of such monolayers, oxygen-dependent biological organisms and plants would be seriously affected. The permeability of monolayers to gases has been described on the basis of a model in which free spaces in the monolayer allow permeation and consequent fluctuating density (Blank, 1964). In a later study, the same model was used to derive monolayer transport properties, in conjunction with transport in membranes (Blank and Britten, 1965).

In Table 4.12, the retardation of water evaporation by different amphiphiles as a function of area/molecule is given. It can be seen that the area covered by an amphiphile in a compact monolayer is the factor that determines the degree to which it can retard the evaporation of water (Baines *et al.*, 1970). The retardation of oxygen penetration into water by monolayers showed the same general trend; i.e., those amphiphiles that do not form highly compact monolayers (i.e., fatty acids and lecithins) give the least resistance to both oxygen and water.

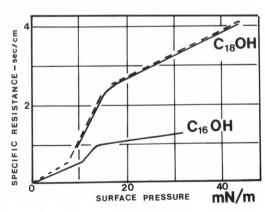

FIGURE 4.32. Retardation of evaporation of water surface by spread monolayers of $C_{16}OH$ and $C_{18}OH$. From O'Brien (1976).

TABLE 4.12. Retardation of Evaporation by Monolayers
of Different Lipids on Water[a]

Lipid	Area/molecule (Å^2)	Retardation (%)
DPL	40.3	20
Soybean lecithin	40	20
Oleic acid	28	5
Stearic acid	24	28
Stearyl alcohol	22	59
Cholesterol	40	0

[a] From O'Brien *et al.* (1976).

It has been suggested that solvation of the polar groups of monolayer molecules is an important factor in evaporation resistance (Trapeznikov and Avetisyan, 1970; Trapeznikov, 1964; Trapeznikov and Lonomosova, 1967).

The water (w) content of a monolayer (M) can be estimated as (Mansfield, 1974)

$$\pi = (RT/a_w) \ln[(\Gamma_w + \Gamma_M)/f_w^s \Gamma_w] \tag{4.123}$$

where a_w is the partial molar area of water at the surface, f_w^s is the activity coefficient of water at the surface, and Γ_i is the surface concentration.

The transport of gases through monolayers was reported in a recent study (Petermann, 1977). Expanded films exhibited very little resistance, while condensed ("solid") films showed great resistance to the adsorption of gases, e.g., CO_2, SO_2, and NH_3. The amphiphiles used were *n*-hexadecanol, *n*-octadecanol, oleic alcohol, and cetyl trimethyl ammonium bromide. It can be concluded that such investigations are urgently needed, in view of the dependence on the kinetics of gas transport through monolayers (which are invariably present) of such important phenomena as CO_2 exchange in the lungs through lecithin films and O_2 adsorption on the surface of seas, lakes, and rivers (which is necessary to support the aquatic life cycle).

EVAPORATION RESISTANCE DATA FOR
ONE-COMPONENT MONOLAYERS

Table 4.13 lists evaporation resistance values for a number of one-component monolayers at three different π values. By introducing the resistance concept (Langmuir and Schaefer, 1943) and using the analogy with electrical resistance and Ohm's law, and also in accord with the

TABLE 4.13. Evaporation Resistance of One-Component Lipid Monolayers at Various Surface Pressures (π)

	Resistance ($r/s\ cm^{-1}$)		
Monolayer	Surface pressure: 10 mN/m	20 mN/m	30 mN/m
n-Eicosanoic acid	5.2	5.2	5.2
n-Nonadecanoic acid	—	3.9	3.9
n-Octadecanoic acid	—	2.3	2.3
n-Heptadecanoic acid	—	1.4	1.4
n-Docosanol	3.9	8.8	14.0
n-Eicsanol	3.2	5.1	6.3
n-Octadecanol	2.5	3.15	3.9
n-Hexadecanol	0.7	1.1	1.4
n-Tetradecanol	0.19	0.26	0.42
n-Docosanoxy ethanol	0.6	3	13.7
n-Octadecanoxy ethanol	0.4	1	3.2
n-Hexadecanoxy ethanol	0.1	0.2	0.3
Methyl octadecanoate	2.3	—	—
Ethyl octadecoanate	1.6	3.6	4.1
Ethyl hexadecanoate	0	0	—
Ethyl 9-octadecenoate	0	—	—
Ethyl 9,12-octadecadienoate	0	0	0
n-Docosyl sulfate	0.1	0.25	0.4
Calcium eicosanoate	8.1	9.3	10.9
Calcium nonadecanoate	8	6.8	8
Calcium octadecanoate	4.2	5	5.6
Cholesterol	—	0.1	0.1
Poly(vinyl octadecanoate)	0.14	0.28	—
1,1,13-Trihydroperfluorotridecanol	0.1	1.7	1.7

concepts of irreversible thermodynamics, one obtains

$$\text{Rate} = \text{driving force}/\text{resistance} \qquad (4.124)$$

APPLICATIONS OF MONOLAYERS TO VARIOUS SYSTEMS

Evaporation Control on Water Reservoirs. For this purpose, alkanols, cetyl alcohol (hexadecanol) and stearyl alcohol (octadecanol), and mixtures of these substances fulfill the requirements.

Fogs and Mists. The processes of droplet formation and growth of condensation are important in such natural phenomena as the formation and dispersal of natural fogs and mists, cloud formation, and rain formation. They are also important in such technical applications as generation of artificial fog to protect orchards and beehives from frost, of foams to protect aircraft from risk of fire and explosion in forced landings, and of aerosol sprays, in general.

4.13. DIVERSE MONOLAYER ANALYSES

4.13.1. DYNAMIC LIPID MONOLAYER MEASUREMENTS

As regards the stability of monolayers, many different kinds of experiments have been carried out. The stability of fatty acid monolayers at the collapse region has been investigated by electron micrographs (of monolayers removed on solid surfaces) (Ries and Kimball, 1957). The compression rates affected some monolayer systems (Rabinotvitch et al., 1960). π vs. A isotherms of hexadecanoic acid ($C_{16}OOH$), octadecanoic ($C_{18}OOH$), eicosanoic acid ($C_{20}OOH$), and docosanoic acid ($C_{22}OOH$) at different monolayer compression rates were investigated (Heikkila et al., 1970). It can be seen from Fig. 4.33 that the compression rate had no effect on transition (T), collapse point (C), or the shape of the collapse region. Very slow compression rates, as obtained by intermittent compression and equilibrium, however, showed profound effects (Fig. 4.34). The postcollapse

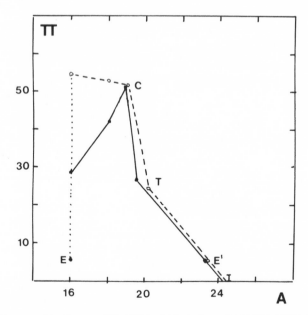

FIGURE 4.33. π vs. A isotherms of $C_{20}OOH$ acid (subphase: 0.1 N HCl, 25°C). Monolayers were compressed at 0.62 Å^2/molecule per min (●) and 5.93 Å^2 per min (○). See the text for details. Redrawn from Heikkila et al. (1970) with changes.

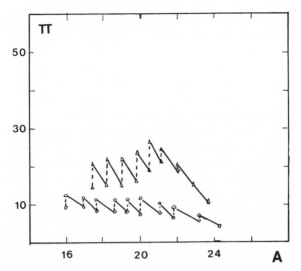

FIGURE 4.34. π vs. A isotherms for $C_{16}OOH$ (O) and $C_{20}OOH$ (●) during intermittent compression and equilibration. Compression rate: 2 min at 0.46 $\text{Å}^2/\text{min}$; equilibration time: 20 min; subphase: 0.1 N HCl, 31°C. See the text for details. Redrawn from Heikkila *et al.* (1970) with changes.

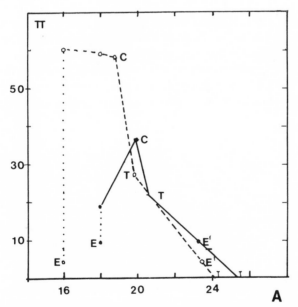

FIGURE 4.35. π vs. A isotherms of $C_{16}OOH$ (●) and $C_{20}OOH$ (O). Compression rates varied from 0.5 to 6.0 $\text{Å}^2/\text{min}$; subphase: 0.1 N HCl, 26°C. See the text for details. Redrawn from Heikkila *et al.* (1970) with changes.

π for $C_{16}OOH$ monolayers decreased rapidly when the film was compressed at either a rapid or a slow rate (Fig. 4.35). The postcollapse π for $C_{20}OOH$ monolayers did not show any changes (Fig. 4.35).

In biological and other systems in which mixed lipids are present, one might expect the existence of domains with varying compositions. A domain could exhibit time-dependent characteristics under any dynamic compression-expansion measurement. This reluctance exhibited by lipid molecules may be quantitated as rheologic hysteresis, in the sense that the molecular arrangement resists deformation and that hysteresis might be considered the basis of function and stability. In some investigations, commercially available monolayer balances have been used for such dynamic studies (Mendenhall and Mendenhall, 1963; Lusted, 1973; Birdi, 1972). In these apparatuses, the areas were compressed from 35 to 16 cm^2 with varying compression-expansion rates (e.g., 0.5 cycle/min). The evolution of hysteresis loop size and shape for a mixed film of 20% lecithin-80% cholesterol is depicted in Fig. 4.36. It is worth noting that the collapse state (C) occurs parallel to the time axis. In Fig. 4.37, various lipid hysteresis π vs. A isotherms are given (Lusted, 1973).

Dynamic isotherms of mixed lipids were also investigated, and the isotherms were described in terms of their van der Waals interactions. These isotherms should be compared with the time-dependent behavior of lipid films (Sims and Zografi, 1972). From these studies, it was concluded that three-dimensional separation occurs when the lipid film is compressed to π approaching the liquid-condensed-solid-condensed transition point. The relative rates of molecular expulsion from the surface, and probably lens formation and growth, could be the determining factors in the rate of π loss when compression is stopped. However, it has also been reported that hysteresis is not observed for such monolayers as oleic acid (which at 24°C

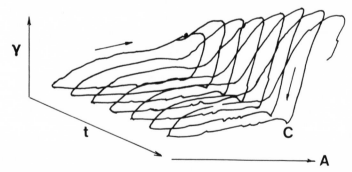

FIGURE 4.36. Evolution of a hysteresis loop of a mixed film of 20% lecithin-80% cholesterol: plot of surface tension γ ($\pi = 72 - \gamma$) vs. area (compressed from 35 to 15 cm^2) compression-expansion isotherms. (C) Collapse state (25°C). Redrawn from Lusted (1973) with changes.

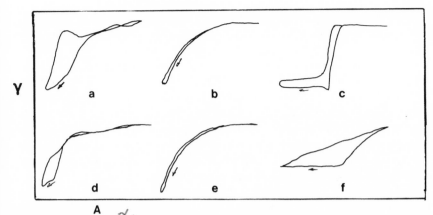

FIGURE 4.37. Dynamic π vs. A isotherms of different lipids on a water surface at 25°C. (a) Sphingomyelin; (b) phosphatidylserine; (c) cholesterol; (d) dipalmitoyl lecithin; (e) phosphatidylinositol; (f) phosphatidylethanolamine. Redrawn from Lusted (1973) with changes.

is only liquid-expanded). These dynamic studies would be of relevance to evolutionary prebiotic systems (Calvin, 1969).

Dynamic monolayers have been investigated by using π and ΔV measurements (Vollhardt and Wuestneck, 1974; Vollhardt *et al.*, 1978). The ΔV showed strong changes in monolayers of n-octadecanol around 28–170 Å^2 and in monolayers of n-octadecanoic acid around 38–190 Å^2 by the dynamic method.

4.13.2. SPECTROSCOPIC ANALYSES OF MONOLAYERS

Understanding of the structure and dynamics of interfaces at the molecular level is limited, as noted above. However, with advances in infrared (IR) technique, this situation might be changing. The main reason for the lack of spectroscopic information has been the weak signal strength due to the low-area, flat surface. Most spectroscopic information on flat surfaces such as the air–water interface of the Langmuir film balance has been obtained by the use of transferred films to combat the sensitivity problems. However, it is pertinent to realize that films adsorbed at solid surfaces may or may not remain identical (see Chapter 9).

Recent advances in spectroscopic measurement techniques for monolayers *in situ* have been made in visible adsorption and fluorescence methods (Kuhn *et al.*, 1972), and resonance Raman dye monolayers at the oil–water interface (Nakanaga and Takenaka, 1977).

In a recent study, an alternative view of phospholipid phase behavior at the air–water interface has been described from studies using the epifluorescence microscope technique (Tscharner and McConnell, 1981). Monolayers of DPL have been analyzed with respect to the liquid-expanded (L_{ex}) ↔ liquid-condensed (L_{co}), and solid-condensed (S_{co}) phase regions. The S_{co} region corresponded to a gel phase of the lipids in which there was no flow in the membrane, i.e., lateral diffusion was low, compressibility was low, and the membrane was optically homogeneous. The L_{co}–L_{ex} region appeared to be a homogeneous membrane in which lateral diffusion and membrane flow were both rapid. This region was found to be of high compressibility. The L_{ex} region was not homogeneous as seen under the microscope, and the flow of the surface layer could be very fast. In the S_{co} region, the compressibility, K, was nearly constant. A sharp increase in K was observed as the L_{co} region was approached. The magnitude of K in the L_{co}–L_{ex} transition region was large and changed only slightly as the area was changed. After the L_{ex} region, the value of K decreased and remained fairly constant. From these data, it was concluded that (1) the L_{ex} region is a two-phase region, part of which is membranelike and the other part is not; (2) the L_{co}–L_{ex} region is a single homogeneous phase; (3) the L_{co} region is a two-phase region; (4) the S_{co} region is a solid, single-phase region.

Fourier transform (FT) IR reflectance spectra were obtained using monolayers of oleic acid (which spreads rapidly to $\pi = 30$ dynes/cm) (Delany and Cornell, 1985). The IR bands are due to symmetric and antisymmetric $-CH_2-$ stretching bands of the saturated hydrocarbon chain at 2489 and 2916 cm^{-1}, respectively. In addition, the band at 3002 cm^{-1} is due to the alkenic $-C-H-$ stretching vibrations of the $=C-H-$ group in the 9-*cis* double bond in oleic acid (Bellamy, 1975). The relatively great intensity at 3002 cm^{-1} could be due to the double bond and water interaction (further studies are needed).

The prominent band at 1708 cm^{-1} is due to the $=CO$ stretching vibration.

EFFECT OF LIGHT ON MONOLAYERS

Several stereoisomers of retinal (vitamin A aldehyde) are known, and their interconversions can easily be promoted by the action of light (Hubbard *et al.*, 1952). The photoconversion of 11-*cis* to all-*trans* retinal takes place with the highest efficiency (Robeson *et al.*, 1955; Brown and Wald, 1963) and is involved in the first step of the visual process (Wald *et al.*, 1963). Studies of vitamin A and its derivatives have suggested that their surface-active properties are closely related to their physiological functions (Weitzel *et al.*, 1952; Bangham *et al.*, 1964; Maeda and Isemura, 1967). π vs. A isotherms of 11-*cis* isomer and all-*trans* give liquid-expanded-type films (Maeda and Isemura, 1967). However, the 11-*cis* isomer has a larger limiting area ($A_{lim} \approx 65 \, \text{Å}^2$) and lower π_{col} ($\approx 14 \, \text{mN/m}$) than the all-*trans* isomer ($A_{lim} \approx 55 \, \text{Å}^2$; $\pi_{col} = 16 \, \text{mN/m}$).

When all-*trans* and 11-*cis* films were irradiated, changes were observed in the surface pressures required to keep the film area constant. The changes were maximal after irradiation for 30 sec.

4.13.3. SPREADING SPEEDS OF LIPID MONOLAYERS AT THE AIR–WATER INTERFACE

A few studies on the spreading speeds of lipid monolayers at the air–water interface have been reported (Gaines, 1966; Davies and Rideal, 1961; Deo *et al.*, 1961; Katti and Sansare, 1970; O'Brien *et al.*, 1976a,b). The velocity of spreading of many polar organic molecules is approximately 10 cm/sec. This can be of interest in emulsion and foam-formation phenomena (Chapter 10).

A suitable procedure for measuring is to place a camera above the trough and measure the speed of spreading by monitoring the movement

TABLE 4.14. Spreading Speeds and π_{eq} of Various Lipids and Surfactants at 21°C[a]

Lipid	Speed (mm/sec)	π_{eq} (mN/m)	Lipid	Speed (mm/sec)	π_{eq} (mN/m)
Alcohols			Ethyl esters		
Dodecanol	308	—	Ethyl dodecanoate	197	—
Tetradecanol	265	46	Ethyl tetradecanoate	195	—
Hexadecanol	155	40	Ethyl hexadecanoate	151	15
Octadecanol	8.5	35	Ethyl octadecanoate	97	—
Eicosanol	8.1	33	Alkyl trimethyl ammonium bromides (TMABs)		
Docosanol	0	28	Dodecyl-TMAB	253	33
Acids			Tetradecyl-TMAB	254	37
Dodecanoic	295	—	Hexadecyl-TMAB	247	34
Tetradecanoic	135	15	Octadecyl-TMAB	260	32
Hexadecanoic	22	9	Amine hydrochlorides		
Octadecanoic	0	2	C_{12}-HCl	300	—
cis-9-Octadecanoic	215	—	C_{18}-HCl	111	—
Methyl (ME) esters			Sodium sulfonates		
ME-C_{12}	218	—	C_8SO_3	132	15
ME-C_{14}	175	—	$C_{10}SO_3$	270	32
ME-C_{16}	30	14	$C_{14}SO_3$	181	18
ME-C_{18}	15	—	$C_{16}SO_3$	159	16
ME-C_{20}	13	—			
ME-C_{22}	0	—			

[a] From O'Brien *et al.* (1976).

of talc particles after the lipid is applied to the surface. In another method, spreading speeds were measured by using a thermistor.

The spreading speeds of various substances are given in Table 4.14. From these data, a linear correlation was found between the entropy ($\propto d\pi_{eq}/dT$) (or enthalpy) difference involved in the bulk-to-monolayer "phase change" and the spreading speeds of the long-chain un-ionized acids. Spreading speed vs. π_{eq} curves were found to indicate the relative lattice energies between homologous series having different polar groups. The data on hydrophobic–hydrophilic balance were found to be inappropriate for predicting spreading speed.

4.14. SURFACE VISCOSITY OF LIPID FILMS

Even though η_s is one of the most important physical properties of monolayers that have been found to be correlated with other properties, e.g., evaporation rates and foam characteristics, η_s remains the least satisfactorily investigated of those physical properties.

4.14.1. SURFACE RHEOLOGY

A fluid interface interacts with liquid in motion because the stress tensor in the interface is not the same as in the bulk fluid phase, as a result of the surface tension. As described above, surface tension is present because of the abrupt change in density or composition near any liquid interface (generally of a thickness of about 1 nm).

At the interface, because of the mechanical equilibrium, the stress-tensor component that is normal to the flat interface will be constant. The tangential force component, on the other hand, changes in going from the bulk to the interface, and surface tension, γ, is defined as the integral of these stress-tensor components (Bakker, 1928; Buff, 1960; Defay et al., 1966; van den Tempel, 1977):

$$\gamma = \int_{z_1}^{z_2} (p_n - p_t)\, dz \qquad (4.125)$$

where the distances z are chosen such that the integral vanishes. In the Laplace equation, the aforementioned mechanical equilibrium is assumed to be valid. In the presence of added substances, i.e., amphiphiles, the Gibbs adsorption equation [equation (2.40)] (Chattoraj and Birdi, 1984) gives the relationship between Γ and the chemical potential in the bulk phase.

A pure liquid with a clean surface therefore has no surface rheology, since the γ is not affected by the motion (or motion within the surface).

The molecular relaxations occur at such fast rates that any mechanical disturbance leads to no measurable effects (Wegener and Parlange, 1964).

4.14.2. SURFACE VISCOSITY OF DIFFERENT LIPID MONOLAYERS

Data on the surface viscosity, η_s (rotational method), of n-alkyl alcohols were reported as a function of π (Fourt and Harkins, 1938). These η_s data showed some peculiarities. At low π (< 3 mN/m), η_s increased with π and with greater chain length. When π was 8 mN/m, the value of η_s was independent of chain length ($\eta_s = 0.015$ surface poise) (Fig. 4.38). These data are qualitatively in agreement with the literature data (Trapeznikov, 1941; Joly, 1952; Rosano and Mer, 1956). The magnitude of η_s increases abruptly in the region where π is between 10 and 16 mN/m for $C_{14}OH$ to $C_{18}OH$. After this region, the value of η_s remains constant for higher π up to 35 mN/m. Close-packed films therefore indicate that η_s is greater for $C_{14}OH$ than for $C_{18}OH$, by a factor of 10. The abrupt change in η_s with π is found to occur where there is a change in the slope of π vs. A isotherms (Table 4.15).

An apparatus was described that measured oscillations of very large amplitude by means of a torsional device for determining η_s (Bragadin, 1972). The main feature of the design was that it allowed measurement of

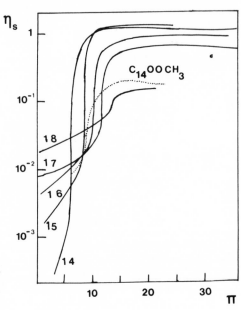

FIGURE 4.38. η_s (by rotational method) vs. π isotherms of a series of n-alkyl alcohols at 25°C. Redrawn from Fourt and Harkins (1960) with changes.

TABLE 4.15. π and A Values of n-Alkyl-Chain
Alcohols at Changes in Slope $(20°C)^a$

Alcohol	A $(\text{Å}^2)^b$	π $(\text{mN/m})^c$
Tetradecanol	21.1	6.2
Pentadecanol	21.0	8.2
Hexadecanol	20.9	10.2
Heptadecanol	20.6	11.8
Octadecanol	20.6	12.9
Eicosanol	20.4	15.0

^{a} From Nutting and Harkins (1939). ^{b} Extrapolated to $\pi = 0$.
^{c} At change in slope.

floating ring angles of oscillations as great as 270°, which allowed study of the same film for the entire range of the elastic and the Newtonian viscous state. The viscosimeter was calibrated by measuring the η_s of palmitic acid monolayers on a subphase of 0.001 N HCl. Mean values of the angular velocities and the corresponding values of the viscosity, η_s, as a function of π are given in Table 4.16. The same apparatus was also used for measuring the η_s of spread monolayers of polymers (Gabrielli et al., 1971).

Metal ions are known to affect π vs. A isotherms radically, which further affects the η_s properties of these films (Enever and Pilpel, 1967a–c). These effects were determined by the metal ion, the pH of the subphase, the temperature, and the concentration of the ion. In the case of stearic acid monolayers, the time effect was slow when the liquid-condensed film was injected with metal ions.

TABLE 4.16. Surface Viscosity, η_s, as a Function of π of
Palmitic Acid Monolayers Spread on a
0.001 N HCl Subphase

Angular velocity (rad/sec)	π (mN/n)	$\eta_s{}^a$	$\eta_s{}^b$
		Surface poise $\times 10^{-3}$	
454	20.0	4.4	4.4
760	18.0	2.6	2.6
858	17.3	2.3	2.3
2472	11.3	0.8	0.8
4338	8.1	0.5	0.5

^{a} From Bragadin (1972). ^{b} From Boyd and Harkins (1939).

The interaction between stearic acid monolayer and Ca^{2+} ion was found to consist of four probable stages:

1. Diffusion of Ca^{2+} ions into the interface.
2. Reaction at the interface between Ca^{2+} ions and the ionized carboxyl groups.
3. Subsequent structural reorientation in the mixed Ca–stearate–fatty acid film, possibly leading to . . .
4. Further reaction between Ca^{2+} ions and stearate ions due to the removal of steric hindrances.

Other investigators have shown that stages (1) and (2) are completed within periods varying from a few tenths of a second to about 2 min, depending on pH, ionic concentration, temperature, and physical state of the monolayer (Langmuir and Schaefer, 1937; Matsubara, 1965). Therefore, stages (3) and (4) involve the rate-determining equilibrium of the soap and the unchanged stearic acid. The order of the reaction was determined by the change in η_s. η_s studies were also reported for mixed films of stearic acid and n-octadecanol (Fig. 4.39).

In a recent study, the η_s of long-chain alcohols was investigated by the canal method (Hühnerfuss, 1985). The motivation for these investigations was that unexpected results were obtained during a large-scale international marine remote sensing experiment using airborne microwave radiometer measurements at 1.43 and 2.65 GHz over a sea surface covered with a monolayer of 9-octadecen-1-ol, Z-isomer (oleyl alcohol). The canal method was used because it is considered to be about two orders of magnitude more sensitive than other methods. η_s was estimated from the following equation:

$$\eta_s = [(\pi_2 - \pi_1) W^3/(12QL)] - (Wn_0/3.14) \qquad (4.126)$$

where W is the width of the canal and L the length, π_2 is larger than π_1 on each side of the canal, and Q is the rate of film flow (in cm^2/sec) (by

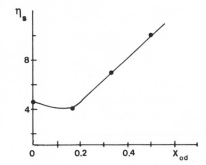

FIGURE 4.39. Surface viscosities (η_s) of mixed n-octadecanol (od)/stearic acid films at 24 $Å^2$ on water (pH = 5.2, 21°C). Redrawn from Enever and Pilpel (1967) with changes.

analogy with the Poiseuille formula for the flow of liquids through capillary tubes). The last term on the right-hand side arises from the drag correction. This relationship is valid when $L \gg W$, if the walls are smooth and parallel with no slip of the film along them, if the flow is Newtonian (independent of shear rate), and if η_s is constant within the canal. The magnitude of π_2 was held constant at either 20 or 3 mN/m. The measurements were carried out as a function of canal width and solvent used for spreading (Table 4.17).

These data suggest that optimum W is 0.15–0.2 cm. The values of η_s were found to be larger for $\pi_2 = 20$ mN/m than for $\pi_2 = 3$ mN/m. It was concluded that η_s increases with increasing chain length in the series dodecanol < tetradecanol < hexadecanol < octadecanol. This finding was in accord with the literature data. It confirms the conclusion that increasing chain length increases the number of water molecules bound in the under layer of water that is dragged along by the film (Suggett, 1977; Hühnerfuss et al., 1984). The solvent effect was not analyzed thoroughly and needs more investigation. It would be disturbing to find that different solvents can give rise to highly variable η_s data.

The η_s of different long-chain alcohols (Birdi et al., unpublished) has been measured by the method based on the decaying surface motion of a circular container of liquid following cessation of steady-state rigid body rotation (Krieg et al., 1981). The method was found to be very useful for the measurement of η_s. At complete monolayer coverage ($A = 20$ Å^2), η_s was found to increase for alcohols in the order $C_{14}OH > C_{16}OH > C_{18}OH$. These data agree with the η_s results reported with the use of a different method (see Fig. 4.4).

Many experimental observations reported in the literature (Blodgett, 1937; Buhaenko et al., 1985) suggest that the rheological properties of spread

TABLE 4.17. η_s Values of Homologous n-Alkyl Alcohol Monolayers Spread from Two Different Spreading Solvents at 20°C[a]

		η_s (mPa/sec per m × 10^{-4})			
Solvent	Lipid	$W = 0.05$ cm	$W = 0.1$ cm	$W = 0.15$ cm	$W = 0.2$ cm
Ethanol	Dodecanol	7	7.8	320	45,100
	Tetradecanol	18	280	640	1,400
	Hexadecanol	22	590	400	3,000
	Octadecanol	80	300	1,000	2,000
Heptane	$C_{12}OH$	3	30	40	120
	$C_{14}OH$	16	64	150	220
	$C_{16}OH$	50	40	2,500	2,600
	$C_{18}OH$	40	66	2,700	5,100

[a] From Hühnerfuss (1985). W = Canal width; $\pi_2 - \pi_1 = 20$ mN/m; $L = 7.5$ cm.

monolayers, such as viscosity, can have important consequences for Langmuir–Blodgett films (see Chapter 9). The η_s of fatty acids, as measured by canal and torsion pendulum methods, was reported in one study (Buhaenko *et al.*, 1985). The η_s values obtained with the rotational viscosimeter were higher by an order of magnitude than those obtained with the canal viscosimeter. This is another instance in which the current literature data on surface viscosity, η_s, are far from satisfactory.

Two techniques for measuring nonlinear interfacial stress-deformation behavior using rotational interfacial viscosimeters have been reported (Jaing *et al.*, 1983): Either a single bob or two bobs of slightly different radii can be used. A series of experiments with an interface between air and aqueous solutions of 0.1 wt.% Na-*n*-dodecyl sulfate were carried out using the disk-knife edge and thin biconical bob interfacial viscosimeters. Agreement among the results supports the validity of the proposed methods. The same system was used in a deep-channel interfacial viscosimeter. It was the first time that consistent measurements of interfacial stress-deformation behavior have been reported using different interfacial viscosimeters.

A new, high-sensitivity surface viscosimeter was described in a recent study (Poskanzer *et al.*, 1974). The η_s's of different proteins (hemoglobin and bovine serum albumin) were investigated. The lipid films were also investigated. The instrument was susceptible to an exact hydrodynamic analysis of the flow fields in the interfacial and substrate phases. The η_s could be detected to as low as 10^{-5} g/sec (surface poise). However, the sensitivity remains to be compared with that of data reported with the use of other apparatuses.

4.14.3. THERMAL-PHASE TRANSITION IN MONOLAYERS: SURFACE VISCOSITY

The conformation and the interfacial interactions of phospholipids in membranes are basically important as regards their interactions with water. However, the dependence of lecithin–water interactions on temperature requires much understanding of the effect of temperature on lipid monolayers. The rheological characteristics of monolayers are strongly dependent on temperature. At low temperatures, the film is comparatively rigid and η_s cannot be detected by the canal method. However, as temperature is raised, the film melts and η_s becomes detectable by most instruments. A procedure used to detect the temperature at which η_s becomes appreciable has been described (Hayashi *et al.*, 1973). The monolayer is initially spread at low temperatures ($\approx 10°C$), where the film is viscous (such as Ca-stearate or protein). The film cannot flow through a canal of 1-mm width. After equilibrium π ($= 10$ mN/m) has been attained, the canal

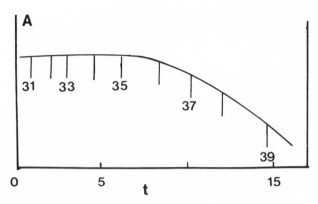

FIGURE 4.40 Rate of flow A (change in area) vs. time, t (and temperature as indicated) of a monolayer of dipalmitoylphosphatidylethanolamine ($\pi = 10 \, \text{mN/m}$). Redrawn from Hayashi *et al.* (1973) with changes.

is opened and the temperature of the subphase is increased. The flow rate of the film shows an abrupt change at the transition temperature (Fig. 4.40).

Mixed-film studies of dipalmitoylphosphatidic acid ($T_{tr} = 45°C$) exhibited a pronounced lowering of T_{tr} on addition of cholesterol. Monolayer viscoelasticity has been investigated by using light-scattering techniques (Langevin, 1981). Monolayers of myristic acid, stearic acid, propylstearate, and polymer films (polyvinylacetate and polymethylmethacrylate) were investigated. The compressibility, shear moduli, and dilational shear viscosity were determined.

Viscoelastic theories of monolayers have been developed. The earliest ones (Lucassen and Lucassen, 1961; Goodrich, 1961, 1962) assumed that a monolayer can be characterized by three elastic moduli—compressibility, shear, and surface pressure—and by two viscosities—dilational and shear. Several more recent treatments (Kramer, 1971; Goodrich, 1962) introduced a third surface viscosity that we will call "transverse," associated with the vertical motion of monolayers (Fig. 4.41). Such properties affect the propagation of surface waves.

A monolayer reduces the surface tension of the liquid on which it is spread. The surface tension, γ, of the interface can be calculated by the Kelvin equation:

$$\gamma = (\rho f^2 \lambda^3)/(2 \times \pi) - [\rho g(\lambda/2 \times \pi)^2] \qquad (4.127)$$

where ρ is the density of the subphase (water), g is 981 dynes/g per sec, f is the frequency, and λ is the wavelength.

The elasticity and viscosity of a monolayer create additional surface forces that also affect the liquid surface motion. In an earlier theoretical

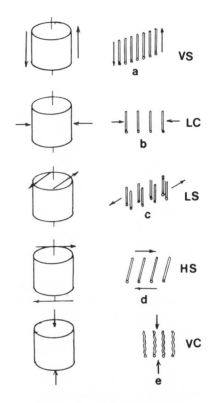

FIGURE 4.41. Various possible types of motion in monolayers. (a) Vertical shear (VS); (b) lateral compression (LC); (c) lateral shear (LS); (d) horizontal shear (HS); (e) vertical compression (VC). Redrawn from Goodrich (1962) with changes.

analysis (Langevin and Bouchiat, 1971), it was shown that the monolayer viscosity or viscoelasticity is expected to modify reflected light. It was thus suggested that the viscoelastic properties could be investigated by surface light-scattering methods, as reported by various investigators (Hard and Lofgren, 1977; Byren and Burnshaw, 1979; Grisner and Langevin, 1978).

The motion of the surface is, however, interconnected with the subphase fluid through the viscosity. The mass of the fluid acts as a reactive element, allowing the fluctuation of surface tension, γ, to propagate on the surface of the liquid as a *wave*—the longitudinal surface wave (Lucassen, 1968; Crone *et al.*, 1980).

A recent study described an instrument with which waves were generated and longitudinal surface waves were detected without touching the surface (Miyano *et al.*, 1983; Ting *et al.*, 1984). This apparatus was used to study dynamic interfacial properties, such as the Gibbs elasticity, the diffusion parameter, the surface (excess) dilational viscosity, and the surface (excess) shear viscosity. From these investigations, it was concluded that parametric study of the dispersion relationship of surface waves and capillary wave characteristics was not sensitive enough for determination of

TABLE 4.18. Temperature Changes in the Water
Surface during the Spread of a Monolayer of
Oleic Acid[a]

Depth of surface (mm)	T °C	
	Calculated from reference index	Thermistor
0.2	−0.19	−0.21
0.5	−0.12	−0.13
2.5	0	−0.02

[a] From O'Brien et al. (1975).

dynamic interfacial properties. A solution of octanoic acid showed surface dilational viscosity of approximately 0.1 surface poise (sp), while the surface shear viscosity was 0.001 sp. The spreading of an alcohol or carboxylic acid monolayer at the air–water interface has been found to give rise to a transient increase in refractive index of subsurface water layers (O'Brien et al., 1975). It was found that this change could be due to the cooling of the water that accompanies the spreading. The variation of temperature during the spreading of the oleic acid monolayer was measured both by thermistor and by changes of refractive index at different depths (Table 4.18).

It has been suggested that an appreciable quantity of water is transported during monolayer spreading (Schulman and Teorell, 1938). Hydrodynamic theories have been used to describe the amount of water that is moved by monolayer and interfacial turbulence (Marangoni effect) and by transport, leading to surface tension gradients and motion of air above the interface (Crisp, 1946; Ahmid and Hansen, 1972; Mansfield, 1959; Sterling and Scriven, 1959; Suciu et al., 1970; Keulegan, 1951). From these studies, it was concluded that water under a spreading monolayer moves at the same speed as the monolayer to a depth of approximately 0.1 mm or less and that some motion is transferred to the water down to greater depths.

The elasticity of stearic acid monolayers on subphases of different compositions has been investigated (Ksenzhek and Gevod, 1975). The degree of repulsion of lipid molecules in a monolayer was found to be insignificant in the subphase with n univalent electrolytes ($n > 1$).

4.15. MONOLAYERS OF COMPLEX MOLECULES

4.15.1. MONOLAYERS OF VITAMINS

The soil-soluble vitamins A, E, and K_1 are all reported to form stable monolayers on the surface of water. In Fig. 4.42, plots of surface pressure,

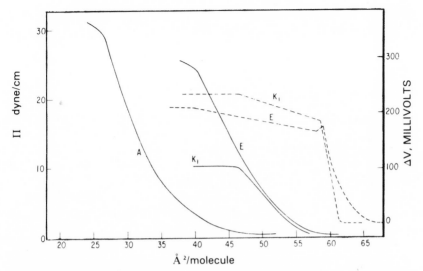

FIGURE 4.42. Plots of surface pressure, π (——), and surface potential, ΔV (- - -), vs. A for spread monolayers of vitamins A, K_1, and E (subphase: 0.1 N HCl, 21°C). Redrawn from Weitzel (1956) with changes.

π, and surface potential, ΔV, vs. A for monolayers of vitamins A, E, and K_1 are given (Weitzel *et al.*, 1952, 1956). It can be seen that the areas at the collapse pressure, π_{col}, are 27, 42, and 47 Å^2 for vitamins A, E, and K_1, respectively. These differences can be ascribed to the structural differences (Fig. 4.43). It can be seen that the explanation of these monolayers is

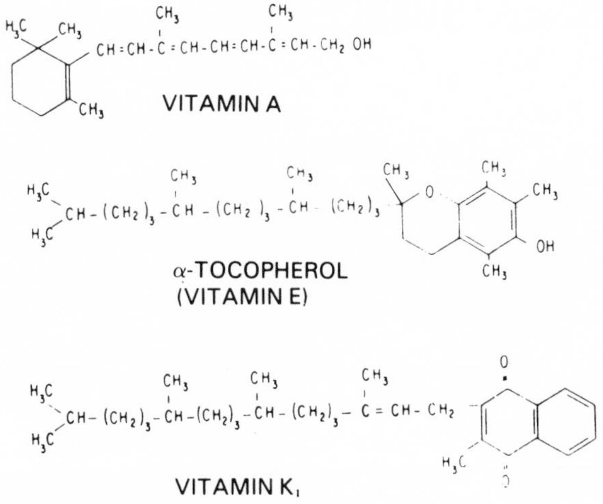

FIGURE 4.43. Structural formulas of vitamins A, E, and K_1.

probably that the molecules of vitamin A pack more tightly than those of vitamins E and K_1. This tighter packing also gives rise to the larger π_{col} for A (30 mN/m) than for E (π_{col} = 23 mN/m) or for K_1 (π_{col} = 10 mN/m).

4.15.2. MONOLAYERS OF STEROIDS

The steroids, and some of the hormones that play an important part in the activity of the reproductive organs, form remarkable surface films, in which a great variety of molecular orientations are found. During the last decade, suggestions concerning the constitution of these substances have undergone extensive changes. Though steroids are known to be of much importance in various biological cell functions and structures, no thorough investigation of the importance of conformation in their monolayers has been undertaken.

Monolayers of cholesterol are very solidlike at 25°C (see Fig. 4.37). The π_{col} is 40 mN/m. The area at collapse is approximately 40 Å^2, which can be used to determine the orientation of molecules at the surface (Adam *et al.*, 1935; Pethica, 1955; Birdi, 1976a). It has been shown that cholesterol monolayers are attacked by oxygen in the air. The monolayers are very compressible; the orientation of the molecules therefores does not change during compression, so there can be little doubt that the molecules are vertically oriented. The measured areas vary from 37 to 44 Å^2.

The surface potentials, ΔV, of these steroids are remarkable, and the differences are probably due to the different angular orientations of the hydroxyl group to the ring system. Cholesterol has ΔV = 400 mV, but epicholesterol has ΔV = almost -100 mV. This drop from a rather high potential to a low value for a single hydroxyl to a negative value is most unusual; it is probably due to the interchange of the positions of the hydrogen and the hydroxyl on carbon number 3.

4.15.3. MONOLAYERS OF QUINONES

In early works on monolayers of p-alkyl phenols, p-alkyl anilines, and oxygenated long-chain derivatives of benzene (Adam, 1923, 1928), it was reported that the limiting areas, A_0, of the *para*-substituted simple benzene derivatives in condensed films generally show a value of approximately 0.24 nm^2/molecule (24 Å^2), which is consistent with close packing of the benzene rings. In a later study, these conclusions were confirmed on the basis of unimolecular film-forming properties of 22 kinds of aromatic azo compounds containing long alkyl chains by measuring π vs. A isotherms at room temperature (Giles and Neustadter, 1952). It was reported that stable condensed films were formed in all the cases in which a hydroxyl group is present on the ring and the alkyl chain has at least 16 carbon atoms. From these studies, it was suggested that the molecules in these condensed monolayers are so oriented that the plane of the aromatic nuclei is vertical to that of the water surface and that the molecules are thus stacked side by side like slices of toast in a rack. Whatever substituent groups may be present, the plane of the ring remains vertical, but the long axis of the aromatic azo group may be tilted in this plane at an angle that depends on the nature and the relative position of the substituent groups.

Monolayers of a series of alcohols, carboxylic acids, and other derivatives containing an anthracene nucleus have been studied (Stuart, 1961). Three types of these derivatives formed fairly stable films, and these films had A_0 in the range of 0.45–0.48 nm^2/molecule (45–48 Å^2). Another study reported the monolayer properties as determined by π vs. A isotherms of several coumarins and 2-f-chromines containing one or two long-chain substituents in relation to their molecular structures (Pommier *et al.*, 1979). Monolayer properties of substances having polycyclic aromatic head groups such as naphthalene, anthracene, pyrene, and pyrelene have been studied in relation to their chemical structures (Stevens *et al.*, 1983).

Monolayers of seven anthraquinone derivatives containing one or two sterol amino groups at different positions of the ring system spread out on the air–water interface (Fukuda *et al.*, 1976). It was concluded that different types of orientation of the anthraquinone could be realized in the surface films of these long-chain dyes, depending on chemical modifications such as variations in the number and relative positions of the long substituent groups.

Monolayers of iris-quinone (2'-*cis*-10-heptadecenyl-6-methoxy-*p*-benzoquinone) were studied in a recent report (Kato *et al.*, 1986). Iris-quinone is a natural substance that is isolated from the seed oil of *Iris pseudacorusl*, a kind of iris indigenous to the area in and around Utsunoniya, Japan; it has a *para*-benzoquinone head group. This substance is of particular interest in that its structure is very similar to those of plastoquinone and ubiquinone, which exhibit biological activities in the chloroplasts of higher plants or in mitochondrial cells. The temperature dependence of π vs. A isotherms of iris-quinone has been measured. Isotherms of other derivatives of iris-quinone, i.e., *trans*-iris-quinone, dihydro-iris-quinone, 3-hydroxy-iris-quinone, and 3-hydroxydihydro-iris quinone, have also been investigated. Their π vs. A transition curves were compared with those of oleoyl alcohol and elaidyl alcohol. It is worth mentioning, before discussing these data, that the original aim of studies on monolayer properties of substances having aromatic head groups was to confirm Bragg's X-ray experiments on the dimensions of naphthalene and anthracene molecules with crystals and other experimental techniques (Adam, 1928). It was reported that long-chain derivatives of benzene such as *p*-alkyl phenols, *p*-alkyl anilines, and *p*-alkyl anisoles form solid films below certain temperatures, and the limiting areas, A_0, of these substances, all centering around the value of 0.238 nm^2/molecule (23.8 $Å^2$), agreed well with the value of 0.233 nm^2/molecule (23.3 $Å^2$) for the cross-sectional area of benzene obtained by Bragg using X-ray analysis. This finding means that the plane of the benzene ring of these substances in condensed monolayers is oriented not parallel but perpendicular to the water surface. The phase-transition data of iris quinones were analyzed according to the thermodynamic equations of surface enthalpy, dH, and entropy, dS.

It has been reported (Jalal and Geografi, 1979) that at least in the case of fatty acids (in the undissociated state), transitions appear to occur only when the temperature is below their melting point (bulk melting point). Furthermore, these authors insisted that there exists no true equilibrium transition between the liquid-expanded and the liquid-condensed monolayer phase, but that at best, change between metastable phases may occur in a range of surface pressures above π_{eq}. In general, however, insoluble monolayers in themselves exist in metastable phases originally, and it has been pointed out that if compressional–decompressional time scales are short compared to any long-term transitions of film to bulk phase, it is acceptable to apply a thermodynamic treatment to the resultant film transitions (Mueller-Landau *et al.*, 1980).

4.16. LATERAL DIFFUSION IN LIPID MONOLAYERS

In recent years, there has been much interest in investigations on the lateral diffusion of lipids and biopolymers in biological membranes. This process is present in the membrane fluidity characteristic, which in turn determines the function and stability of the membrane. Fluidity is a prominent feature of the phospholipid part of biological membranes, as well as of model phospholipid bilayer vesicle and monolayer systems, and it has been measured by different methods (Stoeckenius and Engelman, 1969). In only a few experiments has it been possible to document the correlation between fluidity and function. For example, in some cell membranes, an abrupt increase in the rate of transport of metabolites across the cell wall at temperatures above the melting temperature has been reported (Schirer and Overath, 1969).

The physical properties of lipids extracted from cells have also been correlated with the temperature of bacterial growth (Haest *et al.*, 1969). The extremely slow rate of lateral diffusion across the bilayer implies an almost static bilayer with respect to this process. However, if one considers the rapid chain isomerizations in the hydrophobic interior of the bilayer, which occur more than 10^8 times/sec, it is clear that a bilayer is far from static from this point of view. Further, the Brownian motion (diffusion) of particles in membranes (e.g., lipids and proteins) occurs in a highly anisotropic environment. There has been defined for such molecules a translation mobility (independent of velocity) that takes into account the viscosity of the liquid in which the membrane is embedded. The results of a model calculation based on this definition have been reported (Saffman and Delbruck, 1975). This calculation suggested that for a realistic situation, translational diffusion should be about four times faster in relation to rotational diffusion than in the isotropic case. Lateral diffusion studies have been reported (Edidin, 1974) in which different methods have been used in such measurements: (1) electron spin resonance (Devaux and McConnell, 1972); (2) nuclear magnetic resonance (Lindblom *et al.*, 1976); (3) fluorescence (Galla and Sackmann, 1975; MacGrath *et al.*, 1976; Vanderkooi and Callis, 1974); and (4) a radiochemical method (Stroeve and Miller, 1975).

The translational diffusion of 12-(9-anthroyl)stearic acid in different lipid monolayers has been investigated (Teissie *et al.*, 1978). This study reported the formation of dimers on illumination of the monolayers, with a decrease in fluorescence intensity. The lateral diffusion was described with the help of a mathematical model that was based on less realistic boundary conditions, to allow rapid analysis of the measured data. The diffusion equation can be expressed as

$$dc/dt = D\underline{v}^2 c \qquad (4.128)$$

where $\underline{v}$ is the Laplacian operator, D is the lateral diffusion constant, and c is the concentration of the diffusing species at some distance from the origin at time t.

The reduced fluorescence intensity (total fluorescence intensity divided by the number of probes) of anthroylstearate included in the lipid (DPL), as measured after a 15-sec illumination time interval, was reported to show a small but definite maximum at the beginning of the liquid-to-gel phase transition (Teissie *et al.*, 1976, 1978). The values of the diffusion constant, D, were lower by a factor of 10 after the liquid-to-gel transition region on compression (1.3×10^{-5} to 1.1×10^{-6} cm^2/sec). This result is in accord with those of other studies in which it has been found that fluidity decreases considerably when going from the liquid to the gel phase.

It was concluded that the measured diffusion coefficients of anthroylstearate in DPL monolayers ($1.1 < D < 1 \times 10^{-6}$ cm^2/sec) were similar to the coefficients measured in cholesterol monolayers ($0.1 < D < 1 \times 10^{-6}$ cm^2/sec) (Stroeve and Miller, 1975). On the other hand, these D values differ by two orders of magnitude from the D values measured in vesicles for the same lipid ($1 < D < 6 \times 10^{-8}$ cm^2/sec) (MacGrath *et al.*, 1976).

5

BIOPOLYMER (PROTEIN) AND SYNTHETIC POLYMER MONOLAYERS

The adsorption of an amphiphile molecule at the water interface is ascribed to the alkyl part being attracted to the interface. It therefore becomes imperative to determine what kind of interfacial forces could be present in the case of molecules more complex than those of lipids, such as biopolymers (e.g., proteins, synthetic polyamino acids), synthetic polymers, and other polymers, such as cellulose. In the same context, it is of interest to determine whether any comparisons can be made between simple lipidlike molecules and biopolymers. Because of the great importance of biopolymers in everyday life, there exists a vast literature on the physical properties of biopolymers in solution. On the other hand, the interfacial properties of these biopolymers have not been investigated as extensively as they deserve to be.

To make the analyses clearer, biopolymers are divided for the purposes of this text into two broad categories: proteins and other polymers. Proteins are further separated into globular proteins and membrane proteins. Membrane proteins are discussed in Chapter 8 because some of their properties were found to be so uniquely characteristic that separate treatment is warranted.

The hydrophile–lipophile balance (HLB) of such polymers can be described in terms of the interfacial forces that act on the adsorption process. Before the analysis of biopolymer monolayers is discussed, it is useful to consider very briefly the hydrophobic forces that determine the interaction between nonpolar molecules and water.

5.1. HYDROPHOBIC EFFECT (OR FORCES OR INTERACTIONS)

The ability to predict the effects of even simple structural modifications on the aqueous solubility of an organic molecule could be of great value in the development of new molecules in various fields, e.g., medicine or industry. There exist theoretical procedures to predict solubilities of non-polar molecules in nonpolar solvents (Hildebrand et al., 1970) and of salts or other highly polar solutes in polar solvents such as water or similar solvents (Gurney, 1953). However, predicting the solubility of a nonpolar solute in water has been found to require some different molecular consider-ations. Furthermore, the central problems of living matter comprise the following factors: recognition of molecules, leading to attraction or repul-sion; fluctuations in the force of association and in conformation, leading to active or inactive states; and the influence of electromagnetic or gravita-tional fields and of solvents, including ions, and electron or proton scavengers. In the case of life processes on earth, the main interest is in solubility in aqueous media.

The solubility of semipolar and nonpolar solutes in water has been defined in terms of the "molecular surface area" of the solute and the "interfacial tension" (Langmuir, 1925). This concept was later analyzed in much greater detail by various investigators (Uhlig, 1937; Eley, 1939; Herman, 1972; Amidon et al., 1974; Tanford, 1980; Birdi, 1976a,b; Ben-Naim, 1980; Chattoraj and Birdi, 1984). On the basis of this model, the solubility of a solute, X_{solute}, in water was given by the following equation:

$$RT(\ln X_{solute}) = -(\text{surface area of solute})\gamma \qquad (5.1)$$

where γ is some interfacial tension term at the solute–water (solvent) interface. The quantity (surface area of solute) is the cavity dimension of the solute when placed in the water.

The conformational potential energy of a molecule is, in general, given by (Ramachadran and Sasisekharan, 1968)

$$V = V_{nb} + V_{es} + V_l + V_t + V_\Phi + V_{hb} \qquad (5.2)$$

where the subscript "nb" denotes the nonbonded energy, "es" the electro-static energy, l the strain energy associated with the stretching of bonds, t the strain energy due to bending of bonds, Φ the torsional potential, and "hb" the hydrogen–bond formation energy. The quantity V_{hb} is the sum of two terms, a van der Waals attraction term and a repulsion term. For example, in simple hydrocarbons, since there is not very much stretching or bending deformation, the van der Waals interactions are the most impor-tant. The rotations about near-single bonds, i.e., the nonbonded interactions, make the major contributions to the torsional potential (Hopfinger, 1973). The surface areas of the solute were calculated by computer programs (Amidon et al., 1974).

Data on solubility, total surface area (TSA), and hydrocarbon surface area (HYSA) are given in Table 5.1 for some typical alkanes and alcohols.

TABLE 5.1. Solubility, Surface Areas (TSA and OHSA), Boiling Point (b.p.), and Predicted Solubility [(from Equation (5.6)] of Different Molecules in Water at 25°C[a]

Compound	Solubility[a] (molal)	TSA (Å^2)	OHSA (Å^2)	b.p. (°C)	Predicted solubility (molal)
n-Butane	2.34×10^{-3}	255.2	—	—	1.43×10^{-3}
Isobutane	2.83×10^{-3}	249.1	—	—	1.86×10^{-3}
n-Pentane	5.37×10^{-4}	287	—	—	3.65×10^{-4}
2-Methylbutane	6.61×10^{-4}	274	—	—	6.21×10^{-4}
3-Methylpentane	1.48×10^{-4}	300	—	—	2.08×10^{-4}
Neopentane	7.48×10^{-4}	270	—	—	7.52×10^{-4}
Cyclohexane	6.61×10^{-4}	279	—	—	5.11×10^{-4}
Cycloheptane	3.05×10^{-4}	301.9	—	—	1.92×10^{-4}
Cyclooctane	7.05×10^{-5}	322.6	—	—	7.89×10^{-5}
n-Hexane	1.11×10^{-4}	310	—	—	1.23×10^{-4}
n-Heptane	2.93×10^{-5}	351	—	—	2.33×10^{-5}
n-Octane	5.79×10^{-6}	383	—	—	5.87×10^{-6}
n-Butanol	1.006	272	59	118	0.82
2-Butanol	1.07	264	43	100	1.5
n-Pentanol	0.255	304	59	138	0.21
n-Hexanol	0.06	336	59	157	0.053
Cyclohexanol	0.38	291	50	161	0.43
n-Heptanol	0.016	368	59	176	0.014
1-Octanol	4.5×10^{-3}	399	59	195	3.45×10^{-3}
1-Nonanol	0.001	431	59	213	0.00088
1-Decanol	2.0×10^{-4}	463	59	230	2.24×10^{-4}
1-Dodecanol	2.3×10^{-5}	527	59	—	1.43×10^{-5}
1-Tetradecanol	1.5×10^{-6}	591	59	264	9.4×10^{-7}
1-Pentadecanol	5×10^{-7}	623	59	—	2.4×10^{-7}

[a] From Amidon et al. (1974).

The relationship between solute solubility and surface area of contact between solute and water at their interface was derived from (Amidon et al., 1974)

$$\ln(\text{sol}) = -0.043\text{TSA} + 11.78 \tag{5.3}$$

where (sol) is the molar solubility and TSA is in Å^2. In the case of alcohols, assuming a constant contribution from the hydroxyl group, HYSA = TSA − hydroxyl group surface area (OHSA):

$$\ln(\text{sol}) = -0.0396\text{HYSA} + 8.94 \tag{5.4}$$

However, one can also derive a relationship that includes both HYSA and OHSA:

$$\ln(\text{sol}) = -0.043\text{HYSA} - 0.06\text{OHSA} + 12.41 \qquad (5.5)$$

The relationships as given above did not give satisfactory correlations with the measured data (≈ 0.4–0.978). The following relationship was derived on the basis of solubility data for both alkanes and alcohols (which gave correlations of the order of 0.99):

$$\ln(\text{sol}) = 0.043\text{HYSA} + 8.003\text{IOH} - 0.0586\text{OHSA} + 4.42 \qquad (5.6)$$

where the IOH term equals the number of hydroxyl groups, and 0 if the hydroxyl group is not present.

The term HYSA can thus be assumed to represent the quantity that relates to the effect of the hydrocarbon part on solubility. The effect is negative, and the magnitude of γ is 17.7 ergs/cm^2. The magnitude of OHSA is found to be 59.2 Å^2. As an example, the surface areas of each group containing a carbon atom and of the hydroxyl group in the molecule 1-nonanol were estimated (Table 5.2). It can be seen that the surface area of the terminal methyl group [C-9 (84.92 Å^2)] is about three times greater than that of five of the methylene groups [C-3–7].

The solubility model was tested against the prediction for the complex cholesterol molecule (Amidon et al., 1975). The experimental solubility of cholesterol is reported to be approximately 10^{-7} M. The predicted value

TABLE 5.2. Surface Areas of the Methyl Group, Each Methylene Group, and the Hydroxyl Group in 1-Nonanol[a]

Group:	CH_3	CH_2	CH_2	CH_2	CH_2	CH_2	CH_2	CH_2	CH_2	OH
C-n[b]:	9	8	7	6	5	4	3	2	1	

Surface area (Å^2) at the interface between solute and solvent

OH										59.15
C-1									45.43	
C-2								39.8		
C-3							31.82			
C-4						31.82				
C-5					31.82					
C-6				31.82						
C-7			31.82							
C-8		42.75								
C-9	84.92									

[a] From Amidon et al. (1974).   [b] (C-n) C atom number.

was approximately 10^{-6}. It is obvious that further refinements are needed for predicting the solubilities of such complex organic molecules.

In the following section, this solubility model is applied to the prediction of adsorption energies of such complicated molecules as proteins.

5.2. PROTEIN MONOLAYERS AT THE LIQUID INTERFACE

It is well known that when an aqueous solution of a protein (e.g., of concentration 1 mg/ml) is carefully applied to the surface of an aqueous salt solution, the protein molecules tend to unfold (Birdi et al., 1972; Birdi, 1972, 1973a; Yamashita and Bull, 1967; Joos, 1968; Kretzschmar, 1969; Davies, 1954; Mitchell et al., 1971; Blank et al., 1973; Watterson et al., 1974; Joos, 1975; Yamashita, 1977; Chattoraj and Birdi, 1984). The degree of unfolding would be consistent with the maximum lowering of the surface free energy. Thus, monolayer studies should be useful in providing information about the forces at the surface, especially the forces involved in unfolding–folding phenomena (which can be classified as hydrophobic, electrostatic, and other polar forces, such as hydrogen bonds).

In the case of protein monolayer studies, a few procedures have been described in the literature on the formation of such spread monolayers (Birdi, 1972; Bull, 1947a,b; Joos, 1968; Chattoraj and Birdi, 1984). In some procedures, an additional solvent such as n-propanol was used, to induce unfolding. However, this technique can be less desirable, because the degree of unfolding may thus not be measured. The most useful procedure has therefore been to place the protein solution very carefully on the surface of the bulk phase, thereby allowing the unfolded protein to spread out as it interacts with the interface.

5.2.1. MONOLAYERS OF DIFFERENT POLYMERS

Monolayer films of complex carbohydrates and synthetic polymers and of derivatives of cellulose have been investigated (Katz and Samwel, 1929; Adam, 1933; Harding and Adam, 1933). The ethers of cellulose with methyl, ethyl, and benzyl alcohols spread well. The esters spread less well, but the acetates spread almost completely on the surface of water. The nitrates, although they spread very incompletely on water, can be completely spread on 2 N caustic soda, and these molecules are so rapidly denitrated on this solution that the spread films probably consist of alkali cellulose. Highly nitrated celluloses ($\geqslant 12.6\%$ nitrogen) do not spread completely even on strong soda solutions. All cellulose derivatives show the same general type of π vs. A and ΔV vs. A isotherms. At A of more than about 60 Å^2 per hexose group, π is usually not over 1-2 mN/m, except with considerably depolymerized derivatives such as crystalline trimethyl- and triethylcelluloses; this area is approximately that of a hexose ring lying flat on the surface. The films can be compressed without collapse to about 40 Å^2; during this compression, the value of π falls considerably, indicating a reorientation of the hexose groups, probably a tilting of the plane of the hexose rings toward the water surface. It seems likely that the cellulose chains are stretched out fully on the surface when sufficient area is available. The benzyl ethers require more area, from 65 to 80 Å^2, according to the proportion of the bulky benzyl groups in the molecule.

5.2.2. DEGREE OF UNFOLDING AND AMINO ACID COMPOSITION OF PROTEIN MOLECULES

It is now well established that the folding–unfolding process of proteins in solutions is of considerable importance in understanding the biological reactions. It has long been hypothesized that the information necessary for the folding of a polypeptide chain into a biologically active state is contained within its amino acid sequence (Epstein *et al.*, 1963). Investigation of this hypothesis has led to the reporting of a great number of studies on the prediction of the three-dimensional structure from the amino acid composition or sequence (Davies, 1954; Jones, 1975; Robson and Pain, 1971; Finkelstein and Pititsyn, 1971; Krigbaum and Knuutton, 1973; Palau and Puigdomenech, 1974; Lim, 1974; Burgess *et al.*, 1974; Nagano, 1977; Garnier *et al.*, 1978; Chou and Fasman, 1978; M.-Rias *et al.*, 1987). Different algorithms have been used in these reports. These methods have been used to estimate and predict the three-dimensional structure of a protein molecule from its amino acid sequence. These investigations have been able to predict, with variable success, the α-helix, β-structure, and random coil.

However, the process of unfolding of proteins at interfaces has been analyzed by very few investigators (Birdi, 1973a; Scheller *et al.*, 1975). In an early study, the strong correlation between the amino acid composition or sequence and the degree of unfolding at the air–water interface of various proteins was clearly reported (Birdi, 1973).

The work of compression, W_c, can be estimated from the π vs. A isotherm (Chattoraj and Birdi, 1984):

$$W_c = \int_{\text{initial}}^{\text{final}} \pi dA \qquad (5.7)$$

The integral in equation (5.7) is thus the area under the π vs. A isotherm, where A_{initial} is large ($\pi \approx 10$) and A_{final} is at the collapse point (Fig. 5.1).

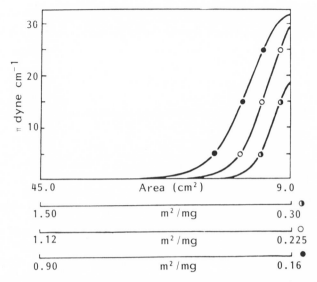

FIGURE 5.1. π vs. A isotherm for ovalbumin. Amount of protein at surface: 3.0 μg (◐); 4.0 μg (○); 5.0 μg (●). Data from Birdi (1972) and Chattoraj and Birdi (1984).

A plot of π vs. surface concentration (C_s) for insulin and insulin–zinc monolayers is given in Fig. 5.2. A linear relationship is observed between W_c and the amount of protein at the surface (Fig. 5.3). The compression rates had no effect on the magnitudes of W_c.

Failure of the plot of a system to cross the 0,0 coordinate merely indicates that some of the protein is lost into the subphase.

Dynamic studies of protein monolayers have been made (Birdi, 1973a). The fact that the difference between the work of compression, W_c, and

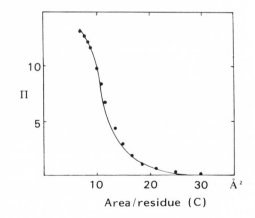

FIGURE 5.2. π vs. C_s [surface concentration (=area/residue)] for insulin and zinc–insulin monolayers at an air–water interface (subphase: phosphate buffer, pH 7.4, 0.43 N NaCl, 25°C). From Birdi (1976b).

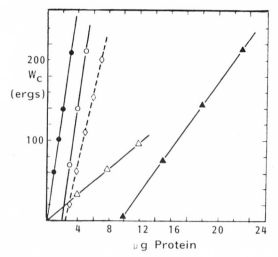

FIGURE 5.3. Plot of W_c as a function of amount of protein added at the surface: bovine serum albumin (●), ovalbumin (○); β-lipoprotein (◇); transferrin (△); myoglobin (▲). From Birdi (1972).

μg Protein

work of expansion, W_e, was not zero indicated that there exists some degree of hysteresis. These data were more complex than the lipid monolayer studies (Chapters 4 and 7). However, the data in Table 5.3 show that hysteresis might be correlated to the degree of unfolding (Table 5.4). Infrared (IR) spectra of spread monolayers of paramyosin and β-lactoglobulin have been investigated (Loeb, 1969). Both proteins gave π vs. A isotherms with a plateau at 13 mN/m for paramyosin and 18 mN/m for β-lactoglobulin. These films were transferred on solid surfaces (as described in Chapter 9). The IR spectra were analyzed for these proteins. The transmission at $1655–1660 \text{ cm}^{-1}$ (amide I region) was less than at $1630–1635 \text{ cm}^{-1}$ when the spectra for transferred protein were compared to those for evaporated protein. These data are given in Table 5.5 for comparison.

The correlation of the IR spectrum and the secondary structure of polypeptide chains has been well established (Fasman, 1967). The major

TABLE 5.3. Dynamic Studies of Protein Monolayers[a,b]

Protein	Hysteresis (cal/residue)[b]	Degree of unfolding	Molecular weight
Bovine serum albumin	95	Complete	68,000
Ovalbumin	87	Complete	40,000
Insulin	197	Complete	6,000
Myoglobin	12	Incomplete	17,000

[a] From Birdi (1973a). Subphase buffer: pH 7.4, 25°C.
[b] Hysteresis = $W_c - W_e$.

TABLE 5.4. Degree of Unfolding of Different Proteins as a Function of Various Molecular Parameters[a]

Protein	Degree of unfolding	Polar ratio	Helix (%)	Number of S–S bridges	Number of chains	Molecular weight
Insulin	Complete	1.08	37	3	2	6,000
Zn–insulin	Complete	1.08	37	3	2	6,000
Hemoglobin	Complete	1.03	75	0	4	68,000
Ovalbumin	Complete	1.26	45	0	1	40,000
Bovine serum albumin	Complete	1.56	50	17	1	68,000
Cytochrome c	Incomplete	2.1	0	0	1	14,000
Lysozyme	Incomplete	1.74	35	1	1	12,000
Transferrin	Incomplete	1.62	—	—	1	90,000
Myoglobin	Incomplete	1.29	75	0	1	17,000

[a] From Birdi (1976b).

absorption band of β-type structures is 1630–1635 cm^{-1}, while α-helical structures correspond to absorption in the range of 1650–1660 cm^{-1}. The ratio of optical dispersion (OD) at 1635 cm^{-1} to that at 1655 cm^{-1} can be taken as an index of conformation (Table 5.5).

Paramyosin is predominantly α-helical and therefore exhibits a lower ratio in Table 5.5. β-Lactoglobulin contains significant amounts of β-structures. These results are compared with those for synthetic polyamino acid monolayers later.

In the adsorption process at the air–water interface (and in most cases also at other interfaces, e.g., oil–water, mercury–water, or solid–water) for monolayer formation, one can safely assume as a first approximation that the energy of adsorption is determined primarily by the nonpolar amino acid residues. One will also expect there to be some adsorption-energy contribution from some of the polar side chains, e.g., Cys (Table 5.6).

TABLE 5.5. IR Spectral Analyses of Transferred Monolayer for Paramyosin and Evaporated and Transferred Monolayer for β-Lactoglobulin[a]

Protein	Ratio[b]
Paramyosin (transferred)	0.6
β-Lactoglobulin (evaporated)	1.1
β-Lactoglobulin (transferred)	0.9

[a] From Loeb (1969).   [b] OD at 1635 cm^{-1}/OD at 1655 cm^{-1}.

TABLE 5.6. Classification of Common Amino Acids on the Basis of Their Electrical Properties

Hydrophobic (Nonpolar) Amino Acids

$$H_2N-\underset{\underset{H}{|}}{\overset{\overset{H}{|}}{C}}-COOH \qquad H_2N-\underset{\underset{CH_3}{|}}{\overset{\overset{H}{|}}{C}}-COOH \qquad H_2N-\underset{\underset{CH}{|}}{\overset{\overset{H}{|}}{C}}-COOH$$

$$\qquad\qquad\qquad\qquad\qquad\qquad\qquad\qquad\qquad H_3C \diagdown\ \diagup CH_3$$

Glycine (Gly) $\qquad\qquad$ Alanine (Ala) $\qquad\qquad$ Valine (Val)

$$H_2N-\underset{\underset{CH_2}{|}}{\overset{\overset{H}{|}}{C}}-COOH \qquad H_2N-\underset{\underset{CH_2}{|}}{\overset{\overset{H}{|}}{C}}-COOH \qquad$$

Leucine (Leu) $\qquad$ Isoleucine (Ile) $\qquad$ Proline (Pro)

Phenylalanine (Phe) $\qquad$ Tryptophan (Trp) $\qquad$ Methionine (Met)

Polar Amino Acids

$$H_2N-\underset{\underset{OH}{\overset{|}{CH_2}}}{\overset{\overset{H}{|}}{C}}-COOH \qquad H_2N-\underset{\underset{CH_3}{\overset{|}{HC-OH}}}{\overset{\overset{H}{|}}{C}}-COOH \qquad H_2N-\underset{\underset{SH}{\overset{|}{CH_2}}}{\overset{\overset{H}{|}}{C}}-COOH$$

Serine (Ser) $\qquad\qquad$ Threonine (Thr) $\qquad\qquad$ Cysteine (Cys)

Tyrosine (Tyr) $\qquad$ Asparagine (Asn) $\qquad$ Glutamine (Gln)

TABLE 5.6 (*continued*)

Charged Amino Acids

$$
\begin{array}{c}
\text{H} \\
\text{H}_2\text{N}-\overset{|}{\underset{|}{\text{C}}}-\text{COOH} \\
\text{CH}_2 \\
\overset{|}{\text{C}} \\
\overset{\diagup}{\text{HO}} \quad \overset{\diagdown\!\!\!\diagdown}{\text{O}}
\end{array}
$$

Aspartic acid (Asp)

$$
\begin{array}{c}
\text{H} \\
\text{H}_2\text{N}-\overset{|}{\underset{|}{\text{C}}}-\text{COOH} \\
\text{CH}_2 \\
\text{CH}_2 \\
\overset{|}{\text{C}} \\
\overset{\diagup}{\text{HO}} \quad \overset{\diagdown\!\!\!\diagdown}{\text{O}}
\end{array}
$$

Glutamic acid (Glu)

$$
\begin{array}{c}
\text{H} \\
\text{H}_2\text{N}-\overset{|}{\underset{|}{\text{C}}}-\text{COOH} \\
\alpha \;\; \text{CH}_2 \\
\beta \;\; \text{CH}_2 \\
\tau \;\; \text{CH}_2 \\
\delta \;\; \text{CH}_2 \\
e \;\; \text{NH}_2
\end{array}
$$

Lysine (Lys)

$$
\begin{array}{c}
\text{H} \\
\text{H}_2\text{N}-\overset{|}{\underset{|}{\text{C}}}-\text{COOH} \\
\text{CH}_2 \\
\text{CH}_2 \\
\text{CH}_2 \\
\text{NH} \\
\text{C}=\text{NH} \\
\text{NH}_2
\end{array}
$$

Arginine (Arg)

$$
\begin{array}{c}
\text{H} \\
\text{H}_2\text{N}-\overset{|}{\underset{|}{\text{C}}}-\text{COOH} \\
\text{CH}_2 \\
\text{C}=\!=\text{CH} \\
\text{HN} \quad \text{N} \\
\text{CH}
\end{array}
$$

Histidine (His)

Analogous to the analysis of the folding phenomena of proteins in the bulk aqueous phase, one can assume that the nonpolar residues are Ala, Leu, Ile, Val, Met, Phe, Trp, and Tyr (Dickerson and Gies, 1969) (Table 5.6). An analysis of the amino acid composition of 48 different proteins (Hatch, 1965) showed that the molar ratio of polar to nonpolar amino acids varied between 1 and 3. There were no proteins for which the ratio was less than 1.

Furthermore, by analogy with the bulk-folding algorithms, it can be argued that short-range interactions are mainly responsible for adsorption (Burgess *et al.*, 1974). Long-range interactions are of much less significance in the unfolded state, as described in detail by various investigators in the case of the bulk phase.

The W_c values for various proteins were found to be different (Birdi, 1972) (Fig. 5.3). The data on W_c in Table 5.7 show some correlation with the HLB of the protein. Of the various algorithms for predicting three-dimensional conformation given in the literature, some have been based on the principle that hydrophobic side chains will be expected to interact if they are present at the i, $i + 1$, $i + 2$, $i + 3$, or $i + 4$ position in the sequence (Palau and Puigdomenech, 1974). This postulate has been used quite successfully in predicting the degree of unfolding at the interface (Birdi, 1976b).

TABLE 5.7. Magnitude of W_c of
Different Protein Monolayers[a]

Protein	W_c/residue (cal)
Bovine serum albumin	182
Ovalbumin	188
Transferrin	20
Myoglobin	36

[a] From Birdi (1972).

The hydrophobic character of the polypeptide chain varies in a sequence for each segment of five amino acids from i to $i + 4$. Table 5.8 lists the number of nonpolar amino acids for each successive segment of five residues for various proteins from their sequence data, as exemplified for insulin in Fig. 5.4. It is obvious that because of the simplicity of this model, one should expect some exceptions. It can also be concluded that merely considering the total number of segments that are nonpolar and that are responsible for the adsorption is not going to provide a completely satisfactory analysis, as can also be seen from the data in Table 5.4. However, it is noteworthy that the distribution of segments with high adsorption energy is a significant characteristic exhibited by these plots. If the nonpolar character in a polypeptide chain is distributed more or less evenly throughout the chain, then the molecule should be expected to adsorb and unfold completely, whereas if this distribution is uneven (patchy) and concentrated in certain parts of the molecule, then the adsorption will be incomplete. It is also useful to compare the plots of two typical proteins, e.g., hemoglobin and cytochrome c. It is found that in hemoglobin, the nonpolar (adsorbing) segments are distributed fairly evenly throughout the chain, thus giving complete unfolding, as has also been found in experiments (area/residue ≈ 17 Å^2). On the other hand, the cytochrome molecule shows predominant

TABLE 5.8. Number of Nonpolar Residues in Each Successive Segment of Five Residues in the Sequences of Various Proteins[a]

Hemoglobin (bovine, α-chain)	41332333413212143333122332433433321
Myglobin (whale)	22443224221223213330132133232122232322
Lysozyme (egg)	224122331110131201123121214430131
Ribonuclease (bovine)	2311212210222312001210231 1414321
Cytochrome c (bovine)	11311002202211223132410431
Insulin	21231212422

[a] From Birdi (1977).

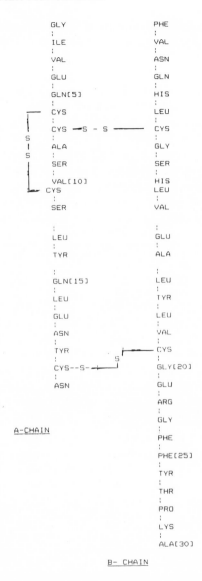

FIGURE 5.4. Primary sequence of amino acids of insulin (monomer). The sequence consists of two polypeptide chains in the insulin that are held together by two disulfide bridges (31% α-helix, mol. wt. 5700).

patches of polar features (determined by the polar amino acid residues), which would be expected to give no (or poor) adsorption energy, and thus to lead to prediction of incomplete unfolding at the interface, as has also been found in experiments (area/residue ≈4 Å^2).

Unfortunately, at this stage, no sequences for other protein molecules are available, so this algorithm cannot be extensively developed at present.

Furthermore, this type of algorithm will be difficult to apply to comparatively small polypeptides, such as insulin, melittin, or valinomycin (Chapter 8).

However, some quantitative dimensions could be added to the aforedescribed model unfolding mechanisms. The hydrophobic energy required to transfer a hydrocarbon chain from water to the interface, e.g., air–water, according to the Traube rule (Traube, 1891) is of the order of 700 cal/mole (2900 J/mole) per $-CH_2-$ group. It is also known that the hydrophobic energy (see above) is most correctly defined by considering the surface area of the alkyl group in contact with water (Hermann, 1972; Amidon *et al.*, 1975; Birdi, 1976b; Anik, 1978). The magnitudes of the surface area of each nonpolar amino acid were estimated by using the data given by various investigators (Amidon *et al.*, 1974). The standard free energy of transfer from water to air, $\Delta G^0_{w \to a}$, can be estimated from the following relationship (Amidon *et al.*, 1975; Birdi, 1976b):

$$\Delta G^0_{w \to a} = 25.5(\text{surface area of molecule}) \quad (\text{cal/mole}) \quad (5.8)$$

$$= 107(\text{SA}) \quad (\text{J/mole}) \quad (5.9)$$

where surface area (SA) is only for the nonpolar alkyl part of the molecule. These values are given in Table 5.9. The hydrophobicity of amino acids has earlier been designated as determined from the free energy of transfer from water to the organic-solvent phase (Tanford, 1963; 1962). The measured values of $\Delta G^0_{w \to a}$ are also given in Table 5.9, for comparison. Note that the

TABLE 5.9. Measured and Calculated Values of Hydrophobic Free Energy of Transfer from Water to Air ($\Delta G_{w \to a}$) or from Water to Oil ($\Delta G_{w \to o}$) for Nonpolar Amino Acid Residues[a]

Amino acid	Area of nonpolar ($Å^2$)[b]	$\Delta G_{w \to a}$ [c]	$\Delta G_{w \to o}$ [d] (kcal/mol)	Ratio $\Delta G_{w \to a}/\Delta G_{w \to o}$
Proline	140	7.1	2.65	3
Phenylalanine	279	3.6	2.6	1.4
Methionine	213	5.4	1.3	4
Valine	166	4.2	1.7	2.5
Isoleucine	253	6.4	2.95	2.2
Leucine	211	5.4	2.4	2.3
Alanine	85	2.2	0.75	3
Tyrosine	278	7.1	2.85	2.5
Tryptophan	294	7.5	3.0	2.5

[a] From Birdi (unpublished).   [b] Calculated from Amidon *et al.* (1975).
[c] Calculated from equation (5.8).   [d] From Tanford (1980).

ratio $\Delta G^0_{w \to a} / \Delta G^0_{w \to o}$ ranges from 1.4 to 4. This variation is not surprising, since the two transfer processes would be expected to be different (mainly due to entropic forces).

In the following discussion, it will be assumed that the energy of transfer from the bulk phase to the interface is approximately the same as that of the transfer of an alkyl group from water to the organic phase. In this process, only the energy of adsorption for the transfer of the nonpolar group of each residue will be calculated, i.e., Ala, Val, Leu, Ile, Met, Pro, Tyr, and Trp. This procedure thus gives the magnitudes of the energy for each five-residue segment from residue i to residue $i + 4$. These calculated energies of various proteins with known sequences are given in Fig. 5.5. The following proteins have been found to unfold partially: cytochrome c, lysozyme, transferrin, and myoglobin (see Table 5.4), and ribonuclease. The data in Fig. 5.5 show that in the cases of ribonuclease, lysozyme, and myoglobin, there are rather extensive patches in the chain in which there are predominantly polar residues; hence, the magnitude of $\Delta G^0_{w \to a}$ is very

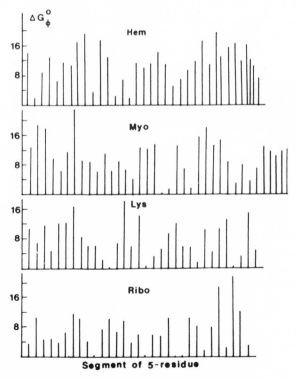

FIGURE 5.5. Magnitudes of adsorption energy, ΔG_{i+4}, per segment of five residues. From Chattoraj and Birdi (1984).

low. A hydrophobic energy value of 0 merely indicates that there are no nonpolar residues in that five-residue segment of the chain. Comparing these proteins with hemoglobin, one finds that in the latter, there are extensive patches of hydrophobic character. This observation would suggest that the hemoglobin molecule should adsorb differently from the other proteins, a prediction that is reported to be in agreement with the experimental data.

It is well known that the hemoglobin and myoglobin proteins differ very little in their folded state; their amino acid compositions are also quite similar. The data in Fig. 5.5 indicate that myoglobin exhibits a number of regions with hydrophobic energy of less than 10 kcal (42 kJ) as compared with the hemoglobin molecule. Another property that is of vital interest in the adsorption process is that the low-energy regions in myoglobin are more concentrated close together, whereas in the hemoglobin molecule, such regions are more evenly distributed near the region of high hydrophobic energy.

An analysis of adsorbed and spread monolayers of lysozyme and a series of its acetyl derivatives at the air–water interface has been reported (Adams *et al.*, 1971) (Fig. 5.6). Adsorption and desorption were carried out by directly monitoring the π and surface radioactivity of films of native and heat-denatured [^{14}C]acetyl lysozyme. Acetylation was reported to be very effective in altering the surface properties of lysozyme. Lysozyme is lost into the subphase, as was also reported for other proteins by different investigators. There was no loss into the subphase on the 3.5 M KCl subphase. On the other hand, as described above, the amount lost can be estimated from W_c plots.

The compressibility of human serum albumin (HSA) monolayers was reported to increase after heat denaturation (Yatsyuk, 1972). The compressibility increased with increasing Cu^{2+} concentration in the subphase.

The kinetics of adsorption from aqueous solutions to the interface (Chapter 6) also supports these conclusions (Chattoraj and Birdi, 1984). Furthermore, the adsorption at interfaces other than air–water, e.g., solid–liquid (Norde, 1976; Andrade, 1985; Birdi, unpublished) and mercury–water (Scheller *et al.*, 1975), has supported the dependence of adsorption at interfaces on the amino acid composition of proteins, as postulated above (Birdi, 1973a).

The thickness of monolayers of lysozyme on water was found to be 2.0 nm (=20 Å) (Kan and Teruk, 1977). This value was twice that of ovalbumin.

In the analyses discussed above, amino acids were characterized only in terms of their hydrophobic–hydrophilic properties. In recent studies, however, other physical characteristics of amino acids have also been used to describe the folding–unfolding processes of protein molecules (Table

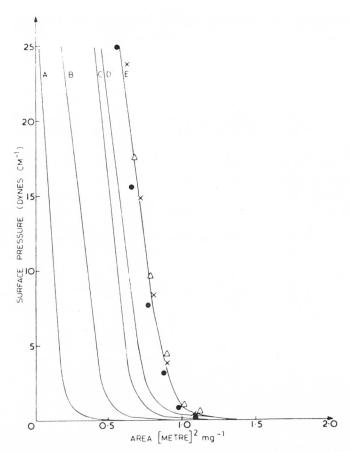

FIGURE 5.6. π vs. A isotherms of monolayers of native and acetylated lysozyme [with various degrees of acetylation (%)] at the air–water (phosphate buffer) interface at 25°C: (A) 0% (native); (B) 20%; (C) 51%; (D) 61%; in subphase of 3.5 M KCl: (E) 0% (native); (●) 20%; (×) 30%; (△) 61%. From Adams *et al.* (1971).

5.10). In an extensive series of studies, amino acids were characterized with respect to a number of physical parameters (Prabhakaran and Ponnuswamy, 1979): hydrophobicity; polarity; acidity; molecular weight; bulkiness; chromatographic index; refractivity; short-, medium-, and long-range energetics; and ability to adopt α-helical, extended, and bent structures. On the basis of analyses of 24 globular protein X-ray structures, the statistical preferences for these proteins in six shells were depicted. The innermost shell showed a preference for nonpolar residues, the outermost shell a preference for both polar and nonpolar residues: Ser, Gly, Ala, and Thr. Almost all the polar residues and Pro were found in shells V and VI. The

TABLE 5.10. Comparative Physical Properties of Amino Acids in Relation to Proteins[a,b]

Amino acid	Total number in proteins	Number buried	Bulkiness	Polarity	Hydrophobicity	Probability of α/β
Val	163	91	21.57	0.13	1.87	1.41/1.65
Ala	183	71	11.5	0.00	0.87	1.45/0.97
Ile	106	69	21.40	0.13	3.15	1.0/1.6
Gly	160	60	3.40	0.00	0.10	0.53/0.81
Leu	138	57	21.40	0.13	2.17	1.34/1.22
Ser	190	46	9.47	1.67	0.07	0.79/0.72
Thr	128	32	15.77	1.66	0.07	0.82/1.2
Phe	60	29	19.80	0.35	2.87	1.12/1.28
Asp	117	17	11.68	49.70	0.66	0.98/0.8
Cys	34	16	13.46	1.48	1.52	0.77/1.3
Pro	67	16	17.43	1.58	2.77	0.59/0.62
Met	28	14	16.25	1.43	1.67	1.2/1.67
Tyr	98	13	18.03	1.61	2.67	0.61/1.29
Glu	65	13	13.57	49.90	0.67	1.53/0.26
Asn	116	12	12.82	3.38	0.09	0.73/0.65
Trp	39	9	21.67	2.16	3.77	1.14/1.19
His	43	8	13.69	51.60	0.87	1.24/0.71
Lys	119	5	15.71	49.50	1.64	1.07/0.74
Gln	80	5	14.45	3.53	0.00	1.17/1.23
Arg	63	0	14.28	52	0.85	0.79/0.9

[a] From Birdi (to be published).   [b] From Prabhakaran and Ponnuswamy, 1979.

preferred spatial arrangement in the six shells is depicted schematically in Fig. 5.7. Future descriptions of protein adsorption energetics should take these various parameters into consideration.

Physical characteristics have also been investigated using data on the thin-film dialysis rates of various amino acids and peptides (Burachik *et al.*, 1970; Birdi and Schack, 1973). The measured half-escape times vs. $M^{1/3}$ (where M is the molecular weight) were found to give linear plots (Birdi and Schack, 1973). However, the slope of Gly, $(Gly)_2$, $(Gly)_3$, $(Gly)_4$, $(Gly)_5$, Ala, $(Glu)_2$, (Tyr-Leu), and (Leu-Ser) was 17. On the other hand, the various Ala peptides, $(Ala)_2$, $(Ala)_3$, and $(Ala)_4$, gave a plot with a slope of 75.

5.2.3. SPECTROSCOPY OF β-LACTOGLOBULIN MONOLAYERS

IR studies of the amide I $(1635\ cm^{-1})$ band of transferred films of β-lactoglobulin spread on 0.5 M KCl were investigated via the multiple

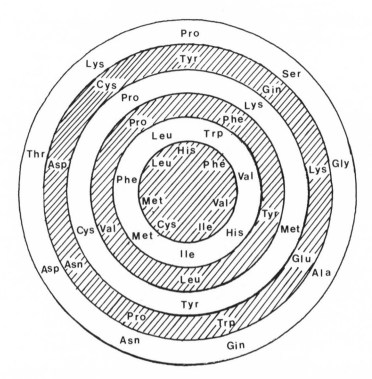

FIGURE 5.7. Schematic representation of preferred residues in successive concentric shells around the centroids of globular protein molecules. The most preferred residue in each shell is underlined, and the rest of the residues are placed clockwise from this residue in order of decreasing preference. Alternate shells are shaded for clarity. From Prabhakaran and Ponnuswamy (1979).

internal reflection technique (Loeb, 1971). The spectra indicated that the conformation of the films changed with π (2–16 mN/m). These data were found to be consistent with a denatured molecule as the predominant species at high area/molecule values and with a species with a secondary structure similar to that of the native molecule at high degrees of compression. The transition was found to be reversed on reexpansion of the film after compression. On the basis of these analyses, it was suggested that approximately equal amounts of native-type and denatured-type spectral components were present when the magnitude of π was 6 mN/m.

The equilibria between adsorbed and displaced segments have been estimated for bovine γ-globulin, catalase, and ferritin monolayers at the air-water interface (MacRitchie, 1981). In these studies, it was assumed that compression of protein monolayers produces molecular configurational changes that are manifested by a falloff of π at constant area or by a reduction in surface area, A, if the monolayers are held at constant π. This phenomenon was earlier studied by Langmuir and Waugh (1940), who referred to it as "pressure displacement." Their procedure was as follows: A few crystals of protein were applied to the surface of a buffer (pH was varied from 3.5 to 7.3) to give $\pi \approx 1$ mN/m. The film was rapidly compressed to $\pi \approx 5$ mN/m. The film was maintained at this π by reducing the area until equilibrium had been established. This protocol was then repeated for the next higher π (10 mN/m), and so on. Two curves were constructed from these data, one of π vs. A at equilibrium and one of π vs. instantaneous area, i.e., the curve that would have been obtained had there been no loss of area. A straight line parallel to the A axis intersecting the π ordinate $(A = 0)$ at $X = 0$, the equilibrium curve at Y, and the instantaneous curve at Z then gave for any value of π

$$\frac{\text{Number of segments at interface } (N_i)}{\text{Number of segments displaced } (N_d)} = \frac{XY}{YZ} \tag{5.10}$$

In this model, it is assumed that only those segments that are adsorbed at the interface contribute to π. One can thus easily derive

$$N_i/N_d = \exp[(\Delta G^0 - \pi A)/kT] \tag{5.11}$$

where ΔG^0 is the standard free energy difference between adsorbed and displaced segments (at $\pi = 0$) and A is the average area per unit segment of the adsorption-displacement process. The plots of $\log(N_i/N_d)$ vs. π were all linear. The data on ΔG^0 and A are given in Table 5.11. The

TABLE 5.11. Values of ΔG ($\pi = 0$) and A for Protein Monolayers at Different Subphase pH and 25°C[a]

Protein	Subphase pH	ΔG (kT)	A (nm^2)[b]
γ-Globulin	7.3	7.3	1.3
	4.4	6.4	1.3
	3.5	5.5	1.4
Catalase	7.3	8.8	1.5
	4.4	8.2	1.6
	3.5	7.5	1.5

[a] From MacRitchie (1981).   [b] 1 nm = 10^{-9} m.

differences in ΔG^0 could be related to the differences in amino acid composition, but this relationship needs to be established.

5.2.4. PROTEIN MONOLAYERS AT THE OIL–WATER INTERFACE

The oil–water interface differs from the air–water interface in that the van der Waals forces, π_{coh}, are negligible. To fully understand the various systems in which an oil–water interface exists, one needs to determine the interfacial forces at such interfaces. However, because of the lack of suitable procedures for obtaining π vs. A isotherms, extensive studies of these systems have not been carried out.

It was suggested that the difficulties in spreading proteins at the air–water interface were obviated by adding propanol acetate (Alexander and Teorell, 1939). The use of mechanical means, such as a glass rod, was also suggested (Trurnit, 1960): Several apparatuses were designed for studying monolayers at the oil–water interface (Askew and Danielli, 1940; Jones *et al.*, 1969; Zilversmit, 1963).

In a recent study (Birdi, 1977), π vs. C_s (surface concentration) plots of protein monolayers at the oil–water interface were investigated (Fig. 5.8). These data are in agreement with earlier results at the oil–water interface. The collapse pressure, π_{col}, of β-lactoglobulin (≈ 15 mN/m) is in agreement with that obtained by the compression method (Zilversmit, 1963). The magnitudes of area/residue at the oil–water interface, approximately 40 Å^2/residue, are somewhat larger than those found at the air–water

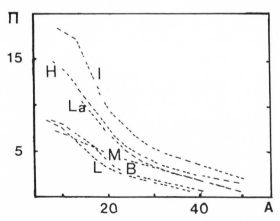

FIGURE 5.8. Surface pressure (π) vs. A (C_s = area/residue) of various protein monolayers at an oil (toluene)–water interface (buffer: ph 7.4, 25°C): Zn–insulin (I); β-lactoglobulin (La); lysozyme (L); hemoglobin (H); myoglobin (M); bovine serum albumin (B). From Birdi (1977).

interface (≈ 25 Å^2/residue). This is the result one would expect, since the absence of the π_{coh} term in equation (4.24) would give more expanded films.

Furthermore, these systems are quite relevant to bacterial adhesion mechanisms (Chapter 7), since the proteins present at the surface are selectively adsorbed. Hence, such investigations can provide much useful information regarding bacterial adhesion mechanisms. In emulsions (i.e., oil–water phases) in which proteins are present, these studies are highly relevant (see Chapter 10). They are also relevant to the interfacial tensions of low-density lipoproteins (from human blood serum) at heptane–water interfaces. The use of polymers in oil recovery would constitute another field to which these investgations are readily applicable (Chapter 10).

5.2.5. RELATIONSHIP BETWEEN HYDROPHOBICITY AND INTERFACIAL TENSION OF PROTEIN SOLUTIONS

It is obvious that the aforedescribed procedure for estimating the hydrophobic properties of protein molecules is not completely satisfactory. For instance, it does not take into consideration the effects from the flexibility and conformation of the protein molecule. This deficiency is particularly important when one attempts to correlate the hydrophobic properties of a protein with its functional properties. This inadequacy is even more apparent when one considers that there are some nonpolar side chains that do remain exposed to the aqueous phase. Furthermore, the nonpolar portion of the charged amino acid lysine, e.g., the four methylene groups, can (and are known to) participate in hydrophobic interactions.

It has been suggested that the interfacial tension of aqueous protein solutions at oil interfaces can be correlated with the protein surface activity (Kinsella, 1976). Data on the interfacial tension of different protein solutions (0.2%) (water–corn oil) are given in Table 5.12 (Keshavaraz and Nakai, 1979). The lower the interfacial tension, the higher the hydrophobicity of the protein, as is known from the behavior of surfactants (Chattoraj and Birdi, 1984). Thus, the hydrophobicity of these proteins is in the order bovine serum albumin (BSA) > ovalbumin > lysozyme > γ-globulin > myoglobin > β-lactoglobulin = trypsin = conalbumin > α-chymotrypsin. This order is in agreement with the degrees of protein hydrophobicity as characterized by the polar/nonpolar amino acid ratio (Section 5.2.2), with some exceptions.

These data on interfacial tension were found to correlate satisfactorily with those on retention coefficients on Sepharose (Pharmacia/Sweden) columns (Keshavaraz and Nakai, 1979), which confirmed the dependence of protein molecules on hydrophobicity. However, more investigations are needed in order to be able to describe these phenomena at a more nearly molecular level.

TABLE 5.12. Interfacial Tension at a 0.2%
Protein Aqueous Solution–Corn Oil
Interface[a]

Protein	Interfacial tension (mN/m)
Bovine serum albumin	6.0
Ovalbumin	6.5
Lysozyme	9.0
γ-Globulin	17.5
Myoglobin	18.0
β-Lactoglobulin	21.5
Trypsin	21.5
Conalbumin	21.5
α-Chymotrypsin	22.5

[a] From Keshavaraz and Nakai (1979).

In another investigation (Neumann et al., 1973), it was reported that the surface tension of poly-L-lysine aqueous solutions changed with temperature and pH. These changes at various temperatures were correlated with changes in the circular dichroic spectra that reflected conformational change. In addition to the major transition at 50°C attributed to the conversion of the α-helical to the β-conformation, two other transitions have been observed, at 30 and 80°C. Furthermore, a minimum in the γ value was observed at pH 10, i.e., near the pK value for polylysine. It was concluded that at this pH, the concentration of hydrophobic side chains at the surface was maximum. Thus, the data from this study support the interfacial tension data described above. In Chapter 10, these physical properties are shown to be related to the formation of emulsions by proteins and to the stability of the emulsions.

The conformation, surface concentration, and thickness of adsorbed protein films have an obvious significance in relation to the structure of biological membranes and the stability of natural colloidal systems. Comparative data on two different proteins, lysozyme and β-casein, have been investigated (Phillips et al., 1975). The values of π at equilibrium (after 16 hr), π_{eq}, of solutions of these two proteins are given in Fig. 5.9. As the bulk protein concentration, C_p, is increased to 10^{-4} g%, adsorption proceeds with a concomitant rise in π to the maximum value of approximately 24 mN/m. However, the π vs. C_p curve for lysozyme is of a different form. Lysozyme, being a rather polar protein, is accordingly found to be less surface-active. The inflection at $\pi = 8$ mN/m is similar to the data on ovalbumin, for which an inflection at $\pi = 12$ mN/m was reported (Bull, 1972). A correlation was found between π vs. C_p curves and Γ (as deter-

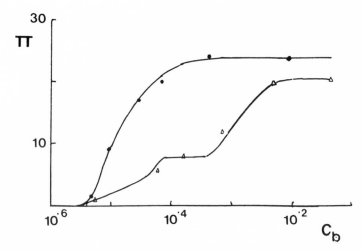

FIGURE 5.9. π_{eq} at the air–water interface as a function of C_p of acetyl-β-casein ($\bullet$) and acetyl-lysozyme ($\triangle$) (pH 7.4, ionic strength 0.1, 22°C). From Phillips *et al.* (1975).

mined by using ^{14}C-labeled proteins) vs. C_p curves. The adsorption isotherms were found to be given by (Phillips *et al.*, 1975)

$$\Gamma_{casein} = 44.36 + 48.73(\log C_p) + 20.44(\log C_p)^2$$
$$+ 3.639(\log C_p)^3 + 0.23(\log C_p)^4 \qquad (5.12)$$
$$\Gamma_{lysozyme} = 99.03 + 93.56(\log C_p) + 34.06(\log C_p)^2$$
$$+ 5.487(\log C_p)^3 + 0.326(\log C_p)^4 \qquad (5.13)$$

where C_p is given in g% and Γ is in mg/m^2.

5.3. MONOLAYERS OF SYNTHETIC POLYAMINO ACIDS

Since synthetic poly-α-amino acids have proven to be most useful models of protein structure, monolayer studies of them are of great interest. π vs. A and ΔV vs. A isotherms of different synthetic polyamino acids were reported (Malcolm, 1971). The π vs. A isotherms indicated a close-packed α-helical arrangement, since a steep rise was observed, and the inflection or plateau arose from the collapse of the monolayer to form a bilayer (Malcolm, 1966, 1968a,b; 1970a,b). However, later studies indicated that through analysis of the collapse state, one could clearly distinguish α-helical polymers from β-polymers (Birdi and Fasman, 1973). The ΔV vs. A isotherms indicated that the potential followed approximately the relationship $\Delta VA = \text{const} = 300$). These data should be better explained by considering the contributions to ΔV that arise from the various parts of the molecule.

The surface viscosities of polyamino acids have been reported by various investigators (Pankhurst, 1958; Joly, 1964). From various π vs. A isotherms, one can estimate the limiting area/residue, A_0, by extrapolation to zero π. Table 5.13 gives these values for different polyamino acids.

As pointed out by some investigators (Birdi and Fasman, 1972, 1973), an important element that had been neglected in most studies was consideration of the conformation of the polypeptides and their monolayer properties. π vs. A isotherms of well-defined polymers, as regards molecular weight and conformation, were investigated (Birdi and Fasman, 1972, 1973). π vs. C_s isotherms of α-helical and β-form polymers are given in Figs. 5.10 and 5.11. The isotherms for the α-helical forms (Fig. 5.10) were reported to give characteristic plateaus at the π_{col} state. The isotherms for the β-forms (Fig. 5.11), on the other hand, showed inflection at the collapse state.

Limiting areas for amino acid residues for different polyamino acids are given in Table 5.14. These data agree with X-ray diffraction data.

The relationship between the magnitude of π_{col} and conformation stability was described in detail (Birdi and Fasman, 1972, 1973). It was reported that π_{col} for the β-form polymers was related to the stabilizing forces in the bulk. However, no such relation was found for α-helical

TABLE 5.13 Comparison of Measured and Calculated Areas per Residue (A_0) for Different Polyamino Acids

Polymer	Nonpolar group in the residue	A_0 Measured	A_0 Calculated	Ref. no.[a]
Poly(alanine) (PLA)	CH_3	13.8	12.8	1
Poly(aminobutyric acid)	CH_2CH_3	15.5	14.5	2
Poly(norvaline)	$(CH_2)_2CH_3$	17.0	16.6	1
Poly(norleucine)	$(CH_2)_3CH_3$	17.3	18.01	1
Poly(leucine$_{11}$)	$CH_2CH(CH_3)_2$	16.6	18.9	2
Poly(γ-methyl-L-glutamate) (PMLG)	$(CH_2)_2COOCH_3$	17.5	17.7	1
Poly(ethylglutamate) (PELG)	$(CH_2)_2COOCH_2CH_3$	19.6	19.7	1
Poly(γ-benzyl-L-glutamate) (PBLG)	$(CH_2)_2COOCH_2C_6H_5$	21.5	21.6	1
Poly(β-benzyl-L-aspartate) (PBLA)	$CH_2CCOCH_2C_6H_5$	20.5	22.0	3
Poly(L-leucine) (PLL)	$(CH_2)_4NHCOOCH_2C_6H_5$	24.00	24.6	4

[a] References: (1) Malcolm (1968a); (2) Hookes (1971); (3) Malcolm (1970a,b, 1976); (4) Malcolm (1968b).

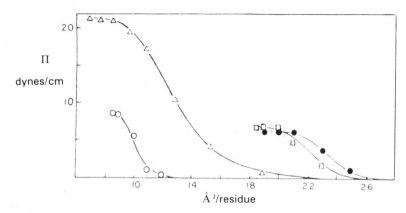

FIGURE 5.10. π vs. C_s isotherms of various α-helical polymers: poly(γ-benzyl-L-glutamate) (□); poly(carbobenzoxy-L-lysine) (●); poly(alanine) (△); poly(leucine) (○). From Birdi and Fasman (1973).

polymers (Table 5.15). The greater stability of poly(γ-benzyl-L-glutamate) (PBLG) as compared with that of poly(γ-methyl-L-glutamate) (PMLG) in solution has been attributed to a difference in side-chain interactions (Fasman, 1967). The efficient shielding of the hydrogen-bonding helical backbone from the solvent by the bulkier benzyl group is undoubtedly one of the factors responsible for the greater stability of PBLG.

Another contribution to stability would be the nonbonded (hydrophobic) interaction between the phenyl moieties, which would be absent

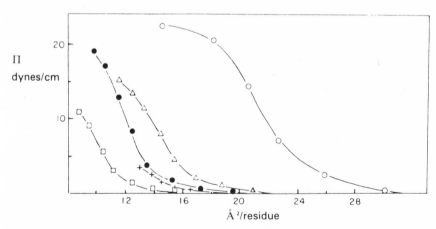

FIGURE 5.11. π vs. C_s isotherms of various β-form polymers: poly(o-butyl-L-serine) (○); poly(benzyl-serine) (△); poly(o-acetyl-L-serine) (●); poly(S-methyl-L-cysteine) (□); poly(carbobenzoxymethylcysteine) (×). From Birdi and Fasman (1973).

TABLE 5.14. Limiting Area per Amino Acid for Different Polyamino Acids (α-Helical or β-Form) Monolayers[a]

Polymer	Conformation in solid state	Å²/residue
PBLG	α-Helix	24
Poly-(carbobenzoxy-L-lysine)	α-Helix	25
Poly(L-alanine) (PLA)	α-Helix	16
PLL	α-Helix	11
Poly(o-acetyl-L-serine) (PALS)	β-Structure	14
Poly(o-benzyl-L-serine) (PBLS)	β-Structure	16
Poly(S-methyl-L-cysteine) (PSMLC)	β-Structure	12
Poly(S-carbobenzoxy-L-cysteine) (PSCC)	β-Structure	16
Poly(o-butoxy-L-serine)	β-Structure	24

[a] From Birdi et al. (to be published).

in PMLG and poly(ethylglutamate) (PELG). Consequently, because of the weak hydrogen bonding at the interface (i.e., water–backbone interaction), when PBLG is in the α-helical conformation, the magnitude of π_{col} would be lower than that for PMLG or PELG. The results in Table 5.15 support this argument, and it is therefore safe to conclude that these polymers retain their α-helical structures at the interface.

It is also possible that in the α-helical conformation, the intramolecular side-chain–side-chain interactions in PBLG, in bulk, contribute to the higher stability, while these interactions would not be expected to contribute to the magnitude of π_{col}. However, intermolecular side-chain–side-chain interactions between helical polymers would be expected to contribute to the magnitude of π_{col}. Thus, PMLG and PELG show higher π_{col} values, since

TABLE 5.15. Comparison of Solvent Composition for Destruction of α-Helixes and the Magnitude of π_{col} for Various α-Helical Polymers[a]

Polypeptide	Solvent composition[b]	π_{col} (mN/m)	Molecular weight
PMLG	60% DCA + 40% CHCl₃	20.0	260,000
PELG	—	22.0	(High)
PBLG	69% DCA + 31% CHCl₃	7.0	75,000

[a] From Birdi and Fasman (1973).   [b] From Fasman (1967). (DCA) dichloroacetic acid.

TABLE 5.16. Comparison of Solvent Composition for
Destruction of β-Structures and the Magnitude
of π_{col} for Various β-Polypeptides[a]

β-Polypeptide	Solvent composition	π_{col} (mN/m)
PALS	40% DCA + 60% CHCl$_3$	19.0
PSCC	3% DCA + 97% CHCl$_3$	3.5
PSMLC	—	11.0
PBLS	—	22.5

[a] From Birdi and Fasman (1972, 1973).

the intermolecular side-chain packing of the methyl and ethyl groups is efficient, while the bulky benzyl groups hinder compact packing.

The analyses summarized in Table 5.16 clearly show the valuable physical information that can be obtained from studies on π_{col} for polymer monolayers. The effect of subphase composition and temperature on α-helical PBLG was investigated (Jaffe et al., 1970). The effect on the α-helix was studied by adding dioxane–dichloroacetic acid (DCA) in subphase. The data were analyzed by the virial equation of state. The effects of dimethyl formamide (DMF) (a helix-inducing solvent), CHCl$_2$COOH and CF$_3$COOH (helix-breaking solvents), and urea (a protein denaturant) on the conformation of poly(β-benzyl-L-aspartate) (PBLA) in monolayer at an air–water interface was investigated by using IR spectroscopy of the collapsed films (Yamashita and Yamashita, 1978). No direct effect of DMF, DCA, or trifluoroacetic acid (TFA) was observed, while urea had some effect on the structural changes of PBLA. The PBLA molecules were probably in the right-handed α-helical conformation irrespective of the subphase composition, while in the solid state, they were in the left-handed β-structure.

It is currently believed that the structure of water at biosurfaces is significantly influenced by constituent macromolecules (Schultz and Asumaa, 1971). In other words, water is considered to affect the properties and structure of biopolymers at interfaces. Recent evidence suggests that the water adjacent to interfaces differs significantly from that in the bulk phase (Drost-Hansen, 1971). The effects of monovalent anions on surface pressure (π) vs. area (A) isotherms of poly(α-benzyloxycarbonyl-L-lysine) were investigated (Yamashita et al., 1978a,b). The magnitude of the collapse pressure, π_{col}, at $A_{col} \approx 20$ Å^2, was found to increase in the order SCN$^-$ > Br$^-$ > Cl$^-$ > F$^-$. This effect is in accord with the effect of this series on the water-structure-breaking effect of biopolymers (Dandliker and de Saussure, 1971; von Hippel and Schleich, 1969). These investigations thus clearly show the usefulness of such monolayer studies at aqueous interfaces.

The mechanism of the collapse process of bidimensional mixtures of PMLG (in α-conformation) and cholesterol has been investigated (Baglioni *et al.*, 1986). It has been shown that the collapse mechanism can be ascribed to a process of nucleation. These monolayers were also subjected to microscopic investigations.

5.3.1. THERMODYNAMICS OF BIOPOLYMER MONOLAYERS

The effect of temperature on monolayers provides the magnitude of surface entropy, S_s, and surface enthalpy, H_s (Harkins *et al.*, 1940; Birdi and Fasman, 1973):

$$\text{Surface free energy} = G_s = dG/dA \qquad (5.14)$$

$$S_s = (dS/dA)_{T,p,n_i}, \qquad H_s = (dH/dA)_{T,p,n_i} \qquad (5.15)$$

where n_i is the number of moles of film-forming component. It can also be written

$$G_s = -\pi \qquad (5.16)$$

$$S_s = (d\pi/dA)_A \qquad (5.17)$$

$$H_s = T(d\pi/dT) - \pi \qquad (5.18)$$

Typical values of S_s and H_s for some biopolymer films are given in Table 5.17.

TABLE 5.17. Surface Entropy, S_s, and Surface Enthalpy, H_s, of Polyamino Acid Monolayers at the Collapse Pressure (25°C)[a]

Polymer	Conformation	S_s(cal/residue)	H_s(cal/residue)
PLA	α-Helix	-7.9	-2529
PBLG	α-Helix	-3.2	-1077
PALS	β-Form	-8.1	-2687
PBLS	β-Form	-7.7	-2283

[a] From Birdi and Fasman (1973).

5.4. EQUATION OF STATE FOR TWO-DIMENSIONAL MONOLAYERS OF BIOPOLYMERS AND SYNTHETIC POLYMERS

The theories on polymer solution in three-dimensional systems are extensively described. In the following sections, two-dimensional monolayers of biopolymers are presented in analogy to these three-dimensional models.

5.4.1. STATISTICAL MECHANICS OF LINEAR MACROMOLECULES IN MONOLAYERS

This section briefly describes some of the most important equations of state, as derived with the help of statistical thermodynamics at the air–water interface, and their application to macromolecules in monolayers. The most fundamental theory was formulated by Singer (1948), and various modifications were later presented by other investigators. The model of the surface at the interface used was a two-dimensional quasi-lattice in which the linear macromolecule may move about, taking up various orientations [analogous to the three-dimensional Flory–Huggins solution theory (Flory, 1942; Huggins, 1942)]. The total surface area, σ, is assumed to be composed of N_0 cells of area σ_0 each. The interaction between the solvent and the macromolecule is assumed to be negligible. This assumption allows application of the model to a majority of systems, as pointed out later. Thus, it makes no difference whether the surface is composed of solvent + macromolecules, holes + macromolecules, or all three. It is assumed that each solvent molecule occupies the same area as the monomer units, s, of the macromolecule, σ_0.

The expression for surface entropy, ΔS_s, is

$$\Delta S_s = R \ln \Phi \tag{5.19}$$

where Φ is the number of different ways in which one can place N macromolecules in the two-dimensional lattice. The following relationship has been obtained (Singer, 1942):

$$(\pi\sigma_0/RT) = -\ln[1 - (sN\sigma_0/\sigma)] + [(s - 1)z/2s]$$
$$\times\ln[1 - (2N\sigma_0/\sigma z)] \tag{5.20}$$

where z is the coordination number of the lattice and $\pi_s = (dG/d\sigma)_{N,T} = -T(\Delta S_s/d\sigma)$, where ΔG_s is the surface free energy and $\Delta H_s \approx 0$ (assumption of this model).

It can be seen that when $(sN\sigma_0/\sigma) \ll 1$, this equation reduces to the ideal gas law, i.e., $\pi A = NkT$. Since $(\sigma/N = A)$, the observed area/molecule, and $(s\sigma_0 = A_0)$, the condensed area/molecule, one obtains

$$(\pi A_0/RT) = -\ln[1 - (A_0/A)] + (s - 1/s)(z/2)\ln[1 - (2A_0/zA)] \quad (5.21)$$

Despite the crude approximation that the enthalpy of surface interface is negligible, this relationship has been found to agree very satisfactorily with the measured π vs. A data, in the low-π region (Birdi, 1972; Birdi et al., 1972a; Chattoraj and Birdi, 1984).

Before a discussion of these analyses, it is worth mentioning the other refinements reported by other investigators. In a later report, Frisch and Simha (1956, 1957) derive the following equation:

$$(\pi A_0/RT) = 1/t + (1/t)^2\{(v'^2/2) - [(v'(v' - 1)/s)$$
$$\times (t - 1/t)(A_0/A) + O(A_0/A)^2]\} \quad (5.22)$$

where t is the number of segments adsorbed and v' is the total number of segments of the polymer. It can be seen that for $s \gg 1$ and $v' \to t$, this relationship reduces to the equation of state as derived by Singer [equation (5.21)].

In a later analysis (Motomura and Matuura, 1963), it was found that the magnitude of the coordination number, z, has generally been evaluated to be too small, i.e., approximately 2. Furthermore, a quantity, $w = z - 2$, was defined (Davies, 1954; Davies and Llopis, 1955) as a measure of flexibility or unfolding as related to the cohesive forces between segments. However, in a careful analysis of Singer's theory, one finds that the connection between flexibility and z is not quite clear. The following analysis which is a development based on the Guggenheim combinatory theory, was given (Motomura and Matuura, 1963):

$$(\pi A_0/RT) = (z/2)\ln\{1 - [2/z(1 - 1/r)A_0/A]\} - \ln(1 - A_0/A)$$
$$-(z/2)\ln\{[1 - 2/z(1 - 1/r)A_0/A]\}(\beta + 1)$$
$$-2[1 - 2/z(1 - 1/r)A_0/A]/[(1 - A_0/A)(\beta + 1)] \quad (5.23)$$

where β is a function of the energy of contact between monomeric units. This relationship is equivalent to Singer's equation, when the segment energy is negligible.

The equations of state based on lattice models (Singer, 1942; Huggins, 1965), as already mentioned above, cannot fit isotherms at the high-π region, nor can they describe the collapse process or predict the collapse point. The collapse process of polymer monlayers has been found to provide much useful information about the interfacial forces (Birdi and Fasman, 1973). In a recent study, a theory based on a simple buckling mechanism was described (Yin and Wu, 1971). This model assumes that the closely packed monomeric units are forced to buckle in successive pairs under high compression. When the protruding angle of the rupture site reaches 90°, the monolayer collapses. This model was found to be able to accommodate the essential features of compression isotherms of polymer monolayers in the condensed region. Beyond the collapse state, the cohesive forces or the adhesive forces determine the stress-bearing capacity of the monolayer. The buckling model thus has much in common with isotherms of α-helical or β-form polyamino acids.

5.4.2. VIRIAL EQUATIONS OF STATE

Equations of state for two-dimensional films and for linear biopolymers (proteins and synthetic polymers) have been reported by different investigators (Huggins, 1965; Gabrielli *et al.*, 1970, 1971, 1972; Birdi *et al.*, 1972), these equations having been made more general by the use of statistical thermodynamics.

The Huggins equation of state of a virial type is applicable for the whole concentration range of biopolymers [in contrast to Singer's equation (5.21)]. From experimental data on π vs. A, equations of state πA vs. A were estimated. The data were fitted to the following polynomial:

$$\pi A = (gRT/M) + b_1\pi + b_2\pi^2 + b_3\pi^3 + \cdots \qquad (5.24)$$

where g is the amount of polymer at the surface, M is the molecular weight of the polymer, and b_i is the virial coefficient. The virial coefficients of Huggins, B_1 and B_2, were estimated as follows:

$$b_1 = B_1/\rho, \qquad b_2 = M(B_2 - B_1^2)/(RT\rho)^2 \qquad (5.25)$$

where ρ is the surface concentration ($M = A_0\rho$), where A_0 is the limiting area deduced by extrapolation of the straight portion in the high-surface-pressure range of the plot π vs. A.

In the case of poly(p-fluorostyrene), which was investigated at the toluene–water interface (Gabrielli and Puggelli, 1972), the following data were found:

$$\pi A = 0.034 + 2.200\pi - 0.524\pi^2 + 0.0490\pi^3 \qquad \text{(at 25°C)} \qquad (5.26)$$

The molecular weights of synthetic polymers were determined. The degree to which the polymer extends into the subphase was estimated as follows:

$$B_1 = (nf_m^2/2)(1 - 2/z) \tag{5.27}$$

$$B_2 = (nf_m^3/3)(1 - 4/z^2) \tag{5.28}$$

where n is the number of polymerization (or the number of amino acid residues in the protein molecule), z is the coordination number (as in Singer's equation (5.21), and f_m is the average fraction of polymer in the surface monolayer.

The values of A_0, B_1, B_2, and f_m deduced from equations (5.27) and (5.28) are given in Table 5.18 for different protein monolayers (Birdi et al., 1972). It can be seen that in all cases, the magnitude of f_m is less than 1, which suggests that all these protein films are such that molecules are partially submerged in the subphase. Furthermore, in the case of synthetic polymers (polyvinylacetate) (Gabrielli et al., 1970), the value of f_m increased as the number of polar groups increased. The degree of anchoring at the surface increases with the subsequent increase in the value of f_m, because of the more regular distribution of the polar groups in these polymers.

In the case of proteins, the distribution (sequence) of polar and apolar amino acids (residues) is complex (see Section 5.2.2), and thus the value of f_m will not be expected to show quite the same relationship to the

TABLE 5.18. Magnitudes of A_0, B_1, B_2, and f_m for Various Proteins and Polyamino Acid Monolayers

Proteins[a]	A_0 (m²/mg)	B_1	B_2	f_m
Proteins				
BSA	1.16	1.178	1.389	0.887
Ovalbumin	1.47	1.179	1.394	0.890
Transferrin	0.206	1.059	1.131	0.802
Myoglobin	0.173	0.351	0.192	0.415
Hemoglobin	0.975	1.264	1.601	0.952
Polyamino acid monolayers[b]				
PLA (α-helix)	1.18	1.45	2.12	1.0
PBLG (α-helix)	0.64	1.14	1.30	0.9
PALS (β-form)	0.87	1.03	1.10	0.8
PBLS (β-form)	0.40	0.89	0.87	0.8

[a] From Birdi et al. (1972). [b] From Birdi and Fasman (1973).

polar-apolar character of the protein molecule. These results can be described as follows:

1. The values of f_m are high when the proteins (BSA, ovalbumin, hemoglobin) are completely unfolded (high A_0 values). The values of f_m are low for incompletely unfolded proteins [transferrin, myoglobin (i.e., low A_0)]. These relationships can be attributed to the degree of protein-subphase interaction, which again will be expected to be related to the degree of unfolding at the surface.

2. Those proteins that give high values of f_m were also found to give values for the work of compression, W_c (Birdi, 1972). This finding suggests that the molecules that are less submerged in the subphase show strong intramolecular cohesion energies. The extent of these cohesive energies can be estimated from z (coordination number of the bidimensional quasi-lattice). However, as can be easily seen from the aforestated relationship, the value of z as defined by equation (5.27) cannot provide any useful information. This is especially the case when the magnitude of n is very large.

3. It was further found that proteins with high f_m values (BSA, hemoglobin, ovalbumin) give collapse surface pressures π_{col} of 20–30 mN/m (Birdi, 1972). On the other hand, protein films with lower f_m values give films with rather high π_{col}. These conclusions are not well understood and need further investigation.

4. The Huggins quantity, f_{mo}, can be estimated as follows:

$$f_{mo} \approx f_m$$
$$= e^{a_{\ddot{e}}}/(1 + e^{a_{\ddot{e}}})$$
$$= \tfrac{1}{2}(1 + a_{\ddot{e}}/2 - a_{\ddot{e}}^3/24 + \cdots) \qquad (5.29)$$

where $a_{\ddot{e}} = (\ddot{e}_b - \ddot{e}_0)/kT$; $\ddot{e}_0$ is the average energy of interaction (per unit outline length) of the monolayer of polymer with its environment, when the surface concentration tends to zero; and $\ddot{e}_b$ is the average energy of interaction of polymer in bulk liquid with its environment, which is assumed to be independent of concentration. The fraction of polymer in the surface layer, f_m, can thus be related to the molecular interaction energies.

The $a_{\ddot{e}}$ values of these protein monolayers were estimated and found to be all negative. This finding suggests that the interaction energy is larger in the monolayer than in the bulk phase ($\ddot{e}_0 > \ddot{e}_b$), and this difference is approximately $4kT$. These values are comparable to those found for synthetic polymers (Gabrielli *et al.*, 1970).

On the basis of the data discussed above, the orientation of biopolymers at the interface is presented schematically in Fig. 5.12.

Investigations were conducted on 2-polyvinyl pyridine (2-PVP), poly-methylacrylic acid (PMA), and a polyampholyte with different contents of

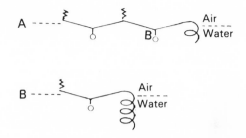

FIGURE 5.12. Schematic representation of the orientation of protein molecules at the air-water interface. (A) Completely unfolded; (B) incompletely unfolded. From Chattoraj and Birdi (1984).

2-PVP and polyacrylic acid (PAA) spread as monolayers at the air-water interface (Jaffe and Ruysschaert, 1964). The effect of ionization on the monlayer properties was analyzed in terms of cohesion and flexibility, as defined by monolayer theories based on statistical thermodynamics (Singer, 1948; Motomura and Matuura, 1963). The magnitudes of z (Singer, 1948) were 2.008 and 2.015 for the monolayers of PMA and PAA, respectively. The term related to the cohesion forces between segments (w/kT) (Motomura and Matuura, 1963) was found to be related to the degree of ionization, i.e., subphase pH (Table 5.19).

5.5. CHARGED POLYMER MONOLAYERS

Spread monolayers of proteins (and of other biopolymers) at the air-water (and the oil-water) interface have been found to be very useful model membrane systems. It is therefore of immediate interest to analyze

TABLE 5.19. Monolayers of 2-PVP at Different Subphase pH[a]

pH	z	w/kT
10.1	>4	0
7.15	3.8	0.9
6.15	2.16	1.05
5.35	2.12	1.1
4.15	2.4	1.05

[a] From Jaffe and Ruysschaert (1964).

the effect of ionization of the amino acid side chains of the spread protein molecule as a function of subphase pH.

As described in Chapter 4, the measured π at the air–water interface is given as

$$\pi_{aw} = \pi_{kin} + \pi_{coh} + \pi_{el} \tag{4.24}$$

The presence of an electrical charge on the film-forming molecule gives rise to an appreciable effect on the π vs. A isotherms, i.e., more expanded films. At low π, the term π_{coh} can be considered to be negligible, and thus one can write the equation of state as (Birdi and Nikolov, 1979)

$$(\pi - \pi_{el})(A - A_0) = kT \tag{5.30}$$

where it is assumed that $\pi - \pi_{kin} = \pi_{el}$. The charged films exhibit larger π, equivalent to π_{el}, at any given area.

The quantitative expression for the term π_{el} is given by the Gouy diffuse double-layer model and the Davies equation

$$\pi_{el} = 6.1c^{1/2}[\cosh \sinh^{-1}(135/A_{el}c^{1/2}) - 1] \tag{4.66}$$

where c is the concentration of the electrolyte in the subphase, A_{el} $(=A/z)$ is the area per charge in the surface of the film-forming protein molecule, and z is the number of charges.

The most general equation of state for monolayers with cohesive, π_{coh}, and electrostatic interactions, π_{el}, can then be written as

$$\{\pi - (C_1 + C_2/A^{2.5}) - 6.1c^{1/2}([\cosh \sinh^{-1}(135/A_{el}c^{1/2})\}(A - A_0) = kT \tag{5.31}$$

This equation is applicable to both charged lipid and biopolymer (and other polymer) monolayers, although there are exceptions, as described for membrane proteins (see Chapter 8). This expression for π_{el} was originally derived for simple lipid molecules. Only recently was it applied to π vs. A isotherms of charged protein monolayers (Birdi and Nikolov, 1979) for the first time. The effect of pH on β-casein monolayers was analyzed by equation (5.31) (Evans et al., 1970). The monolayer data of casein as analyzed by this equation gave a satisfactory number of charges.

The π of protein monolayers was measured by the Wilhelmy method with a sensitivity of ± 0.001 mN/m (dynes/cm). The protein was spread from its aqueous solution by using a microsyringe. πA vs. π plots as a function of subphase pH are given in Fig. 5.13 for insulin, hemoglobin, and BSA.

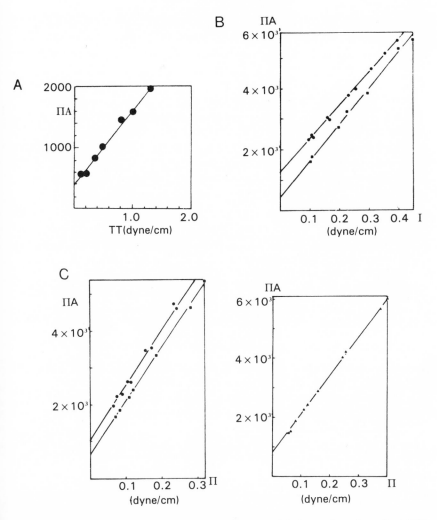

FIGURE 5.13. πA vs. π plots for protein monolayers on subphases of varying pH at 25°C: (A) insulin, pH 7.4; (B) hemoglobin: (■) pH 7.4, (●) pH 5.1; (C) BSA: (●) pH 7.4, (■) pH 5.1, (▲) pH 3.95. From Birdi and Nikolov (1979).

Insulin Monolayer. The πA vs. π plot for an insulin monlayer spread on a subphase of pH 7.4 (=isoelectric pH) shows that $(\pi A) = kT$, as $\pi \to 0$, when a molecular weight of 6000 (i.e., monomeric insulin) is used.

Hemoglobin Monolayer. The protein is reported to be present as the tetramer at this pH range. Using a tetramer molecular weight of 68,000, the plot gives πA vs. π, as $\pi \to 0$, on a subphase of pH 7.4 (=isoelectric pH).

This result thus indicates that the effect of subphase pH on the number of charges can be analyzed as described below. The slope of this plot gives the value of A_0 as $11,500 \, \text{Å}^2/$molecule ($=16.9 \, \text{Å}^2/$residue). This value is comparable to the value of approximately $15 \, \text{Å}^2/$residue determined from X-ray diffraction data for various proteins. The literature value of $13,200 \, \text{Å}^2/$molecule also compares satisfactorily (Kubicki *et al.*, 1976).

BSA Monolayer. The pH effect is described below. The slope of the plot is $14,500 \, \text{Å}^2/$molecule ($=21.3 \, \text{Å}^2/$residue), which agrees with the literature data (Bull, 1945, 1947a,b; Chen, 1976). From these data, it can be convincingly seen that when the film-forming protein molecule bears no net charge (i.e., when subphase pH = isoelectric pH), equation (5.30) is valid, since the data give $(\pi A)_{\pi \to 0} = kT$, using the values of molecular weight for the respective proteins as reported from bulk studies at these pH values. On the other hand, for charged films, i.e., when subphase pH is not equal to isoelectric pH, appreciable deviation from the relationship given in equation (5.30) is observed, this deviation arising from the term π_{el}.

Since it is reasonable to expect the variation in subphase pH to affect mainly the magnitude of the term that arises from the charge repulsion, i.e., π_{el}, one would expect equation (5.30) to be valid for these charged films. The data for BSA were analyzed under these assumptions. The

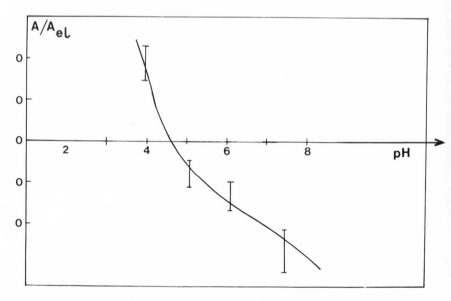

FIGURE 5.14. Plot of A/A_{el} [$=z$ (number of charges per molecule)] vs. subphase pH for monolayers of BSA. From Birdi and Nikolov (1979).

TABLE 5.20. Variation of Charges per Molecule, z, of BSA with pH, as Determined from Bulk Studies and from Monolayer Studies[a]

| pH | Values of z | |
	Bulk	Monolayer
4	25	15–25
5.5	0	0
7.4	10	15–25

[a] From Birdi and Nikolov (1979).

magnitude of π_{el} was estimated from equation (5.30) and the magnitude of A_{el} then estimated from equation (4.66). The number of charges per molecule, z, can thus be estimated $(=A/A_{el})$. A plot of z vs. subphase pH showed that the isoelectric point from monolayer studies was approximately 5, which agrees with bulk data from the literature (Fig. 5.14).

Table 5.20 lists the values of z from bulk studies and compares them with the z values calcuated from the monolayer data for BSA. The data on z in this table indicated that the agreement is fairly acceptable. Further, the monolayer data of hemoglobin at subphase pH 5.1 gave a magnitude of $z = 10$, an acceptable value. These results indicate that Davies's equation provides useful information for polyelectrolytes, even though it was originally derived from simple long-chain charged lipids.

These data indicated that at low π, the following relationship is valid:

$$\left(\frac{\pi A}{kT}\right)_{\pi \to 0} = 1 + 0.00413 z^2 \qquad (5.32)$$

The effect of z on the intercept for πA vs. π plots is thus significant, even for low values of z.

In contrast, the current literature studies reported on such charged protein films, at the air-water interface, do not take into account the repulsion term, π_{el} (Chen, 1976; Kubicki et al., 1976). Accordingly, these investigators reported the molecular weight of BSA to be 16,300, on a subphase of pH 3.9. Since the BSA molecule is not dissociated at this pH, these analyses must be considered as incorrect because of the neglect of the π_{el} term (equation 5.30).

The effect of the specificity of the counterion on the interaction between quaternary ammonium surfactants in monolayers and bilayers was reported in a recent study (Mara, 1986a,b).

5.5.1. ADSORPTION OF BIOPOLYMERS AT THE WATER–MERCURY INTERFACE

Besides the air–water and oil–water interfaces, the mercury (Hg) electrode is of particular interest for studying the properties of proteins or biopolymers adsorbed at the interface. Its advantage arises from the ready accessibility of the thermodynamic parameters—e.g., interfacial tension, charge density, surface concentration (Γ), and potential—to direct measurement. In addition, the field strength in the interface can be varied from zero to approximately 10^4 V/cm at constant ionic strength of the aqueous phase. Therefore, the influence of high electrical fields on biopolymers adsorbed at the interface can be investigated. The mercury–water interface is widely used for characterizing biological redox reactions (Theorell, 1938; Weitzman et al., 1971; Betso et al., 1972; Berg et al., 1972). In this case, electron transfer between the adsorbed biomolecules and the electrode is accomplished by a heterogeneous process. Mercury electrodes are also found to be useful for analytical problems, e.g., estimating small amounts of protein, and for detecting denaturation–renaturation processes (Palecek and Pechan, 1971; Ruttkay-Nedecky and Bezuch, 1971; Müller, 1963).

In earlier studies (Betso et al., 1972; Ruttkay-Nedecky and Bezuch, 1971; Behr et al., 1973), it was incorrectly assumed that proteins have identical structure in solution and when adsorbed on electrodes. However, it is now well established that the proetin molecule undergoes structural changes when adsorbed at any interface. Evidence for structural changes of adsorbed proteins has been reported from polarographic catalytic currents of seven globular proteins, as well as from capacitance–time curves at the hanging-drop mercury electrode (Kuznetsov, 1971; Brdicka, 1933).

It is safe to assume that strongly surface-active substances would adsorb instantaneously at the interface and that desorption would be negligible (Jehring, 1969; Pavlovic and Miller, 1971; Koryta, 1953). Under these assumptions, the surface concentration, Γ, is given as (see Chapter 6)

$$\Gamma = 7.36 \times 10^{-4}(\sqrt{D}C\sqrt{t}) \qquad (5.33)$$

where D is the diffusion coefficient, C is the bulk concentration of protein, and t is the time. Provided that (1) the relative decrease of capacitance, $\Delta C_{ap}/C_{ap}^0$, is proportional to the degree of coverage of the electrode; (2) the surface area, S, occupied by the protein molecule at the interface does not depend on the degree of surface coverage; and (3) only one monolayer is formed, one can derive the following relationship for the capacitance–time curve (Damaskin et al., 1968; Jehring and Horn, 1968):

$$\Delta C_{ap}/C_{ap}^0 = (7.36 \times 10^{-4})N_A(S\sqrt{D}C\sqrt{t_m})(\bar{C}_{ap}^s/\bar{C}_{ap}^0) \qquad (5.34)$$

where $\bar{C}_{ap}^s$ corresponds to the full coverage of the electrode. The surface area of the adsorbed molecule can be estimated from AC polarograms representing capacitance average values. Integration over the drop time can be carried out by introducing a factor of 0.77 into equation (5.34) (Jehring and Horn, 1968):

$$\Delta \bar{C}_{ap}/ \bar{C}^0_{ap} = 0.77(7.36 \times 10^{-4}) N_A (S\sqrt{D} C\sqrt{t_m})(\bar{C}^s_{ap}/ \bar{C}^0_{ap}) \quad (5.35)$$

where t_m is the drop time. The concentration (insulin) dependence of the relative decrease of differential capacity, $\Delta \bar{C}/ \bar{C}_0$, is given in Fig. 5.15 (Scheller et al., 1975). The relative lowering of the capacitance at constant potential in the adsorption region depends linearly on both protein concentration, c, and $\sqrt{t_m}$. The surface areas, S, of the different adsorbed proteins calculated from equation (5.35) are given in Table 5.21.

The values for the surface areas, S, of different proteins determined by the various methods are the same. The electrode surface areas, S, of these proteins are greater than the biggest cross sections of the molecules in the crystalline state. This difference could be explained as being related to the unfolding at the interfacial energy, the degree of unfolding being consistent with the maximum lowering of the surface free energy. The electrical field present across the interface may also be expected to induce unfolding.

At the air–water interface, the value of the area/amino acid has been reported to be approximately 17 Å^2 for different globular proteins, as determined from the limiting slopes of the two-dimensional state equation (Birdi, 1972, 1973a, b; Birdi et al., 1972; Birdi and Fasman, 1972). This

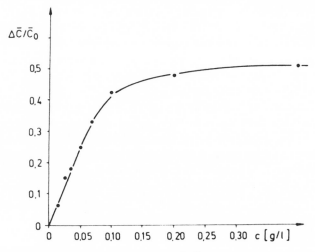

FIGURE 5.15. Variation of differential capacity ($\Delta \bar{C}/ \bar{C}_0$) with insulin concentration in bulk [c (g/liter)] (pH 3, -800 mV, $t_m = 3$ sec). From Scheller et al. (1975).

TABLE 5.21. Surface Areas of Different Globular Proteins Adsorbed at the Mercury–Water Interface (in Comparison with the Air–Water Interface)

Protein	Molecular weight	Surface area, S (Å^2/molecule)					Thickness $(d)^c$
		$S_{aw}{}^a$	$S_{dcp}{}^b$	$S_{acp}{}^b$	$S_{D\text{-}t}{}^b$	S/residue^b	
Insulin	6,000	870	870	800	1,000	16.5	8.5
Cytochrome c	13,300	—	—	2,100	2,200	20.5	7.7
RNAse	13,700	2,600	2,400	2,400	3,200	19.3	7.2
Lysozyme	15,000	—	—	1,800	2,800	14.0	10.5
Met-myoglobin	16,000	—	—	3,500	3,600	23.5	5.7
Met-hemoglobin	32,000	5,200	—	5,700	—	19.7	7.1
Ovalbumin	40,000	9,800	—	7,300	6,400	19.0	7.2
BSA	68,000	12,000	9,200	10,100	13,000	13.5	8.5
Glycogen	180,000	—	—	7,300	—	—	32.0

^{a} From Birdi (1972). ^{b} From Scheller *et al.* (1975). $^c d = M/(0.6 \times S \times 1.3)$.

value corresponds roughly to the surface occupied by one amino acid of the β-sheet structure, as established by X-ray studies (Miller and Bach, 1973). The data on protein adsorption at the mercury–water interface are in agreement with those on the degree of unfolding at the air–water interface (Birdi, 1973a). The results at the mercury electrode gave evidence that in the adsorbed layer, these globular proteins are unfolded and helical structures and intra chain bonds are intact to some extent.

The interaction of nucleic acids with a charged mercury surface has been reported (Miller, 1961). These studies were carried out by measuring the effect of the nucleic acids on the differential capacity of the electrical double layer between a polarized mercury surface and a 0.1 N NaCl solution containing various concentrations of nucleic acid. The lowering of the capacity by the adsorbed nucleic acids was caused by the different molecular residues (e.g., sugar, purine, ionized groups) of the nucleic acids (DNA and RNA) adhering to the surface.

Furthermore, unfolding at the interface was proposed for DNA. The unfolding of the helix starts at that part of the molecule that reaches the surface first and proceeds by rotation about their axes of the two double helical cylinders that protrude into the solution.

5.6. DETERMINATION OF MOLECULAR WEIGHT OF POLYMERS (BIOPOLYMERS) FROM MONOLAYER STUDIES

Early references to the possibility of determining the molecular weights of biopolymers from π vs. A isotherms were not extensive and systematic,

TABLE 5.22. Isoelectric pH for
Different Proteins

Protein	pH
Serum albumin (BSA/HSA)	4.9
Ovalbumin (egg albumin)	4.6
β-Lactoglobulin	5.2
Chymotrypsinogen	9.5
Cytochrome c	10.7
Hemoglobin	6.8
Myoglobin	7.0
Pepsin	1.0
Insulin	7.0

as one would have expected. It must be mentioned that molecular weights could be easily measured by using only 10–50 μg of biopolymer. This method provides an enormous advantage over other known methods, such as osmometer, light-scattering, and sedimentation, for which something like a few hundred milligrams of material is needed to measure the molecular weight.

The molecular weights of proteins as measured from monolayers, e.g., gliadin, egg albumin (ovalbumin), and hemoglobin, were reported half a century ago (Guastala, 1939). Later, data for ovalbumin and β-lactoglobulin (spread on ammonium sulfate aqueous solution) were reported (Bull, 1945, 1947a,b). The range of π was 0.01–1.0 mN/m.

The molecular weights of synthetic polyvinylacetate polymers (of molecular weights ranging from 2×10^3 to 8.5×10^5) were investigated (Benson and McIntosh, 1948). The molecular weights obtained by extrapolation of πA to $\pi = 0$ were found to be satisfactory in the case of lower-molecular-weight polymers. However, as delineated above, molecular weights estimated from π vs. A data are correct only when the term π_{el} is negligible, which is at subphase pH = isoelectric pH (see Table 5.22).

5.7. OTHER PROPERTIES OF BIOPOLYMER MONOLAYERS

The electrochemical properties of rhodopsin at the air–water interface have been reported (Korenbrot and Pramik, 1977; Korenbrot, 1977; Hwang et al., 1977). An analysis of dichroic absorption was recently reported for the light-absorbing pigment of the rod cell (Korenbrot and Jonrs, 1979). The properties reported were of the same character as those reported in the intact disk membrane: transduction of light energy into chemical energy as mediated by the only protein of the purple membrane, bacteriorhodopsin. In the light, bacteriorhodopsin acts as a proton pump and thus generates a transmembrane electrochemical gradient of protons.

Monolayers of a chloroform extract of hydrophobic polypeptide from *Rhodospirillum rubrum* at the air–water interface were investigated (Kopp *et al.*, 1979). Langmuir–Blodgett films were also investigated with the help of IR spectroscopy. These spectral studies showed amide I and amide II bands, which are typical of α-helical and random conformations.

The surface chemical properties of macromolecules of tear film components under static and dynamic conditions were investigated (Holly, 1974). Monolayers of mucin, albumin, globulin, and lysozyme were also investigated, since these proteins are found in the tear film, fluid that covers the cornea. These investigations showed that the most surface-active material of tears is mucin. It was concluded that mucin performs an important role at the air–water interface of the tear film, affecting both formation and stability of the tear film.

5.8. ENZYME ACTIVITY IN MONOLAYERS

Pancreatic lipase (PAL) catalyzes the hydrolysis of fatty acid esters of glycerol and other simple alcohols (Desnuelle, 1961) and acts readily on emulsions of lipids. It is not quite clear whether phase heterogeneity is a necessary activity requirement. The monolayer is the most useful model for determining this kind of mechanism. When 1,2-dioctanoin was spread on a subphase containing PAL, the π showed a substantial time-dependent decrease (Lagocki *et al.*, 1970). Since both the hydrolysis products are soluble, the surface concentrations of the unreacted diester can be estimated from π vs. A isotherms. These studies showed that PAL was active at monolayers without the phases present in emulsions. The reaction was first-order with respect to the enzyme and substrate and independent of the compression in the π range investigated. Thus, the rate was found to be proportional, not to the concentration of surface molecules, but to the total number of substrate molecules at the interface.

On the other hand, the reaction of phospholipase c (from *Clostridium perfringens*) with lecithin monolayers was markedly dependent on π (Miller and Ruysschaert, 1971). The same was observed for porcine lipase with mono-, di-, and triglycerides of short-chain fatty acids (Brockman *et al.*, 1975; Esposito *et al.*, 1973).

Monolayers of dioctanoyl lecithin under hydrolysis with phospholipase a_2 (of *Crotalus atrox* venom) were investigated (Cohen *et al.*, 1976). When the subphase contained phospholipase, a rapid decrease in the π of the lipid monolayer was observed immediately following the spreading of the monolayer and evaporation of the spreading solvent. The kinetics were analyzed at constant area and variable π. The reaction was found to obey the following law:

$$(-d[\text{N}]/dt) = V$$

$$= k_E E_0 [N] \qquad (5.36)$$

where E_0 is the molarity of the enzyme in the subphase and $[N]$ is the surface concentration (estimated from π). At pH 6.3 and $[\text{Ca}^{2+}] = 6.7 \times 10^{-3}$ M, the value of k_E was 1.6×10^6 M^{-1}/sec in the higher π range and 1.1×10^6 M^{-1}/sec in the lower π range. These experimental data were analyzed according to a model in which diffusion of the enzyme to the surface from a stirred subphase is balanced by surface denaturation, thereby creating a steady-state concentration of active enzyme in the proximity of the substrate monolayer.

The dependence on π was also reported for the system porcine pancreatic phospholipase a_2 with lecithin monolayers (Zografi et al., 1971).

Phospholipases from different sources have been used to reveal information regarding the localization of the lipids within erythrocyte membranes (Verkley et al., 1973; Zwaal et al., 1973, 1975; Kahlenberg et al., 1974). These phospholipases show a great variety of behavior against biological membranes and model systems. Some phospholipases can hydrolyze the phospholipids of the intact erythrocyte membrane, which can lead to hemolysis of the cell. Other phospholipases cannot hydrolyze the phospholipids of the intact cell, but can hydrolyze the phospholipids of the ghost membranes (Zwaal et al., 1975). Questions as to why only some phospholipases can degrade the phospholipids of intact cells are difficult to answer from studies on biological membranes. It would be of interest to know whether the phospholipases have different abilities to penetrate the membrane or whether the lipid composition could play a role.

Monolayers have been used for investigating enzymatic hydrolysis (Bangham and Dawson, 1962; Colacicco and Rapport, 1965; Miller and Ruysschaert, 1971; Verger and de Haas, 1973). In the monolayer model system, it is easy to change the molecular packing of the lipids by varying the compression of the monolayer. Monolayer composition can be greatly varied using synthetic and isolated lipids.

The action of phospholipases on lipids was followed by measuring the surface pressure, π, and surface radioactivity (of the ^{14}C-labeled methyl group in lipids) (Demel et al., 1975). Pure phospholipases showed no surface activity at the concentrations used. The effect of the molecular packing of ^{14}C-labeled palmitoyloleoylphosphatidylcholine (POPC) on phospholipase c degradation is shown in Fig. 5.16. At a surface pressure, π, of 29.4 mN/m, a rapid release of the phosphate-choline moiety is measured by the decrease

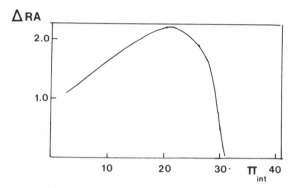

FIGURE 5.16. Relationship between the change in surface radioactivity (ΔRA) and the initial π (π$_{int}$) of POPC monolayers on hydrolysis by phospholipase. From Demel (1975).

in radioactivity, after an inactive period of 4 min. The degradation is completed in about 12 min. The simultaneous measurement of the π shows a decrease due to the transformation of phosphatidylcholine to diacylglycerol. At an initial surface pressure of 34.8 mN/m, no POPC degradation was measured even when additional amounts of enzyme were injected.

Phospholipase c from *Bacillus cereus* is known not to degrade the phospholipids of the intact cell. Monolayer studies have shown that POPC with a π of more than 31 mN/m is not degraded. The maximum enzymatic activity was observed between π values of 27 and 12 mN/m. The reduced activity at π values below 10 mN/m was suggested to be related to surface denaturation as was found for phospholipase from *Clostridium welchii*

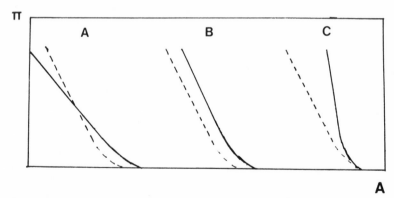

FIGURE 5.17. Typical examples of the three general types of change of slope of π vs. *A* isotherms for polymer monolayers when a solute is introduced into the subphase. (– – – –) Monolayer on water. See the text for details. Redrawn from Giles and McIver (1975) with changes.

(Miller and Ruysschaert, 1971). For POPC, the critical interfacial pressure of 31 mN/m corresponds to a molecular area of 60 Å^2/molecule.

Many biological processes that occur at the lipid–water interface involve soluble enzyme or a soluble enzyme adsorbed onto a solid surface and acting on an insoluble substrate. The kinetic and catalytic characteristics of these two-dimensional enzymatic reactions are prime factors in membrane processes and in the action of insolubilized enzymes, and for this reason they have been the center of many investigations (Dawson, 1968; Katachalski *et al.*, 1971; Brown and Hasselberger, 1971; Laidler and Sundaram, 1971; Miller, 1971; Brockman *et al.*, 1973). The reaction of porcine lipase with the soluble triglyceride tripropionin was reported to exhibit substantial stimulation in the presence of hydrophobic surfaces (such as siliconized glass).

5.9. POLYMER–SOLUTE INTERACTIONS IN MONOLAYERS

The change in slope of π vs. A isotherms for a polymer monolayer film on water when a solute is introduced into the aqueous subphase has been extensively described by three general types of effects (Giles and McIver, 1975) (Fig. 5.17 and Table 5.23). These three general effects on the slopes were identified and interpreted as follows:

Type A. The slope is decreased and a weak complex is formed between the monolayer and a surface-active solute, or a more compressible conformation of the polymer is caused by the solute binding.

Type B. The slope is unchanged and the curve may or may not be displaced to the right. The solute lies beneath and parallel to the polymer monolayer chains, or it forms weak acid–base-type cross-links with the polymer at the interface.

Type C. The slope is increased, the curve is sometimes vertical in the upper portion, and a rigid film with extremely high viscosity is formed by strong polymer–solute bonds, often with cross-linking of the polymer chains by multiple bonds with the solute. In some special cases, the effect can be produced by solutes that appear to become entangled with protein monolayers by hydrophobic forces. These types of changes in slope are illustrated in Fig. 5.17. Interactions in cellulose monolayers with various solutes were studied (Giles and McIver, 1975).

The hydrolytic stability of polymers of polydimethylsiloxane (PDMS) has been investigated by various workers (Noll, 1966; Noll *et al.*, 1972; Fox *et al.*, 1950; Rudol and Ogarev, 1978). Linear PDMS polymers with molecular weights of 5.7×10^4 and 1.25×10^6 were investigated (Rudol and Ogarev, 1978). π vs. A isotherms did not show any changes (after 10 min)

TABLE 5.23. Examples of Changes in Slopes of Surface Pressure vs. Area Isotherms[a]

Polymer	System/solute	Reference
	Type A Changes	
Cellulose	Nonionic/cationic dyes	Agnihotri and Giles (1972)
Cellulose triacetate	Urea; quinol; dye	Cameron et al. (1957)
Nylon-4	Urea; guanadine	Clark et al. (1957)
Protein	Alkyl sulfate	Cameron et al. (1958)
Protein	1:2, Metal dye	Giles and MacEwan (1959)
	Dye	Giles et al. (1974)
	Urea	Giles and McIver (1974)
	Type B Changes	
Cellulose	Azo dyes	Agnihotri and Giles (1972)
	Bisazo dyes	Agnihotri and Giles (1972)
	Urea	Giles and Agnihotri (1967)
Cellulose triacetate	Ionic dyes	Giles et al. (1974)
Chitin	Phenol	Giles and Agnihotri (1969)
Protein	Glucose; mesoinositol	Cameron et al. (1958)
	Type C Changes	
Cellulose	Mono- or bifunctional dyes	Giles and McIver (1974)
Methoxymethyl nylon	1:2, Metal–dye complex	Giles and MacEwan (1959)
Protein	Alkyl sulfate	Giles and MacEwan (1959)
Nylon 66/6	Silicic acid	Clark et al. (1957)
Protein	Monosulfonated dye	Giles et al. (1974)
	Reactive dyes	Giles and McIver (1974)
	Silicic acid	Clark et al. (1957)
	Tannic acid	Ellis and Pankhurst (1954), Lanham and Pankhurst (1956)

[a] From Giles and McIven (1975).

when subphase pH was varied from 6.3 to 11.0. However, a significant increase in π was observed when subphase pH was 1.0 or 1.7. The PDMS monolayer becomes more expanded with decreasing pH. From the kinetics of these hydrolyses in monolayers, the rate constant was found to be 7.2×10^{-3} min^{-1}. This value compares with ester hydrolysis rates in monolayers of 0.05–0.005 min^{-1} (Davis, 1956).

6

KINETICS OF ADSORPTION
AND DESORPTION

The surface tension of water decreases when an amphiphile molecule, such as alcohol or protein, adsorbs at the air–water or oil–water interface. The adsorption is determined by the energy barriers that oppose both the removal of the polar hydroxyl group from the water phase and the entry of the hydrophobic alkyl group into the aqueous medium. The kinetics by which the alcohol molecule comes close to the interface are determined by the diffusion process alone. On the other hand, the kinetics of desorption of the alcohol monolayer are related to quite a high energy barrier. The same holds true for polymers, although in this case conformation (kinetics of conformation changes) and molecular weight are added parameters.

The kinetics of such adsorption and desorption processes are obviously of much importance in many different interfacial phenomena, and extensive studies concerning the mechanisms and energetics have been reported in the literature.

6.1. THEORY OF ADSORPTION
KINETICS AT INTERFACES

Because of their hydrophile-lipophile balance (HLB), amphiphile molecules, such as lipids and proteins, are at a higher energy state in the bulk phase than when they are adsorbed at any interface (e.g., air–liquid, liquid$_1$–liquid$_2$, liquid–solid). This circumstance leads to the observation that when a fresh surface is created, the Gibbs surface excess, Γ_i, of a solute (i) attains an equilibrium value only after some definite time has elapsed (Chattoraj and Birdi, 1984). The adsorption kinetics from a high-energy state (bulk) to a lower-energy state (interface) are analogous in some degree to the flow of heat, as regards the diffusion properties. Under these conditions, Fick's laws of diffusion would apply for describing the kinetics of adsorption. The net rate of transfer of a substrate from a concentrated to

197

a more dilute solution is proportional to the concentration gradient, to the area of the plane through which the transfer takes place, and to some coefficient related to the system under consideration. This process can be described by applying Fick's first law of diffusion (Scatchard, 1976):

$$\text{Flux} = J_1 = dm/dt \tag{6.1}$$

$$= -DA(dC/dx) \tag{6.2}$$

where m is the mass transferred, t is the time, x is the distance in the direction in which diffusion occurs across an area A, D is the diffusion coefficient of the molecule or a proportionality coefficient, and C is the concentration. Fick's second law of diffusion is given as

$$dC/dt = D(d^2C/dx^2) \tag{6.3}$$

This relationship specifies that the rate of change of concentration at any plane perpendicular to the direction of diffusion is proportional to the rate of change of the concentration gradient in the plane. Under some constraints, the solution to this equation is

$$dc/dx = C/2(\pi Dt)^{1/2}\{[\exp(-x^2/4Dt)]\} \tag{6.4}$$

where C is the concentration difference between the two states. The kinetics of diffusion to a flat surface, like those of adsorption to the surface, can be deduced from equation (6.4) as follows: Since $x = 0$ for adsorption at a flat surface, one obtains

$$dC/dx = C/2(\pi Dt)^{1/2} \tag{6.5}$$

On substituting Fick's first-law relationship (neglecting the negative sign), one obtains

$$dm/dt = AC/2(D\pi t)^{1/2} \tag{6.6}$$

On integration, this leads to

$$m = AC(Dt/\pi)^{1/2} \tag{6.7}$$

Thus, in an unstirred and unrestricted diffusion process, when a molecule reaches the plane at which $x = 0$, the chances that the molecule will go forward or backward are each one half. However, if all the molecules arriving at the plane pass through it or are irreversibly adsorbed on it, there is no opportunity for the molecules to flow backward, and accordingly, under these conditions, the right-hand side of equation (6.7) should be multiplied by 2, and one obtains

$$m/A = 2C(Dt/\pi)^{1/2} \qquad (6.8)$$

or, in an appropriate form for surface adsorption:

$$\Gamma = 2C_b(Dt/\pi)^{1/2} \qquad (6.9)$$

where C_b is the concentration in bulk in moles/liter and Γ is the number of adsorbed molecules/cm^2 at time t sec from the beginning of the adsorption experiment. This relationship was derived only in the case of irreversible processes, which one would expect to be valid only in the initial stages of adsorption (Ward and Tordai, 1946; Hansen, 1960, 1961; Delahay and Trachtenberg, 1957; Delahay and Fike, 1958; Knotecky and Brdicka, 1947; Davies and Rideal, 1963).

For example, in the case of a small amphiphile molecule, such as $C_{14}H_{29}OH$, $D \approx 6 \times 10^{-8}$ cm^2/sec, if $\Gamma = 2.14$ molecules/cm^2, for $C_b = 0.0015$ mole/liter, one obtains a value for t of 0.0064 sec. In the case of a protein solution of a macromolecular bovine serum albumin (BSA), with $D = 10^{-7}$ cm^2/sec, the adsorption time can be as long as 10–100 sec. This therefore allows the investigation of such adsorption kinetics by simple methods.

The rate of adsorption of molecules from the bulk phase can also be written as

$$dm/dt = B_1 C_b(1 - \Theta) \qquad (6.10)$$

where B_1 is a constant and Θ is the fraction of the surface actually covered by molecules. In the early stages of adsorption, before Θ and the desorption rate are appreciable, dm/dt is proportional to C_b. The desorption rate is given as (Davies, 1952)

$$-(dm/dt) = B_2 m \, e^{(ze\Phi_0 - W)kT} \qquad (6.11)$$

where B_2, like B_1, is a constant that depends on the diffusion process, W is the energy of adsorption of the hydrocarbon chain, z is the valency of the molecule, and Φ_0 is the surface potential. For small intervals of time t

$$\log[-1/m(dm/dt)] = ze\Phi_0/2.3kT + \text{const} \qquad (6.12)$$

or

$$\log[-d\pi/dt] = ze\Phi_0/2.3kT + \text{const} \qquad (6.13)$$

The constant includes W and is actually constant for films at the air–water interface only when the area/molecule A is constant. At the oil–water interface, W is independent of A because cohesive forces are absent.

Let us now consider the adsorption process as it approaches equilibrium, that is, as the rate of adsoprtion tends to become equal to that of desorption. The following relationship has been found to describe the process when desorption becomes appreciable (Ward and Tordai, 1946; Davies and Rideal, 1963):

$$\Gamma = 2(D/\pi)^{0.5}[(C_0 t)^{0.5} \int \Phi(z)d(t - z)]^{0.5} \qquad (6.14)$$

where z is the arbitrary time between zero and Γ, where equilibrium is attained, and $\Phi(z)$ is the surface concentration at time z after surface adsorption. There remains the difficulty as to how to estimate the value of the integral term on the right-hand side, since $\Phi(z)$ is not known as a function of time, t. A procedure used for this has been to plot the equilibrium surface tension as a function of concentration, and from comparison of the plotted values with dynamic values, the surface concentration $\Phi(z)$ can be estimated:

$$\pi(z) = RT\Gamma_\infty \ln(1 + \Phi)z/a \qquad (6.15)$$

where $\pi(z)$ is the surface pressure at time z, a is the adsorption coefficient (moles/cm^3), and Γ_∞ is the saturation adsorption (moles/cm^2). The relationship between π and Γ is given as

$$\pi = -RT(\Gamma_\infty) \ln(1 - \Gamma_\infty/\Gamma) \qquad (6.16)$$

Typical values of D have been found to range from 0.4 to 1.69×10^6 cm^2 sec^{-1} (Joos, 1968).

It has been pointed out (Fordham, 1954), however, that the Gibbs adsorption model cannot be easily applied to a system in which adsorption at the surface is taking place as a function of time. According to the Gibbs dividing plane (Chattoraj and Birdi, 1984), the surface excess, Γ_2, would also include the excess in the diffusion zone. Under these conditions, one will be faced with the question whether to consider the bulk concentration far away from or close to the diffusion zone. However, more experiments are needed before these criticisms can be verified.

A theoretical model for adsorption based on the kinetic viewpoint has been described (Baret, 1969). In this model, two opposing concepts were used to describe the adsorbed layer: first, that the adsorbed layer is localized; second, that it is mobile. This mixed model corresponds to an important aspect of physical actuality of adsorption; i.e., the adsorbed molecule may have different degrees of interaction with the interface.

6.1.1. RATES OF DESORPTION OF LIPIDS

The desorption kinetics of slightly soluble lipid present as monolayers have been investigated by different workers (Saraga, 1955; Davies, 1951a,b, 1952; Brooks and Alexander, 1960; Gaines, 1966; Gonzales and MacRitchie, 1970; Patlak and Gershfeld, 1967; Shapiro, 1975; MacRitchie, 1985). Desorption mechanisms were investigated for monolayers of phosphatidyl serine, DMPA (dimyristoyl phosphatidic acid), DPPA (dipalmitoyl phosphatidic acid), DSPA (distearyl phosphatidic acid) (Shapiro, 1975). The kinetics were determined from π measurements. From the effect of temperature on the rate constant, the activation energy was estimated. The measured activation energies were suggested to be related to both the polar and the nonpolar molecular interactions of lipid films (see Chapter 4).

Few data have been reported for desorption of mixed monolayers, even though such mixed monolayer desorption is of major importance in industrial and biological phenomena.

6.1.2. RATES OF ADSORPTION OF BIOPOLYMERS

Studies of adsorption of proteins from their dilute aqueous solutions as a function of different parameters have been reported by many investigators (MacRitchie and Alexander, 1963; Bull, 1972; Ghosh and Bull, 1963; Blank and Britten, 1965; Muramatsu, 1959; Archer and Shank, 1967; McCrakin et al., 1963; Gonzalez and MacRitchie, 1970; Asakura et al., 1974; Birdi, unpublished; Chattoraj and Birdi, 1984).

The most convincing demonstration that the relationship in equation (6.9) describes the state of kinetics for surface absorption can be seen from the data in Table 6.1. The agreement is reasonable considering the experimental reproducibility and difficulty with experimentation under these very low concentrations, where even very small contaminations would jeopardize the data. The value of $n = 0.7 \times 10^{-7}$ g cm^2 at $\pi = 0.1$ dyne/cm is found from the π vs. n curve. This gives the value of πA as

$$\pi A = 0.1[10^{16}/(7 \times 10^{-8}/68,000) \times 6 \times 10^{23}] \qquad (6.17)$$

$$= 1600$$

which shows that $\pi A \gg 411$. In this connection, recent π vs. A (or n) data on different spread monolayers of proteins at low π values must be considered.

TABLE 6.1. Experimental and Calculated [from Equation (6.9)] Times (t) for BSA Films to Give 0.1 dyne/cm Surface Pressure (π)[a]

Concentration of BSA (mg/liter) (0.1 ionic strength)	$t_{measured}$	$t_{predicted}$
30	6 sec	7 sec
20	15 sec	16 sec
10	55 sec	64 sec
5	3.5 min	4.2 min
3	9 min	12 min
2	19 min	27 min
1	42 min	107 min

[a] From MacRitchie and Alexander (1963). $D = 6 \times 10^{-7}$ cm^2/sec $= 6 \times 10^{-11}$ m^2/sec.

The adsorption of β-casein and κ-casein at the air–water interface from aqueous solutions containing concentrations ranging from 10^{-5} to 10^{-1} wt% was studied using both radiotracer and ellipsometric techniques (Benjamins *et al.*, 1975). The surface properties of 1-[^{14}C]acetyl-β-casein and the unmodified protein were found to be identical. The surface concentration Γ, as a function of $\sqrt{t}$, was reported to be linear, in the case of β-casein, when the initial protein concentration was $C_{bulk} = 10^{-4}$ wt%. At high bulk concentrations, it was not possible to measure the Γ, in contrast to the case of the data on π vs. $\sqrt{t}$ for other proteins described herein. The thickness of β-casein and κ-casein was estimated at final adsorption states to be of the same values, though related to the bulk concentration, ranging from 25 to 50 to 120 Å^2, for C_{bulk} values of 2×10^{-5} to 2×10^{-4} to 2×10^{-2} wt%, respectively.

The various adsorption kinetic data are given in Table 6.2. The difference in D as measured by the adsorption method (D_{ads}) and as reported from sedimentation techniques (D_{bulk}) is clearly evident. The plausible reasons can be as follows:

1. Convection Effects. These effects can be serious, but would, however, lead to excessive experimental scatter in the data. In the author's laboratory, it has been observed that the experimental reproducibility is rather high, which therefore rules out this effect.

2. Effect of Unfolding at the Interface. In a conventional sedimentation experiment, when D_{bulk} is determined, the protein molecule moves under a homogeneous phase. On the other hand, in an adsorption experiment, the protein molecule diffuses to the interface and unfolds (in most cases irreversibly). This might be a two-step diffusion process, which may or may

TABLE 6.2. Diffusion Constants as Determined from Surface Adsorption Kinetics

Protein	Method	$D_{surface}$ $(m^2/sec \times 10^{10})$	D_{bulk} $(m^2/sec \times 10^{10})$	Ref. no.[a]
β-Casein	Radiotracer	0.033	0.01	1
κ-Casein	Radiotracer	0.015	0.01	1
BSA	$\pi - t^{1/2}$	60	60	2, 3
Hemoglobin	$\pi - t^{1/2}$ ⎱	Ratio = 1/900	60	3
Myoglobin	$\pi - t^{1/2}$ ⎰		107	3
Ovalbumin	$\pi - t^{1/2}$	$\approx$80	80	4

[a] References: (1) Benjamins *et al.* (1975); (2) MacRitchie and Alexander (1963); (3) Birdi (unpublished), Chattoraj and Birdi (1984); (4) Blank and Britten (1965).

not give rise to differences between the diffusion constants measured by the two methods.

The physical factor of greatest interest in studies on the mechanism of protein adsorption kinetics is that the *folding–unfolding* of these biopolymers is of biological significance (Chapter 5). In this adsorption process, the molecular conformation in the bulk phase undergoes a change on adsorption at the interface. The degree of unfolding is obviously determined by the HLB, which is related to the amino acid sequence (as described in Chapter 5).

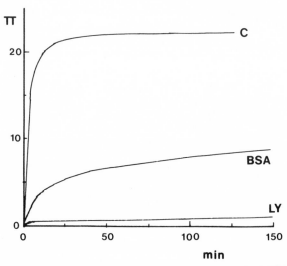

FIGURE 6.1. π vs. time (min) of β-casein (C), BSA, and lysozyme (LY) (protein concentration 0.001% in bulk phase). From Phillips *et al.* (1975).

The amounts adsorbed at the surface of water, Γ, of the proteins [^{14}C]acetyl-β-casein and [^{14}C]acetyl-lysozyme were investigated (Phillips *et al.*, 1975). The adsorption kinetics are given in Fig. 6.1. It can be seen that the most polar protein, lysozyme, exhibits the slowest kinetics. This suggests that the polar–nonpolar character of a protein molecule can be easily estimated from such investigations.

In a recent study (Chattoraj and Birdi, 1984; Birdi, unpublished), the comparative kinetics of two related but functionally important proteins, hemoglobin and myoglobin, were reported. It is known that myoglobin shows no aggregation under normal physiological conditions (i.e., pH 7.4, ionic strength, temperature), while hemoglobin aggregates to dimer-tetramer size as a function of pH and ionic strength. The $d\gamma/d\sqrt{t}$ vs. protein bulk concentrations are given in Fig. 6.2. It is found from equation (6.9) that the term $n/\sqrt{t}$ vs. C_{bulk} should be linear, and the slope should be related to the diffusion constant, D. These data clearly show that the kinetics of myoglobin are different from those of hemoglobin, as one would expect. Since the data when t is small fit very well with the relationship given in equation (6.9), the following virial-type equation was derived to fit these data (Chattoraj and Birdi, 1984):

$$\gamma(t) = a_1\sqrt{t} + a_2t^4 + a_3t^3 + a_4t^2 + a_5t + a_6 \qquad (6.18)$$

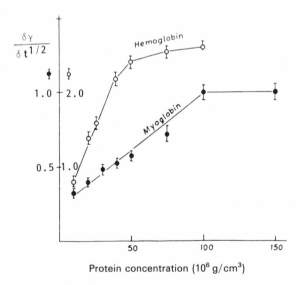

Protein concentration (10^6 g/cm^3)

FIGURE 6.2. Plot of $d\gamma/d\sqrt{t}$ vs. C_b of protein for hemoglobin and myoglobin solutions (25°C, pH 7.4). From Chattoraj and Birdi (1984).

At the value of $d\gamma/d \sqrt{t} = 1$, we find the following:

$$d\gamma/d \sqrt{t} = (2C_{bulk}/\pi) \sqrt{D} \qquad (6.19)$$

$$= 1$$

$$= (2 \times 100/\pi \times 17,000) \sqrt{D}_{myo} \qquad (6.20)$$

$$= (2 \times 15/\pi \times 68,000) \sqrt{D}_{hem} \qquad (6.21)$$

or

$$\sqrt{D}_{myo}/\sqrt{D}_{hemo} = 1/28 \qquad (6.22)$$

$$\approx 1/30$$

$$D_{myo}/D_{hemo} = 1/900, \qquad D_{myo}^{bulk}/D_{hemo}^{bulk} \approx 2 \qquad (6.23)$$

It can thus be seen that the ratio of diffusion constants is approximately 900. To explain these data, we need to consider that:

1. The molecular weights of myoglobin and hemoglobin (as a tetramer) are 17,000 and 68,000, under the experimental conditions.
2. The association of hemoglobin, i.e., chains 2α and 2β, is known to be stabilized through hydrophobic interactions.
3. The amino acid sequences of these two hemoproteins differ somewhat, as regards the degree of distribution of the polar and nonpolar amino acids (Chapter 5).
4. In biological function, hemoglobin carries O_2, while myoglobin is able to complex O_2 strongly in muscles.

The surface adsorption kinetics of sickle hemoglobin (HbS) and normal hemoglobin (HbA) were investigated (Asakura *et al.*, 1974). Sickle cell disease is known to be due to an abnormal hemoglobin with valine (Val) substituted for glutamic acid (Glu) at the sixth position of the β-chains. Because of enhanced polymerization properties of HbS, deoxygenation of erythrocytes takes place, causing the sickling phenomenon.

It was found that on shaking, HbS precipitated from aqueous solutions approximately 10 times faster than HbA. The precipitation phenomenon may be ascribed mainly to surface denaturation. The adsorption–desorption processes must be different in these two proteins. However, it might be of interest to consider that in designing chemicals for treatment, the effect on the surface property of HbS should be kept in mind.

We can thus conclude that the surface adsorption kinetic differences at the air–water interface are related to the amino acid sequence differences. This leads to the further conclusion that the inherent HLB of proteins is of much importance in interfacial biological reactions.

A recent study using the drop-volume method (Tornberg, 1978) investigated the adsorption behavior of three different food proteins, a soya protein isolate, a Na-caseinate, and a whey protein concentrate, at the air–water interface. The kinetics of surface tension decay were investigated in terms of different rate-determining steps at different ionic strengths and bulk protein concentrations. The soya protein was found to diffuse slowly to the interface as compared to the other proteins, probably because of the large particle size of the association complex of soya proteins. The diffusion of soya protein was slower in distilled water than in 0.2 N NaCl solution. Although the caseinate has a complex quaternary structure, like the soya proteins, its surface adsorption behavior is very different. The diffusion step is rapid at concentrations above 0.001 wt% and contributes to a large extent to the interfacial tension decay, especially when it is dispersed in 0.2 M NaCl solution. These results were correlated with the structural differences of these protein molecules, especially the observation that the hydrophobicity was related to the adsorption behavior (Chapter 5) (Birdi, 1973a).

6.2. DESORPTION KINETICS OF LIPID MONOLAYERS

The desorption of slightly soluble components from pure monolayers has been studied quantitatively by many workers (Heikkila *et al.*, 1950; Saraga, 1955; Davies, 1951a,b, 1952; Brooks and Alexander, 1960; Gaines, 1966; Gonzales and MacRitchie, 1970; Gershfeld, 1964; Petlak and Gershfeld, 1967; Patil *et al.*, 1973; Baret *et al.*, 1975). The kinetics of adsorption are best followed by measurements of the decrease of surface area, A, as a function of time, t, at constant surface pressure, π. At a given π, the monolayer establishes rapid equilibrium with thin layers (several molecular diameters in thickness) adjacent to the surface. The desorption rate [given by $\ln(A/t)$] is then equal to the rate of diffusion from these layers into the bulk phases, which are at zero concentration. Little work appears to have been done on desorption of soluble components from multicomponent monolayers, which is often the more practical situation in industrial processes as well as biological systems.

The general behavior of desorption is best illustrated by a monolayer of *n*-octadecanol (nondesorbable) and *n*-dodecanol (desorbable), spread from petroleum ether solution (MacRitchie, 1985). Plots of log A vs. time, t, for $\pi = 5$–30 mN/m are shown in Fig. 6.3. For comparison, the log A vs. t plot for a dodecanol monolayer at 10 mN/m is also shown. The analysis of these results is summarized in Table 6.3. It can be seen that the rates of desorption of *n*-dodecanol from the monolayer and the mixed monolayers (linear region) are the same within experimental error at any π. This finding suggests that the rate of desorption depends only on π and is independent

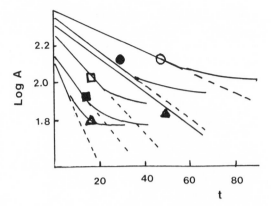

FIGURE 6.3. Log(A) vs. t plot for desorption of n-dodecanol from mixed monolayers with n-octadecanol as a function of π: (○) 5 mN/m; (●) 10 mN/m; (□) 15 mN/m; (■) 20 mN/m; (▲) 30 mN/m. From MacRitchie (1985).

of the surface composition over a wide range, a result relevant to transfer processes across interfaces in the presence of strongly adsorbed surfactants. If the rate of desorption is given by the following equation:

$$\text{Rate} = K \exp(\pi A / kt) \qquad (6.24)$$

where K is constant, then a plot of log(rate) vs. π should be linear, allowing A, area/molecule, to be calculated. The plot is linear and the value of A is 22 Å^2 from the slope. This value agrees with the previous value for linear-chain amphiphiles.

An important element was considered by one of the investigators (Baret *et al.*, 1975). In all the desorption studies, the basic interfacial parameter is the surface pressure, π, of the film. Therefore, any theoretical analysis based on constant area would have to take into account that none of these parameters is constant, since the magnitude of π is variable. At present, there is no theory in the literature that takes this point into consideration.

TABLE 6.3. Rates of Desorption of n-Dodecanol from Pure and Mixed Monolayers at Different Surface Pressures, π (at 20°C)[a]

π (mN/m)	Rates of desorption (l/min)		Dodecanol desorbed (%)
	Dodecanol	Dodecanol + octadecanol	
5	0.016	0.016	88
15	0.032	0.03	82
30	0.062	0.062	74

[a] From MacRitchie (1985).

Mixed films of monooctadecyl phosphate and stearyl alcohol or methyl stearate were used to determine the effect of charge separation on the rate of desorption of soluble monolayers from the air–water interface (Gershfeld, 1964). It was found that (1) at large separations of charge, desorption is primarily influenced by coulombic repulsions, and (2) short-range attractive forces operate between the close-packed phosphate groups. The π dependence of the desorption rate was attributed to the influence of the charged groups on the structure of the water layer in the interface.

6.2.1. DESORPTION OF BIOPOLYMER (PROTEIN) MONOLAYERS

As regards the desorption of protein monolayers at the air–water interface, the adsorbed parts of the polymer must desorb from the interface. The desorption is generally not much at low π, but becomes significant at higher π values (Gonzalez and MacRitchie, 1970; Langmuir and Waugh, 1940; Adams *et al.*, 1971; MacRitchie, 1977).

The energy barriers for desorption were estimated as a function of π (MacRitchie, 1977).

7

LIPID–PROTEIN MONOLAYERS

The importance of studying mixtures of lipids and proteins in mono-molecular films is obvious when one considers the currently accepted fluid mosaic model of membranes (Fig. 7.1) (Singer, 1971, 1974; Singer and Nicolson, 1972). The classic Danielli–Davson lipid bilayer (Gorter and Grendel, 1925; Danielli and Davson, 1935) functions in its liquid crystalline phase transition region. Furthermore, it was realized many decades ago that the diversity in chemical composition and functional specialization of membranes presents a formidable obstacle in assigning a specific functional role to each of the membrane components, e.g., lipids and proteins. It has been suggested that the lipid bilayer functions as a protein solvent with solvent domains in equilibrium with lipid domains (Papahadjopoulos *et al.*, 1973). Phase or domain fluctuations, possibly regulating membrane function, can be related to localized energy fluctuations of approximately kT (≈ 2400 J $\approx$ 600 cal) or to localized protein perturbations resulting from a change in protein conformation.

Transport of a wide variety of solute molecules and ions is one of the most fundamental properties of biological membranes, yet there has been little success in isolating ionophoric species that could begin to account for the broad spectrum of biological transport capabilities (Green *et al.*, 1980).

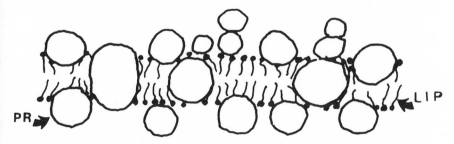

FIGURE 7.1. Schematic representation of membrane structure. Filiform alkyl chains, polar groups (●), and protein molecules (large open circles) are depicted as either attached to the bilipid layer or extending across the membrane.

209

Phospholipids, especially phosphatidic acids, have been suggested as possible transport species. The studies that led to this suggestion employed oil–water partitioning of complexes of lecithin and phosphatidic acid with calcium and a variety of solute molecules (Green *et al.*, 1980; Putney *et al.*, 1980).

The best argument in favor of using a monolayer as a model system of membranes is that it provides more quantitative information than is available from other model systems (e.g., bilayers or vesicles). One can inject protein molecules into a lipid monolayer and determine the changes in π, ΔV, η, or some other monolayer property in order to estimate the interaction energies in lipid–protein mixtures. Kinetic studies of such experiments would provide information about the relaxation mechanisms and the activation energy.

7.1. LIPID–PROTEIN MONOLAYERS AS MODELS FOR CELL MEMBRANE STRUCTURE AND FUNCTION

The early studies investigating π measurements in mixed lipid (cardiolipin, cholesterol, cephalin) and protein (albumin, globulin) monolayers indicated that the interactions were dependent on subphase pH (Doty and Schulman, 1949). Irreversible adsorption of proteins (hemoglobin, albumin, globulin) on lipid (cardiolipin, stearylcholine, behenyl sulfate) monolayers was reported (Matalon and Schulman, 1949). Interactions between basic proteins (cytochrome c, lysozyme, human serum albumin) showed that these proteins could penetrate negatively charged lipid monolayers (phosphatidic acid, cardiolipin) (Quinn and Dawson, 1969). Several studies concerned with the interaction between cytochrome c (CYTc) and phospholipids have been carried out (Quinn and Dawson, 1969, 1970; Fromherz, 1971; Nicholls *et al.*, 1967; Morse and Deamer, 1973).

TABLE 7.1. Cytochrome c–Lipid Complex Formation[a]

Lipid	Final π (mNm)	π (mN/m)	$d\pi/dt$
None	16	16	0.4
Stearylamine	18	12	4
Stearyl alcohol	16	10	7
Stearic acid (pH 6.8)	18	12	9
Stearyl sulfate	28	22	>17
Stearyl phosphate	30	24	33
Oleic acid (pH 6.8)	23	17	15

[a] Morse and Deamer (1973).

The most useful parameter besides π is the ΔV that follows on injection of protein under the lipid monolayer. A second parameter of protein–lipid interaction in the monolayer is the rate of π change, $d\pi/dt$. The rate of change of π on injection of CYTc at the air–water interface in the absence of lipid was 0.37 dyne/cm per min. A $\Delta\pi$ of approximately 15 mN/m was obtained after several hours. The presence of the lipid monolayer was found to greatly increase the magnitude of $d\pi/dt$ (Morse and Deamer, 1973) (Table 7.1).

From these studies, it was concluded that some portion of CYTc molecules might penetrate the lipid monolayer and that as more CYTc interacts in this process, π increases. These investigations established two relationships between the lipid and the protein:

1. The extent of interaction, π, is principally controlled by the expanded or condensed character of the lipid film. A more expanded lipid film generally produced greater $\Delta\pi$ values in interacting with CYTc. The expanded nature of the lipid film may be controlled either by π, by surface charge, by unsaturation of the hydrocarbon chain, or by temperature.
2. The rate of interaction, $d\pi/dt$, was mainly determined by favorable electrostatic interaction between CYTc and the lipid. Thus, CYTc had much greater rates of interaction with negatively charged film than with neutral or positively charged lipid film.

The interaction of β-lactoglobulin with phospholipids in monolayers at varying pH has been reported (Cornell, 1982). Typical π vs. A curves for egg yolk phosphatidic acid (EYPA), β-lactoglobulin, and a mixture of the two at pH 4 are given in Fig. 7.2. The film of pure β-lactoglobulin gave an area/residue of 21 Å^2 at $\pi = 5$ mN/m. The calculated surface area at $\pi = 5$ mN/m was less than the measured area in mixed films. Table 7.2 lists the average values of the area per molecule of lipid or residue of protein

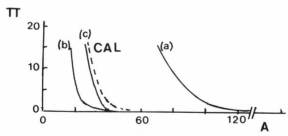

FIGURE 7.2. π vs. A isotherms of EYPA, β-lactoglobulin, and a mixture of the two on pH 4 subphase at 20°C. (a) EYPA; (b) β-lactoglobulin; (c) EYPA-β-lactoglobulin (80 mole% β-lactoglobulin): (———) experimental; (- - - -) calculated from equation (4.104). From Cornell (1982).

TABLE 7.2. Area per Molecule of Lipid or Residue of Protein (A) Obtained
from π vs. A Curves of Pure Components[a]

| | | | Area (Å^2/molecule or residue) | | |
| | | | π (mN/m) | | |
Component[b]	pH	[Ca^{2+}]	3	5	17
EYPA	6	0	104.8	97	73
EYPC	6	0	109.9	103	—
β-Lactoglobulin	6	0	23	21	—
EYPA	4	0	96.6	89.9	68.5
EYPC	4	0	103	96	—
β-Lactoglobulin	4	0	22.7	20.6	—
EYPA	4	1 mM	84.4	78.6	—
β-Lactoglobulin	4	1 mM	22	20	—
EYPA	1.3	0	91	85	66
EYPC	1.3	0	108	101	—
β-Lactoglobulin	1.3	0	25	23	—
EYPA	1.3	1 mM	91	84	—
β-Lactoglobulin	1.3	1 mM	24.4	22	—

[a] From Cornell (1982).
[b] (EYPA) Egg yolk phosphatidic acid; (EYPC) Egg yolk phosphatidyl choline.

at different π values. The effect of calcium ion added to the subphase is also shown. These monolayer studies indicated that EYPA interacted with β-lactoglobulin on a subphase of pH 4, but not on one of pH 6.

A similar result was reported for the pH-dependent interaction between the phosphatides of milk and β-lactoglobulin (Payens, 1960). The isoionic pH of β-lactoglobulin is 5.3 at 0.1 ionic strength, as used in these experiments. The pK_1 for EYPA monolayers has been reported to be 3.5 at high π (Patil et al., 1979). Thus, at pH 4, β-lactoglobulin would be expected to be positively charged, while EYPA would be negatively charged, and electrostatic attraction between the lipid and the protein in mixed monolayers would be expected. This agrees with the observation that the measured A is less than that calculated from the ideal equation (4.104). Calcium ion is known to bind both to phosphatidic acid (PA) (Abramson et al., 1964) and to β-lactoglobulin (Zittle et al., 1957; Carr, 1953).

Some studies have focused on characterization of the interaction between gelatin and octadecanoic acid monolayers in terms of surface viscosity, η_s, and surface elasticity (Wuestneck and Zastrow, 1985; Zastrow et al., 1985).

The interactions between lipids and synthetic polypeptides have been reported by a number of investigators (Shah, 1969; Hammermeister and

Barnett, 1974; Shafer, 1974; Chatelain *et al.*, 1975; Hookes, 1971; Gabrielli, 1975; Kretzschmar, 1969; Subramanian *et al.*, 1978; Ralston and Healy, 1973; Sears, 1969; Yamashita *et al.*, 1978a,b). It was reported that for a poly(lysine)-stearic acid system, the maximum interaction occurred at a subphase pH value at which the polypeptide is in the α-helical conformation, and the α-helical conformation was found to be stabilized through electrostatic interactions. On the other hand, hydrophobic interactions were found for systems composed of phosphatidylserine (DPS) or DL-α-dipalmitoyl-lecithin (DPL) and poly(L-aspartic acid) (PLASP), poly(L-glutamic acid) (PGA), or poly(L-lysine) (PLL).

Investigations in which mixed polypeptide and lipid monolayers were used have also been reported (Hookes, 1971; Gabrielli, 1975; Kretzschmar, 1969; Malcolm, 1968a,b, 1973; Hatefi and Hanstein, 1969). One of these studies reported complexes of 5:1 and 1:1 (residue/lipid molecule) for a poly(DL-leucine)-sodium octadecyl sulfate system and 1:1 for a poly(DL-phenylalanine)-sodium octadecyl sulfate system (Kretzschmar, 1969). The interactions of polypeptide (benzoyloxycarbonyl) derivatives of basic poly-α-amino acids (ornithine), benzyl esters of acidic poly-α-amino acids, and poly(leucine) and lipid (fatty acids, fatty alcohols, and cholesterol) at the air–water interface have been investigated (Yamashita *et al.*, 1978). In the

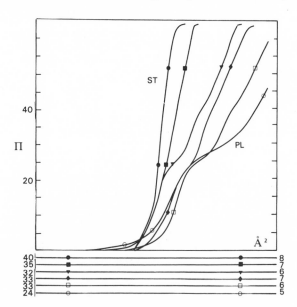

FIGURE 7.3. π vs. A isotherms of mixed stearyl alcohol (ST)–poly(alanine) (PL) monolayers on a water subphase (25°C) as a function of molar ratio (with the alanine residue as the unit in PL): (●) ST; ST/PL: (■) 85:15, (▼) 50:50, (♦) 30:70, (□) 15:85; (○) PL. From Birdi and Sørensen (1979).

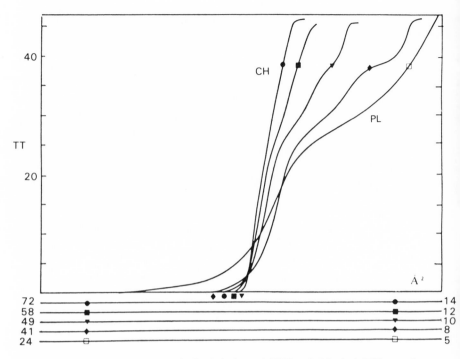

FIGURE 7.4. π vs. A isotherms of mixed cholesterol (CH)–poly(alanine) (PL) monolayers: (●) CH; CH/PL: (■) 70:30; (▼) 50:50; (♦) 30:70; (□) PL. From Birdi and Sørensen (1979).

case of ornithine and C_{14} (tetradecanoic acid) or cholesterol mixed films, it was found that these data are of the same nature as those reported for poly(alanine) and lipids (Birdi and Sørsensen, 1979). The two transition-state π values were clearly evident. The lower collapse π (≈ 12 mN/m), which corresponds to pure ornithine, increases as the C_{14} content increases. On the other hand, the second collapse π remains almost unchanged, i.e., approximately 22 mN/m, corresponding to the pure C_{14}. However, a third collapse π, intermediate between these two, appears about the same way as reported for poly(alanine)–lipids (Fig. 7.3) (Birdi and Sørensen, 1979).

It is of interest to mention that no intermediate collapse states were found for poly-benzyl-L-glutamate (PBLG) and C_{14} (Yamashita *et al.*, 1978). Mixed monolayers of poly(alanine)–stearyl alcohol (Fig. 7.3) and poly(alanine)–cholesterol (Fig. 7.4) were investigated as a function of composition (Birdi and Sørensen, 1979). Plots of mean area/molecule vs. composition in mixed monolayers of these films were linear in both mixed monolayer systems (Fig. 7.5). To determine the usefulness of the mixed monolayer equations (see Section 4.10), the area/molecule was calculated

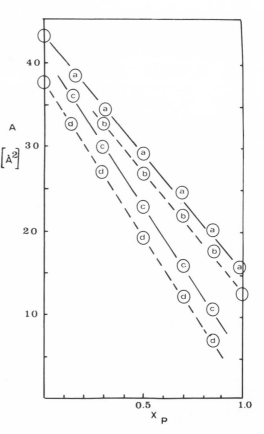

FIGURE 7.5. Plots of mean area/ molecule (A) vs. composition of mixed poly(alanine) (PL)– cholesterol (CH) monolayers calculated at four collapse states: (a) limiting π; (b) first collapse state (corresponding to that of PL); (c) second collapse state; (d) third collapse state. (x_p) Molar fraction of PL. See the text for details. From Birdi and Sørensen (1979).

at the following four collapse states (Fig. 7.5):

   (a) Limiting π
   (b) First collapse [corresponding to that of poly(alanine)]
   (c) Second collapse state
   (d) Third collapse state

In both mixed monolayer systems, the magnitudes of the areas at states (a)–(d) were linear. This finding would suggest that such mixed monolayers behave as "ideal" mixed films. If the isotherms of the two systems are compared, however, it is found that in both systems, a new collapse state, intermediate between that of poly(alanine) and lipid, is measured. If these were ideal films, then no such new collapse state ought to have been observed. Furthermore, the magnitude of the intermediate collapse pressure, π_{int}, is determined by the characteristic of the lipid. The magnitude of π_{int} is ca. 40 mN/m in the poly(alanine)–stearyl alcohol system, while it is 35 mN/m in the poly(alanine)–cholesterol system. The dependence of collapse pressure in the mixed films on the character of the lipid suggests that the new collapse state arises from complex formation and could be related to the magnitudes of the van der Waals force (since the systems consist of neutral molecules). It was suggested that since the magnitude of van der Waals force [as calculated from equation (4.39)] in stearyl alcohol is larger than in cholesterol, the magnitude of π_{int} was accordingly different.

In a later study (Gabrielli *et al.*, 1981) (L-, DL-, L-) mixed poly(alanine) and arachidic acid (with subphase K^+, Na^+, and Ca^{2+}) monolayers were reported. The poly(alanine) monolayers were analyzed with the help of Huggins's theory for two-dimensional monolayers and found to exhibit negative deviation from the ideal equation. These findings agreed with other reports for different polypeptide–lipid mixtures (Hookes, 1971).

Mixed poly-methyl-glutamate (PMG)–cholesterol monolayers were investigated, and negative deviation from the additivity rule was reported (Baglioni *et al.*, 1986) (Fig. 7.6).

Diverse surfactant–protein monolayers were reported in very early data (Schulman and Hughes, 1935; Schulman and Rideal, 1937; Doty and Schulman, 1949; Chen and Rosano, 1977).

These studies show that very specific intermediate complexes are easily observed by the monolayer method in lipid–polymer systems.

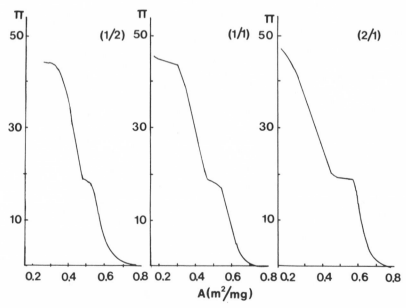

FIGURE 7.6. π vs. A isotherms of mixed monolayers at 25°C. (1/2), (1/1), (2/1) PMG/ cholesterol molar ratio. Redrawn from Baglioni *et al.* (1986) with changes.

7.1.1. BACTERIA–LIPID INTERACTIONS IN MONOLAYERS

The interaction of lipid surface films of bacteria with free fatty acids and phospholipids at the air–water interface has been studied by various investigators. The surface microlayers at the air–water interface of the sea have been studied extensively with respect to both chemistry and microbiology and have been found to consist of three upper strata: the lipid film, the polysaccharide–protein complex, and the bacterial layer (Kjelleberg *et al.*, 1979; MacIntyre, 1974; Norkrans, 1979). Evidence from field and model system experiments suggests a positive correlation between the number of microorganisms and the amount of lipids on the surface (Norkrans and Sorensson, 1975). The nature of the microbial interaction, which is of general interest for most interfaces, has not been fully revealed. This interaction can be conveniently studied with the surface Langmuir film balance technique (Kjelleberg *et al.*, 1976; Kjelleberg and Stenström, 1980). Addition of bacteria or molecules to the surface is indicated by an increase in the π at a given area per molecule, A. On the other hand, dissolution of the film (e.g., as a result of enzymatic activity) decreases the film π.

The interaction between bacteria and palmitic acid, stearic acid, and oleic acid has been investigated at pH 7.4, a pH at which bacteria have negative charges. The data in Fig. 7.7 illustrate the interaction of bacteria with a palmitic acid monolayer. Both the expanded film and the condensed monolayer were affected. This finding allows one to conclude that the forces binding bacteria to free fatty acid films are not strong enough to penetrate the condensed lipid monolayer. These findings are consistent with other protein–lipid monolayer studies (Birdi, 1976b). The degrees of interaction between bacteria and lipid monolayer are given in Table 7.3.

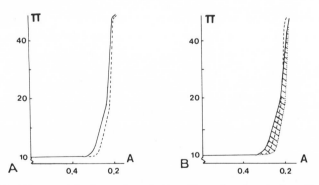

FIGURE 7.7. (A) π vs. A isotherms for palmitic acid on phosphate buffer with (——) and without (– – – –) cells of *S. marinorubra*. (B) π vs. A isotherms with (——) and without (– – –) cells of *P. halocrenaea*. The hatched area represents overall adhesion of cells after compensating for the film loss at the point of monolayer collapse. Redrawn from Kjelleberg and Stenström (1980) with changes.

TABLE 7.3. Degrees of Interaction between Test Bacteria and Lipid Monolayers at the Air–Water Interface[a]

Test organism	Change in area (A) as a percentage of the lipid monolayer[b]					
	$C_{16:0}$	$C_{18:0}$	$C_{18:1}$	ODA	EPC	DOPC
Pseudomonas halocrenaea	14	27	<5	43	40	27
Serratia marinorubra	37	35	21	90	29	33

[a] From Kjelleberg and Stenström (1980).
[b] ($C_{16:0}$) Palmitic acid; ($C_{18:0}$) stearic acid; ($C_{18:1}$) oleic acid; (ODA) octadecylamine; (EPC) egg phosphatidylcholine; (DOPC) dioctanoylphosphatidylcholine.

The bacterial interaction with lipid films, as revealed by the surface balance, is influenced by several factors, partly resembling the mechanisms of bacterial attraction to interfaces (reviewed by Marshall, 1975, 1976).

These results can be described as follows: If the surface molecules are sufficiently separated to prevent interchain reactions, the possibility for bacterial interaction is greatly enhanced, as can be seen from the experiments on the spreading of C_{16} (myristic acid). Similar reasoning explains the stronger bacterial influence on the condensed C_{16} and C_{18} (stearic acid) films than on the $C_{18:1}$ (oleic acid) film. Furthermore, bacteria carry a characteristic negative charge that varies from species to species (Longten, 1975). This charge provides an explanation for the following phenomena:

1. The large increase in π caused by changing from negatively charged fatty acid films to octadecylamine films (which are positively charged) at pH 7.4.
2. The higher overall addition of different bacteria to stearic acid and palmitic acid monolayers.

Specific surface molecules such as pyrol-containing prodigiosins, which are pigments characteristic of some bacteria, may also play a role in these interactions.

The surface charge, ψ_0, however, cannot be wholly responsible for the interaction of the bacteria, since strong effects are observed between bacteria and zwitterionic phospholipids. This phenomenon indicates that a third factor of importance, arising from hydrophobic interaction (Chapter 5), is present. Since part of the outer surface of some bacteria is hydrophobic, it is reasonable to suppose that such bacteria are rejected from the aqueous phase and attracted toward any nonaqueous phase (e.g., air–water, oil–water, or solid–water). At low π, bacterial surface groups are probably mixed with the nonpolar part of the surface film.

The chemical composition and microflora of the interface film of fresh water and seawater seem to reflect the properties of the underlying aqueous media. For example, in lakes with a high degree of pollution, these surface films may consist of several layers of hydrophobic molecules. Surface films of aquatic environments and their lipid composition were investigated (Sodergren, 1978). Enzyme activity is manifested in the reduced π [as reported from different enzyme–lipid systems (Shah and Shulman, 1967; Verger *et al.*, 1976; Zografi *et al.*, 1971)]. The film loss observed as free fatty acids, using phosphate buffer solutions, has been attributed to β-oxidative activity.

7.1.2. CELL SURFACE HYDROPHOBICITY AND THE ORIENTATION OF CERTAIN BACTERIA AT INTERFACES

The sorption of microorganims to surfaces is an important factor in microbial behavior in natural habitats (Corpe, 1970; Marshall, 1971; Hsu, 1987). It has been shown that an initial reversal phase of sorption of bacteria to solid surfaces is followed by a time-dependent, and possibly growth-dependent, irreversible phase of sorption of some bacteria (Marshall, 1971). It has also been suggested that the irreversible phase may be associated with the production of extracellular fibrils that serve to anchor the cell to the solid surface. It has been shown that individual cells orient themselves perpendicular to the interface in air–water, oil–water, and solid–water systems (Marshall and Cruickshank, 1973). Electrostatic phenomena are probably not involved in this orientation, since no evidence was found of any localized distribution of positively charged ionogenic groups on the bacterial surface. It was therefore suggested that the orientation results from a relatively hydrophobic portion of each cell being ejected from the aqueous phase of the system. This property may also be related to the formation of rosettes by these bacteria. Electron micrographs of thin sections of cells sorbed to araldite blocks showed that the cell proper is not in contact with the solid surface, but is achored to it by extracellular adhesive material. The extracellular material has been suggested to be of a polysaccharide nature. A theoretical model was used to describe the kinetics of bacterial adhesion (Hsu and Wang, 1987; Hsu, 1987).

The attraction of a specific region of the cell to the oil rather than the aqueous phase of an oil–water system suggests that this portion of the cell is relatively hydrophobic. This possibility is supported by the observations that a surface-active agent (Tween 80) prevents this distinct orientation at the oil–water interface and, as well, prevents sorption to a solid surface. The adsorption mechanism would account for the preferred orientation at the interface of a two-phase system regardless of the nature of the nonaqueous phase (air–water, water–oil, or solid–liquid).

The adhesion of cells to solids has been correlated with the interfacial forces, as determined from contact-angle measurements (Birdi, 1981).

7.2. DETERMINATION OF VAN DER WAALS FORCES IN LIPID-PROTEIN MONOLAYERS

The structure of biological membranes and the role of lipids and proteins in determining that structure are the object of much intensive biophysical investigation. In these structures, the lipid-lipid, lipid-protein, and protein-protein interactions are therefore expected to be of major interest in determining the structural stability and function of biological membranes. In the monolayer method, protein-lipid complex formation has been investigated by means of the changes in surface pressure ($\Delta \pi$) that follow the injection of protein under the lipid film at some well-defined π. Generally, the magnitude of π for lipids varies from 2 to 10 mN/m.

The interaction between cholesterol and gliadin was investigated by this method. The results indicated that the protein was ejected from the lipid-protein mixed monolayer at high π. In other studies (Eley and Hedge, 1956, 1957; Chattoraj and Birdi, 1984), it was proposed that a layer of spread protein was formed beneath the lipid film, followed by a second layer of unspread protein (i.e., in native conformation) beneath the first layer. However, these interpretations have been questioned by later studies reported by other investigators, the results of which were interpreted to show that protein instead unfolded at the interface and formed a mixed film with the lipid film, giving rise to a change in surface pressure, $\Delta \pi$. It has been reported that the addition of proteins to charged lipids indicated that a primary association between the protein and the lipid was taking place.

Insulin and various other proteins (Birdi, 1976, 1977) have been reported to be able to penetrate the monolayers of various lipid films. At a certain lipid film pressure, called the "limiting π," no further penetration by the protein molecules is observed.

Even though a considerable number of such investigations by the monolayer method have been reported in the literature, only in a few cases has a more quantitative description been presented as regards the van der Waals forces in such lipid-protein mixed monolayers (Birdi, 1976).

Van der Waals Forces in Lipid-Protein Monolayers. As described in Chapter 4, in lipid monolayers, the magnitude of van der Waals forces, which are known to play a very important role in the stability of such films, can be estimated by the relationship

$$W_{dis} = -1.24 \times 10^3 \, (N_c/D^5) \tag{4.39}$$

It is reasonable to expect that when the value of lipid π is equal to the limiting π, the van der Waals interactions between the alkyl chains (groups) are of such magnitude that no further penetration by the protein molecule is possible.

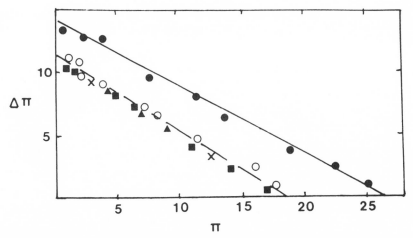

FIGURE 7.8. $\Delta\pi$ vs. π_{lipid} isotherms for several films as a function of the lipid component. (■) Stearyl alcohol; (□) stearic acid; (×) DPL; (●) cholesterol. Amount of insulin added: 10 μg. From Birdi (1976).

Data on π_{lipid} vs. $\Delta\pi$ are given in Fig. 7.8 for the interaction between insulin and various lipids, e.g., stearyl alcohol, stearic acid, DPL, and cholesterol (Birdi, 1976; Chattoraj and Birdi, 1984). These results are given for the addition of 10 μg insulin, which corresponded to 14.7 $\mathring{A}^2$/residue. It can be seen that in all cases, $\Delta\pi$ decreased linearly over the entire range of π_{lipid}. It has been found (Birdi, unpublished), however, that at very low π_{lipid} values, the plots of π_{lipid} vs. $\Delta\pi$ are nonlinear. The magnitude of π_{lim} for stearyl alcohol, stearic acid, and DPL is of the same magnitude, i.e., 19.0 mN/m. On the other hand, the magnitude of π_{lim} in the case of cholesterol film is much larger, i.e., 27.0 mN/m. Various data reported in the literature support this finding that the value of π_{lim} is larger for cholesterol than for other lipids or other proteins. The magnitudes of π_{lim} for cytochrome and stearyl alcohol or stearic acid films were also found to be 17.00 mN/m (Birdi, unpublished).

These studies indicate that the degree of penetration of insulin (or other proteins) is different for different lipid films, as one would expect. These differences arise from the structural differences among these lipids, as has also been reported by many different investigators. In only one case in the current literature has a quantitative description been given for the significance of the limiting π, π_{lim}, in these $\Delta\pi$ vs. π_{lipid} plots (Birdi, 1976). In lipid films, the magnitude of van der Waals forces, π_{coh}, can be estimated by equation (4.39). The magnitude of π_{lim} can be described as follows: It is reasonable to expect that when the value of lipid film π is equal to π_{lim}, the van der Waals interactions between the hydrocarbon chains are of such

TABLE 7.4. Magnitude of van der Waals Forces, $\dot{W}_{dis}$ [Calculated from equation (4.39)], at π_{lim} for Lipid and Insulin Films[a]

Lipid	π_{lim} (mN/m)	A_{lim} (Å^2)	N_c	D (Å)	W_{dis} (kcal/mole)
n-Stearyl alcohol	19.0	34	18	6.3	−2.30
n-Stearic acid	19.0	34	18	6.3	−2.3
DPL[b]	19.0	58	15	5.8	−2.84
Cholesterol	27.0	40	32	5.2	−2.3

[a] From Birdi (1976).   [b] There are two alkyl chains in DPL.

magnitude that no penetration by protein molecules is possible, especially in the case of stearyl alcohol and cholesterol. The value of $W_{dis,SA}$ is found to be −2.3 kcal/mole by using equation (4.39) ($A = 34$ Å^2, $D = 6.3$ Å, $N_c = 18$). A straight-chain molecule, such as stearyl alcohol, will pack better than the nonlinear-chain cholesterol, and one thus cannot use the relationship in equation (4.39) directly. However, one finds that $W_{dis,ST} \approx W_{dis,CH}$ at the respective π_{lim} if one uses $N_c = 32$ for cholesterol, when actually there are 27 carbon atoms in the nonlinear-chain cholesterol molecule (Table 7.4).

Under the experimental conditions reported, the electrostatic interactions in the case of stearic acid and DPL would be rather weak in comparison to the van der Waals forces. It is thus in agreement with the results given in Fig. 7.8 that the magnitude of π_{lim} for stearic acid is the same as that for stearyl alcohol, i.e., approximately −2.3 kcal/mole. At the π_{lim} for stearic acid, W_{dis} is the same as that of stearyl alcohol. At the π_{lim} for DPL, one finds that $A = 58$ Å^2, i.e., 29 Å^2 for each alkyl chain. Using this value and since $N_c = 15$ for DPL, one finds that W_{dis} for DPL is −2.84 kcal/mole (Table 7.4).

7.2.1. ELECTROSTATIC INTERACTIONS AT LIPID–PROTEIN MONOLAYERS

Monolayers of lipid (DPL, DPS) spread at the air–water interface were investigated with regard to their interaction with polypeptides (PGA, PLASP, PLL) (Chatelain et al., 1975). The interactions were measured by using radioactivity and π measurements. The adsorption was studied on monolayers of DPL, with a zero net charge, and DPS, with a negative net charge. The data in Table 7.5 show the influence of the subphase pH on the polypeptide surface concentration. These data show that adsorption of polypeptide is low when the lipid and the polypeptide have the same charge.

TABLE 7.5. Polypeptide Adsorption at the Lipid–Water Interface[a]

Subphase pH	PGA (mg/m^2)		PLL (mg/m^2)	
	DPL	DPS	DPL	DPS
4.1	0.11	0.041	—	—
4.7	0.10	0.040	—	—
5.15	0.09	0.041	—	—
5.90	0.07	0.032	—	—
8.5	0.07	0.032	0.10	2.5
10	—	—	0.1	2.4
11.5	—	—	0.1	0.1

[a] From Chatelain *et al.* (1975).

7.3. LUNG SURFACTANTS

The alveoli of the lung are lined with a highly surface-active, phopholipid-rich material, the so-called "pulmonary surfactant" which prevents their collapse on expiration. The existence of surfactant was first suggested decades ago (von Neergard, 1929), but not until more than 25 years later was its presence actually demonstrated (Pattle, 1955). It was shown that remarkably stable bubbles could be squeezed from a lung cut under water. Later, it was reported that lung extracts lowered the surface tension at the air–water interface (Clements, 1956, 1957). It was thereafter found that the lungs of infants who died from the respiratory distress syndrome (RDS) were deficient in surfactant. The synthesis of phospholipids that serve as surfactants is an essential step in the development of fetal lungs.

Lung surfactants from different animals were analyzed and found to be surface-active and to consist of lipid (79–90%) and protein (18–28%) (Harwood *et al.*, 1975). The lung surfactants of dogs have a similar composition. Lipids from rabbit lung lavage consist of approximately 85% phospholipids, 10% glycolipids, and 5% neutral lipids (Rooney *et al.*, 1974). Over half the phosphatidylcholine (PC) is disaturated, and palmitic acid accounts for 90% of the saturated fatty acids (Rooney *et al.*, 1974). The most significant finding is that dipalmitoyl-PC is the major component of the pulmonary surfactant. The fetal lung produces surfactant in increasing quantity toward the end of gestation. In the rabbit, there is a 10-fold increase in the amount of PC and disaturated PC in lung lavage between 27 and 31 (full term) days' gestation. During the same period, the composition of the phospholipids in lung lavage also changes. PC increases while sphingomyelin decreases. This results in a dramatic increase in the PC-

TABLE 7.6. Development Changes in the Phospholipid Content and
Composition of Rabbit Lung Lavage[a]

	PC		PC (%)	
Gestation days	Total	Disaturated	Lecithin/sphingomyelin	LS ratio
27	2.6	1.3	29/38	0.8
29	7.4	3.4	50/11	5
31	25.4	13.4	68/7	10
+1	274	161	79/2.6	31
Adult	264	143	86/1.2	>50

[a] Rooney *et al.* (1977).

lecithin/sphingomyelin ratio (the so-called "LS ratio"). Since lung fluid contributes to amniotic fluid (Adams *et al.*, 1967), measurement of the LS ratio in amniotic fluid can be used to predict the degree of maturity of the human fetal lung (Gluck *et al.*, 1971; Kulovich and Gluck, 1979; Kulovich *et al.*, 1972) (see Table 7.6).

The surface-active material isolated from the lungs of a man who developed adult RDS after massive trauma showed lipid/protein ratios different from those found in normal lungs (Petty *et al.*, 1977, 1979; von Wichert and Kohl, 1977). The lung fluids exhibited different surface compressibilities in this patient with RDS. This finding was associated with the decreased levels of dipalmitoylphosphatidylcholine in the patient.

During surface-tension measurements of mixed DPL–cholesterol films, extreme differences in the kinetics of film-forming became evident (Reifenrath *et al.*, 1981). Similar observations were made with lung-surfactant dynamic studies.

When pulmonary surfactant is repeatedly and gently compressed and expanded, in a manner that might imitate the breathing movements of the lungs, its chemical composition at the surface changes. The suggestion that it changes from a liquidlike to a solidlike monolayer was made (Bangham *et al.*, 1979; Colin and Bangham, 1981). This change was suggested to have taken place due to the squeezing out of the asymmetric lipids (Figure 7.9).

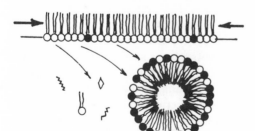

FIGURE 7.9. Schematic representation of the manner in which compression of a mixed lipid monolayer is suggested to give rise to squeezing out of some lipids into the bulk phase (forming vesicles). Redrawn from Colin and Bangham (1981) with changes.

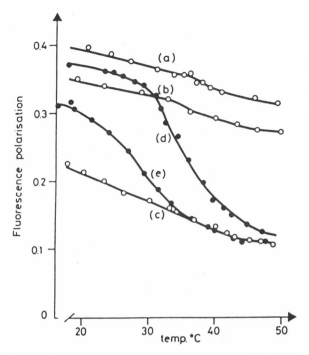

FIGURE 7.10. Variation of fluorescence polarization of human amniotic fluid at different ages of gestation: (a) 14–15 weeks; (b) 19th week; (c) 40th week. Lipid vesicles: (d) sphingomyelin; (e) 2:2:1 mixture of DPL–egg lecithin–sphingomyelin. From Petersen and Birdi (1983).

The role of protein in the LS ratio was reported in recent monolayer studies (Mutafchieva *et al.*, 1984).

The fluorescence probe hexatriene was used to determine the lipid-phase behavior of human amniotic fluid (Petersen and Birdi, 1983). The variations of fluorescence polarization of different mixed lipid vesicles and amniotic fluids were compared (Fig. 7.10).

7.3.1. DYNAMIC LIPID MONOLAYER STUDIES

As described in Chapter 4, dynamic lipid monolayers have been investigated in order to determine the composition domains, if present, in mixed lipid systems.

The periodic expansion–compression of the lungs gives rise to an important property as regards the interfacial behavior of lipids under these dynamic conditions. Since π can affect the intermolecular configurations of the surface molecules, the cyclic changes of π could be expected to give rise to changes in other interfacial properties such as η_s. The hydrophobic-hydrophilic balance is suitable to register a total force that results from the kinetic interaction of film components, whether these be the hydrocarbon chain or a large domain. If a cyclic study is to be done in a larger body, the period of the cycle needs to be longer.

In dynamic compression–expansion studies, one might expect the composition of the mixed films to be changing. It was demonstrated (Ries and Walker, 1961) that on collapse, there could be extrusion of one of the molecular species, which would continue to sustain the π. Furthermore, a change in film composition could occur by means of the mechanism whereby complete or irreversible loss of the extruded species has taken place. Structural formation as a function of cycling rates has been investigated (Bienkowski and Skolnick, 1972; Lusted, 1973a,b).

On compression, the maximum surface pressure, π, is 52 mN/m, while it is independent of the amplitude of area oscillation. The minimum π on expansion is dependent on the amplitude of oscillation and is reported to become larger with increasing amplitude. The dependence of this property on lung function thus raises the question of how to interpret this dynamic behavior in terms of lipid composition and monolayer characteristics. Mixed lipids of DPL and cholesterol in ratios of 1:2 and 8:1 were studied (Reifenrath and Zimmermann, 1976). The maximum π was beween 49 and 54 mN/m. The dependence of π on amplitude of compression–expansion was also investigated. The maximum π was found to be dependent on both the amplitude of area oscillation and the DPL/cholesterol ratio. An increase in the amplitude of area oscillation gave a decrease in minimum π. The role of cholesterol in these fluids remains to be investigated. The film area between barriers was changed from 35 to 16 cm^2, with a rate of 0.5 cycle/min (area reduction of 19 cm^2/min) (Lusted, 1973a,b).

In a recent study, a leakproof automated apparatus for reproducible determination of dynamic interfacial properties of lipid monolayers was described (Somasundaran *et al.*, 1974). Dynamic surface properties of alveolar surfactants have been investigated (Mateeva *et al.*, 1975). Dynamic film compression of phosphatidylcholine–phosphatidyglycerol was studied, since this process is relevant to lung surfactants (Wojciak *et al.*, 1985).

7.4. CANCER AND INTERFACIAL INTERACTIONS AT CELL SURFACES

The role of the cholesterol molecule in cell membranes has been given much attention, even though it is not quite clearly established. It is reasonable to assume that in the evolutionary process, the first unicellular organisms divided and became separated. The role of any surface-active material, which must be present at the surface of membranes, can be ascribed to cell adhesion in colonies or to cell separation. It has been postulated that this controlled-release mechanism in cells, if it were understood, could provide a means of cancer treatment (Sebba, 1972a,b). This postulate is based on the report that in biliquid foams, a small amount of surface-active agent in the oil phase was shown to modify the adhesive properties to such an extent that the cells separated (Sebba, 1972a,b).

Furthermore, it is well established that cancer can be induced by several means, e.g., by radiation, by certain chemicals, or by viruses. These processes lead to the damage of a given gene. Suppose, however, that this particular gene is not essential for cell growth or reproduction, but does affect cell-adhesion properties, such that the loosened cells move away to induce cancer tissue elsewhere (metastasis).

Surface-active agents, such as fatty acids, cetyl alcohol, and unsaturated fatty acids, could be expected to be effective in inducing cell separation. Oxidation in these molecules, however, would have to proceed in several stages before the surface activity would be lost. The cholesterol molecule, on the other hand, is known to lose its surface activity in only one oxidation step.

It is not known clearly why cells manufacture cholesterol or why this lipid molecule is present so universally in the biological world. Although cholesterol is necessary to the manufacture of steroid hormones, it is known to be unable to form vesicles or bilipid membranes (though it loses this ability only after oxidation). Furthermore, the nonideal mixing behavior of cholesterol–lecithin monolayers (see Fig. 4.23) also needs further investigation to determine the biological role of the cholesterol molecule.

On the basis of these findings, it was postulated that if cancer cells do not degrade cholesterol and given that normal cells are known to do so, effective cancer treatment may be achieved by bathing cancerous tissues with excess cholesterol. This treatment would be expected to lead to the rupture of cancer cells, since they would be unable to destroy the cholesterol molecule. Normal cells would be expected to survive, since they can destroy cholesterol.

8

MONOLAYERS OF MEMBRANE PROTEINS

In biological membranes, among many different transport phenomena, ion transport is one of the major processes that are vital for all kinds of cell function. The extensive studies over the past few decades of the mechanisms of ion transport across artificial membranes have revealed a great deal both about ionophores themselves and about the properties lipid membranes must have for the ionophores to function efficiently. In the case of a carrier molecule, for example, it is of little value for it to have a high ion selectivity if it does not combine with the membrane or, if it does combine, if it does not move or bind ions in the membrane. In the case of pore-forming molecules, binding to different membranes may be very similar, but the efficiency of the transport process may depend critically on the state of ionization of the lipid or on the membrane thickness.

Ion transport has to take place from an aqueous medium (high dielectric constant, $\varepsilon \approx 78$) through a "lipid medium" of bilayer with a low dielectric constant ($\varepsilon \approx 2$). This process can be considered in terms of the energy of interaction between ions and membrane. It is known that a membrane exhibits an energetic barrier to the ions in the adjacent aqueous solution. It is also known that the modulation of this barrier is the primary means by which a membrane controls the ionic flow. However, one generally tends to think of this interaction only as one imposed by the membrane on the ion, rather than as a mutual interaction that might act to perturb the structure of the membrane.

Biological membranes bear a net negative charge, which gives rise to an electrostatic potential at the membrane-solution interface. The Gouy-Chapman theory of the diffuse double layer predicts that at the surface of a negatively charged membrane, there is produced an electrical field that attracts cations and repels anions. Consequently, the concentration of cations at the membrane surface is higher than in the bulk aqueous phase, whereas for anions the reverse is true. The concentration difference will depend on the surface potential.

The monolayer system (at the air–water or oil–water interface) has been found to be a useful model system for studying the state of charges in globular proteins, as described in Chapter 5. Since the charges of globular spread proteins can be satisfactorily analyzed, it is obvious that similar analyses of membrane proteins should be carried out. Among the proteins that have been recognized to be the best candidates for such analyses are melittin, valinomycin, and gramicidin. Such studies could focus on the process whereby ions interact with the membrane through a coulombic field that emanates from the ionic charge. This field acts to polarize the surrounding medium, which gives rise to the attraction of oppositely charged ions and the repulsion of like charges, with the result that the ionic energy is reduced.

Furthermore, ions passing through a bilayer structure must surmount an energy barrier. This energy barrier has been separated into *electrostatic* and *nonelectrostatic* components (Ketterer *et al.*, 1971). The adsorption of hydrophobic ions or dipolar molecules or both on the membrane surface influences mainly the electrostatic barrier encountered by the ions. The electrostatic barrier can be studied by using as probes such hydrophobic ions as tetraphenylborate (Andersen and Fuchs, 1975; Melnik *et al.*, 1977; Szabo, 1974; Bruner, 1975; Wulf *et al.*, 1977).

Many integral or intrinsic membrane proteins are anchored in the membrane (see Fig. 7.1) by hydrophobic interactions between the interior of the lipid bilayer and sequences of hydrophobic amino acids (Capaldi, 1982; Low *et al.*, 1986). In some cases, these hydrophobic polypeptide sequences are located at the $-COOH$ or NH_2 terminus of the protein; in others, the polypeptide chain is believed to traverse the membrane one or several times, with substantial domains on both sides of the bilayer. In the effort to understand the mechanisms of attachment of membrane proteins to bilayers, various polypeptides of membranes have been extensively studied in the current literature.

8.1. MELITTIN MONOLAYERS

Melittin is a peptide of much interest in the current literature, its strong interaction with different cell membranes, vesicles (Haberman, 1972), and bilayers (Schoch and Sargent, 1980; Tosteson and Tosteson, 1981) having been reported. Because melittin can be incorporated into the lipid bilayer structure, its monolayer at the air–water interface needs detailed investigation analogous to that made of globular spread protein monolayers (Birdi, 1972, 1973a, 1977; Birdi *et al.*, 1972; Birdi and Nikolov, 1979; Chattoraj and Birdi, 1984). This peptide is the main component (50% of dry weight) of bee venom (Haberman, 1972). Furthermore, melittin affords an advantage

(+)GLY-ILE-GLY-ALA-VAL-LEU-LYS(+)-VAL-LEU-THR-THR-GLY-LEU-PRO-

ALA-LEU-ILE-SER-TRP-ILE-LYS(+)-ARG(+)-LYS(+)-ARG(+)-GLN-GLN(NH₂)

FIGURE 8.1. Primary sequence of melittin.

in such investigations in that its amino acid sequence (Haberman and Jentsch, 1976) and X-ray diffraction three-dimensional structure (Eisenberg *et al.*, 1980) are now known. The primary structure sequence of melittin is shown in Fig. 8.1. This sequence clearly reveals the hydrophobic–hydrophilic characteristic of this peptide, due to the presence of many nonpolar amino acids (see Chapter 5). It can be seen that except for the positive charges at the N terminus and at Lys-7, the first 20 residues are mainly nonpolar. In contrast, the last 6 amino acid residues comprise 4 charged and 2 polar amino acids. Furthermore, all the hydrophilic residues are ionized and positively charged when melittin is dissolved in water at neutral pH (Schoch and Sargent, 1980; Tosteson and Tosteson, 1981).

The unusual sequence of the amino acid residues in the melittin molecule is therefore of interest both for its biological activity and for the physical properties of its monolayers. Melittin exhibits very strong surface activity and forms stable monolayers at the air–water interface (Sessa *et al.*, 1969; Degrado *et al.*, 1981; Birdi *et al.*, 1983; Gevod and Birdi, 1984; Birdi and Gevod, 1987). The amphiphile nature of its structure also gives rise to its strong binding properties to all kinds of cell membranes (Haberman, 1972). The latter property further leads to the disorganization of the hydrophobic matrix of membranes in bilayers.

Melittin is reported to exist in aqueous solution in two states that differ in both secondary and quaternary structure. In solutions of low ionic strength and at low peptide concentrations ($\approx 1.9 \mu M$), it behaves as a monomer (mol. wt. 2840). On the other hand, under conditions of high peptide concentrations or in solutions of high ionic strength, or both, a monomer-tetramer equilibrium exists (Schoch and Sargent, 1980). Exposure of polarized bilayers to melittin at concentrations of about 10^{-9} M gives rise to the formation of single voltage-dependent ion-selective channels, organized from tetramers (Tosteson and Tosteson, 1980). No lytic activity has been observed up to peptide concentrations of 10^{-8} M in the aqueous phase. However, when the concentration is higher than 10^{-7} M, a rapid rupture of bilayers has been reported independent of the polarized state of membranes (Schoch and Sargent, 1980; Tosteson and Tosteson, 1981).

Furthermore, the time-course of melittin-induced conductance in bilayers showed two distinct steady-state regions that differed both in kinetics of activation and in their voltage dependence (Tosteson and Tosteson, 1984). One region activated rapidly in response to changes in voltage with a gating charge of 1, while the other region showed a gating charge of 4. The magnitude of the steady-state conductance (at a fixed voltage) in both regions was reported to be proportional to the 4th power of the melittin concentration (as a monomer).

Systematic studies of melittin monolayers have only recently been conducted under different experimental conditions. These analyses are described below.

π VS. A AND π VS. C_s

π vs. A and π vs. C_s data are given in Fig. 8.2. The magnitudes of π_{col} and A_0 on pure water are 18 mN/m and 145 Å^2, respectively. The magnitudes

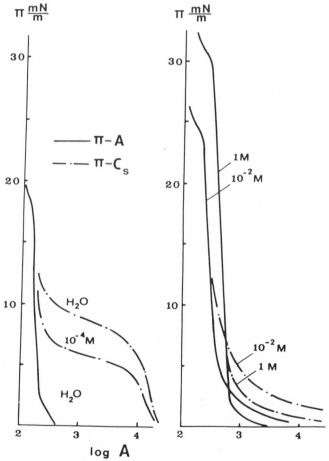

FIGURE 8.2. Surface pressure (π) vs. log(A) of melittin monolayers (spread by the π vs. A and π vs. C_s methods) on subphases with different KCl concentrations at 26.4°C. Units: π = mN/m (=dynes/cm); $A = 10^{-20}$ m^2 (=Å^2). From Birdi and Gevod (1984).

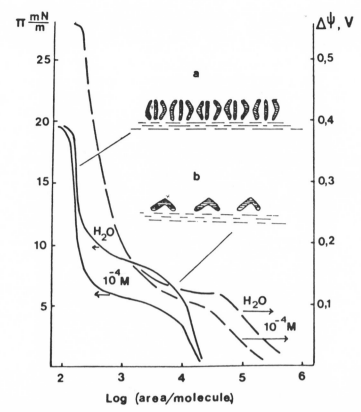

FIGURE 8.3. π and $\Delta\psi$ vs. C_s isotherms of melittin monolayers at the air–water interface. (a, b) Schematic representations of the orientation of melittin molecules in concentrated (a) and dilute (b) monolayers. From Birdi *et al.* (1983).

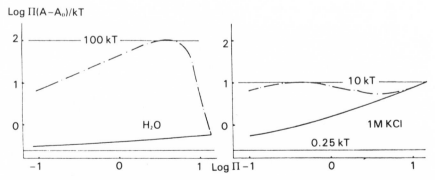

FIGURE 8.4. Plots of $\log[\pi(A - A_0)]/kT$ vs. π for melittin monolayers on subphases of water and 1.0 M KCl ($\pi A = kT = 411$, if $\pi = $ mN/m and $A = 10^{-20}$ m^2 = Å^2). From Birdi and Gevod (1984).

of A_0 on subphases of 10^{-2} and 1.0 M KCl are 208 and 275 Å^2, respectively. The values of π_{col} are 23 and 30 mN/m, respectively. Very dilute monolayers were studied by the π vs. C_s method (Fig. 8.3). It can be seen that when $C_s \approx 40,000$ Å^2/molecule, the data are very different from those measured by the π vs. A method. However, when the subphase contained a high concentration of electrolyte (KCl), the two methods gave the same kinds of isotherms. These data were analyzed using equation (4.29). The analysis shows that on a pure water subphase, the value of $[\pi(A - A_0)]_{\pi \to 0} = kT/4$ (Fig. 8.4). This result indicates that melittin is present as a tetramer when compressed and that it exhibits no charges, i.e., $\pi_{el} = 0$. The π vs. C_s data, on the other hand, showed very large deviations, i.e., $\pi A \gg kT$.

These results mean only that larger π values are measured due to strong electrostatic interactions, since at high C_s values, the van der Waals forces

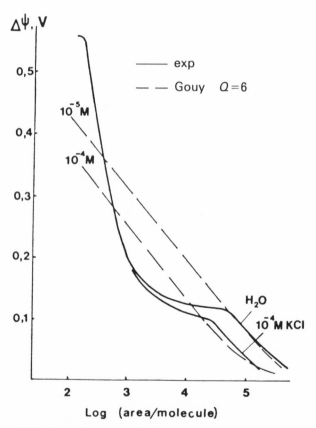

FIGURE 8.5. Experimental and calculated values of $\Delta\psi$ for melittin monolayers. Calculated values were derived by the Gouy–Chapman diffuse charge model with the number of charges per melittin molecule (Q) as 6. From Birdi *et al.* (1983).

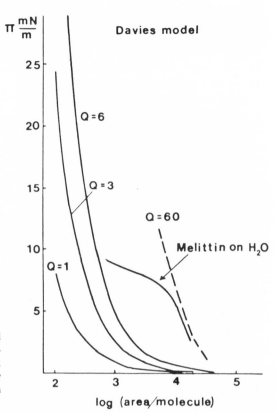

FIGURE 8.6. Experimental and calculated π for melittin monolayers for different values of Q. Calculated values were derived by the Davies equation (4.66). From Birdi *et al.* (1983).

would be negligible. Accordingly, the Davies relationship for (diffuse) charged monolayers [equation (4.66)] was applied to these data. It was found that the number of charges, Q, necessary to fit the data would be around 50–100 (Figs. 8.5 and 8.6). Since melittin can exhibit a maximum of only six charges (see the sequence in Fig. 8.1), it was concluded that melittin monolayers exhibit charges of a different kind (discrete charges) from those reported for charged globular protein monolayers (Birdi and Nikolov, 1979).

8.2. DISCRETE-CHARGE BEHAVIOR OF MEMBRANE PROTEIN MONOLAYERS

8.2.1. SURFACE POTENTIAL AND ION TRANSPORT

The relationship of the ion permeability of a lipid membrane to the change in the electrostatic potential across its surface has been discussed by many investigators (Finkelstein and Holzer, 1973; Haydon and Hladky, 1972; Szabo *et al.*, 1973; Chandler *et al.*, 1965; Gilbert and Ehrenstein, 1969; Neumcke, 1970; McLaughlin *et al.*, 1970; Hladky and Haydon, 1973). In earlier studies, only the diffuse and Stern layer potentials were considered, but it was later realized that the potential change arising from the oriented dipoles of the polar head groups may be of similar or even greater importance. This importance is clear for carrier transport from the fact that either the ion or the ion–carrier complex must pass through all the potential changes due to the presence of the polar groups (Fig. 8.7).

For the classic pore, in which, as far as the permeating ions are concerned, the lipid polar groups are considered to be shunted, the position is less obvious. The Debye–Hückel reciprocal length is likely to be sufficiently large that the diffuse layer potential will affect the ion concentration near the entrance to a pore. However, the range of influence of the lipid polar group dipoles is much less, and the permeating ions could be unaffected by their presence.

Recently (Ohki, 1971, 1973), a theory was proposed whereby the membrane potential would arise from the difference between two surface potentials, one on either side of the membrane, produced by the fixed charges, as well as from the polarization of the molecules at the membrane surfaces and the surrounding electrolyte solutions. Asymmetrical phospholipid bilayers (i.e., a phosphatidylcholine monolayer on one side and a phosphatidylserine on the other) were found to give rise to a potential difference when both sides were surrounded by the same NaCl solution at pH 7.0.

Furthermore, it is well established that divalent cations, in particular Ca^{2+}, are of much importance to the chemistry of excitable cells (Ringer, 1883; Triggle, 1972; Manery, 1966; Poste and Allison, 1973). The subsequent discovery that numerous bioelectrical, secretory, and contractile phenomena are Ca^{2+}-dependent substantiated the early opinion that divalent cations

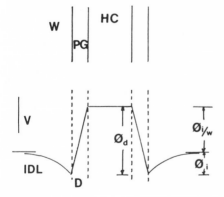

FIGURE 8.7. Schematic representation of electrostatic potential across the lipid-bilayer membrane. (W) Water phase; (*V*) potential change; (IDL) ionic double layer; (PG) polar groups; (D) dipoles; (HC) hydrocarbon; ($\phi_{i/w}$) total potential difference [$= \phi_i$ (ionic double layer) $+ \phi_d$ (dipole)]. From Haydon (1975).

are essential to the physicochemical integrity of biological systems. For instance, it is now well established in the case of the neuromuscular system that an elevated Ca^{2+} concentration in the external medium affects the cell resting potential, enhances the membrane resistance, and raises the threshold for excitation (Weidmann, 1955). A reduced external concentration of Ca^{2+} was found, in contrast, to decrease the membrane potential and induce spontaneous firing (Hodgkin et al., 1952; Bülbring and Tomita, 1970). An increase in the amount of Ca^{2+} in the bathing solution of nerves was shown, for instance, to produce a shift in the positive direction along the voltage axis of the Ni^+ and K^+ conduction voltage curves, without apparently affecting the maximum amplitude of conductivity (Hille, 1968). Subsequent studies on the effect of Ni^{2+}, Cd^{2+}, Ba^{2+}, and Co^{2+} on the excitation process in the lobster axon revealed that most divalent cations are more or less equivalent, as far as the Na^+ conduction shift is concerned.

At present, there is no model that can describe at a molecular level the effects of divalent cations on the physicochemical properties of excitable membranes (Adam, 1970; Jakobsson, 1971; Hamel and Zimmerman, 1970). It is therefore apparent that our understanding of the physicochemical basis of these Ca^{2+}-mediated events depends largely on our understanding of the interaction between divalent cations, such as Ca^{2+}, and certain membrane-forming molecules. There is a great deal of evidence that shows that phospholipids constitute a very important binding locus for bivalent cations. The orientation in the membrane matrix, and the electrochemical characteristics, undoubtedly favor an interaction of this kind (Tien, 1974; Dawson and Hauser, 1970). Studies of the interaction of Ca^{2+} with lipid monolayers are discussed in Chapter 4.

8.2.2. EQUATION OF STATE FOR CHARGED MELITTIN AND VALINOMYCIN MONOLAYERS (DISCRETE-CHARGE MODEL)

As already described in Chapter 4, the term that arises from the electrostatic interaction, π_{el}, in monomolecular films has been analyzed by various models in the current literature. However, only in some recent studies (Chattoraj and Birdi, 1984) has a modified equation of state for charged monolayers been reported. According to this analysis, the fluctuation pressure arises from the following factors:

1. The discreteness of surface charge.
2. The forces due to the formation of an electrical field at the interface.
3. The screening of the double layer in the presence of neutral salts.

These modifications are not sufficient to describe the state in which the charges in the monolayer could be further removed from the bulk phase, i.e., water with a high dielectric constant. This state is implicit in that the depth of the screening layer is much less than the Debye length. Thus, in such systems, the coulombic repulsions between the charged film-forming molecules become appreciable in the region with lower dielectric constant (e.g., air or the hydrophobic part of peptide away from the bulk aqueous phase), even if the distances D ($=2\sqrt{A_0}/\sqrt{3}$ for hexagonal molecular packing) between charges are large [i.e., A (area/molecule) $= 10^5\text{-}10^2 \text{ Å}^2$]. The discrete-charge model as described in recent reports for charged membrane proteins (Birdi *et al.*, 1983; Gevod and Birdi, 1984; Birdi and Gevod, 1984, 1987) differs from earlier theories in that the latter were described for simple amphiphiles, while the former was developed for membrane protein monolayers. The adsorbed peptide molecule could differ in more than one way from simple amphiphiles, possible ways being that

1. Conformation could be α-helix or β-form.
2. The peptide could exhibit channel-forming properties.

In other words, the state of ions near a film with peptide at an interface could be expected to be different under these conditions from when the simple amphiphiles are situated at an interface.

In bilayers, certain peptides can induce gating in an electrically inert bilayer (Stein, 1967; Tien, 1974; Christensen, 1970; Kotyk and Janacek, 1975). Furthermore, it has been shown that channels exhibit discrete units of conductance with amplitudes too large to be explicable in terms of single carrier ions. It is safe to assume that a monolayer of peptide is analogous to a bilayer structure and that the same kind of model can be used to describe charged monolayers.

As regards the magnitude of π_{el}, it is known that apart from having an inverse dependence on distances of various orders, all kinds of electrostatic interactions depend on the increases in the dielectric constant of the medium (Moelwyn-Hughes, 1974). Furthermore, ignoring solvation effects for simplicity, the potential energy of a "naked" ion in a hydrocarbon medium ($\varepsilon \approx 2$) would be some 40 times greater than in bulk water ($\varepsilon \approx 80$). In other words, the exceptionally large values of π_{el} as measured for melittin and valinomycin charged films can be explained only by theories based on discrete charged repulsions.

In the case of melittin monolayers, it was postulated that the peptide could be oriented in such a way that parts of the molecule are inside the aqueous medium, while other parts are pointed away from the aqueous phase (Fig. 8.8). Amino acid residue Lys-7, in particular, could be situated in such a way that it could not be fully surrounded by the aqueous medium, possibly because Pro-14 would give rise to a "bend" in the peptide, as suggested by many investigators.

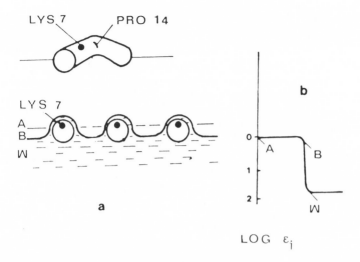

FIGURE 8.8. (a) Schematic representation of melittin molecules in a dilute monolayer. The amino acids Lys-7 (+) and Pro-14 (at the bend) are indicated. (b) Schematic representation of variation of the dielectric constant (ε) in the air (A)-boundary (B)-water (W) system. From Birdi *et al.* (1983).

To estimate the magnitude of π_{el}, a square lattice arrangement of peptide molecules in a monolayer was used (Fig. 8.9). Analogous to the Langmuir bidimensional pressure model for such monolayers, the repulsion force vectors between molecule 0 and all the surrounding molecules were estimated as shown in Fig. 8.9. The magnitude of π_{el} was found to be

$$\pi_{el} = (10^8/D) \times 10^5 \times \sum_{n=1}^{N} \sum_{-N}^{N} (1/D^2)[n/(i^2 + n^2)^{3/2}F] \qquad (8.1)$$

where F is the electrical repulsion force, for a general case:

$$F = \sum_{j=1}^{2} \alpha_j \{(Qe_0^2(1 + k_jD) \exp(k_j\alpha)/[8\pi D^2 \varepsilon_0 \varepsilon_j(1 + k_j\alpha)]\}$$
$$\times [\exp(-k_jD)] \qquad (8.2)$$

where Q is the number of charges; D is the distance between charges; $1/k_j$ is the Debye-Hückel length; the dielectric constants, ε, for water and air are denoted by the summation limits 1 and 2; the factor 10^5 converts the

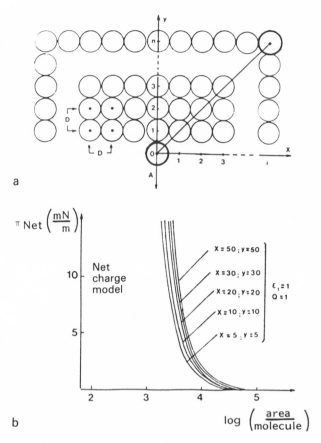

a

b

FIGURE 8.9. (a) Schematic representation of the square lattice model used for deriving the discrete charge equation about the reference molecule (0) at the interface. (D) Distance between centers of melittin molecules; (x_i, y_n) number of neighboring molecules taken into account in each direction in the two-dimensional monolayer lattice. The resultant force vector $(0 \rightarrow A)$ is the sum of vectors due to interaction with all the molecules situated from x_1 to x_i and y_1 to y_n. (b) Values of π_{net} calculated with the help of the discrete-charge equation, as a function of A, and (x_i, y_n) (for $\varepsilon = 1$, $Q = 1$) in vacuum. From Birdi *et al.* (1983).

result to mN/m units; n and i are numbers of molecules in each direction of the model lattice; and N is the limiting size of the lattice.

Under greatly simplified conditions, this equation becomes

$$\pi_{el} = [(1.15 \times 10^5 \times Q)^2/(\varepsilon_2 D)^3][2\ln(\sqrt{2}N)] \tag{8.3}$$

for $N > 10$ and, assuming that interactions take place only through the medium of low dielectric constant, ε_2. This equation was used to analyze

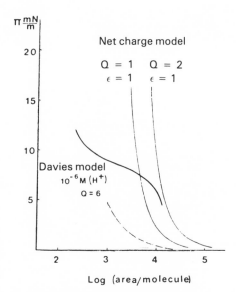

FIGURE 8.10. Comparison of experimental values of π (——) and values calculated by the Davies diffuse-charge model and by the net (discrete)-charge model, assuming interactions between N (=50) molecules and interactions in the aqueous phase. From Birdi *et al.* (1983).

the measured π vs. A isotherms of melittin and valinomycin monolayers, as described below.

MELITTIN (CHARGED) MONOLAYERS

The values of π_{el} calculated by equation (8.3) are given in Fig. 8.10. It was concluded that the discrete-charge equation of state gave a satisfactory fit to the measured π vs. A data, when the magnitude of A was large. This was the region in which the diffuse-charge model and the Davies model [equation (4.66)] gave Q values (>50) much larger than the six charges.

COMPARATIVE MELITTIN AND MELITTIN_{8-26} MONOLAYERS

Since the biological activities of melittin and the 8-26 fragment were reported to be different, monolayers were investigated to determine the mechanism of these differences (Gevod and Birdi, 1984). The differing biological properties were reported especially in the presence of lipids (i.e., in vesicles and membranes). Monolayer studies can be carried out both with and without the presence of lipids, which is advantageous in determining the physical properties at interfaces of biologically important molecules. π vs. A and $\Delta\psi$ vs. A isotherms for both melittin and the 8-26 fragment are given in Fig. 8.11. The data for both melittin and the 8-26 fragment gave plots in which $[\pi(A - A_0)]_{\pi \to 0} = kT/4$. From this result, it was concluded

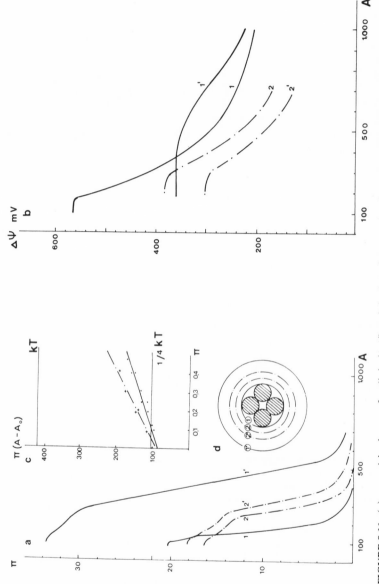

FIGURE 8.11. (a) π vs. A isotherms of melittin $(1, 1')$ and the 8–26 fragment $(2, 2')$ on subphases of water $(1, 2)$ and 1 M KCl $(1', 2')$ at 25°C. (b) $\Delta\psi$ vs. A isotherms. See (a) for details. (c) $\pi(A_0 - A)$ vs. π plots of melittin (———) and fragment 8–26 (—•—). (d) Relative area per tetramer for melittin $(1, 1')$ and fragment 8–26 $(2, 2')$ at a surface pressure of 11 mN/m. From Gevod and Birdi (1984).

that these peptides are present as neutral tetramers on the surface of pure water.

The limiting values of $\Delta\psi$ for melittin and the 8–26 fragment are 560 and 380 mV, respectively. Since in monolayers these peptides differ from other globular proteins in their orientation (i.e., dipoles along the vertically packed α-helix), the large magnitudes of $\Delta\psi$ were ascribed to these differences. Furthermore, the peptides would be expected to be oriented parallel with the polar end (i.e., 21–26 residues) inside the aqueous phase, and the predominantly nonpolar part to be pointing away from the aqueous phase, such that the system arrives at a minimum surface free energy. Hence, the ratio of the limiting $\Delta\psi$ on a pure water subphase is proportional to the number of amino acids in the peptide: $560/26 = 22$ for melittin and $380/19 = 20$ for the 8–26 fragment. These results allow one to conclude that the potential gradient exists along the channels in the case of both native melittin tetramers formed by the 8–26 fragment. They also support the postulate

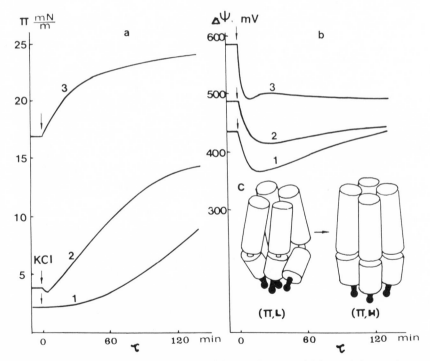

FIGURE 8.12. Plots of change in π vs. time, t (a), and of $\Delta\psi$ vs. t (b) for melittin monolayers at different initial π values. The subphase concentration was changed by injection of an appropriate amount of KCl (in solid form), as indicated by the arrows. (c) Schematic drawing showing the expected changes in conformation of the tetramers at low (L) and high (H) π. From Gevod and Birdi (1984).

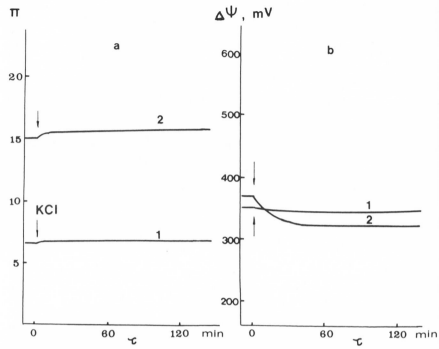

FIGURE 8.13. Plots of change in π with t (a) and of $\Delta\psi$ with t (b) for fragment 8-26 monolayers at 25°C under the experimental conditions described in the Fig. 8.12 caption. From Gevod and Birdi (1984).

that melittin retains the α-helical structure at the interface, since the dipoles of these structures would be approximately proportional to the number of amino acid residues.

The absence of Lys-7 in the 8-26 fragment is found to give much lower Cl^- interaction than observed with melittin. This may be ascribed to the fact that Cl^- fixation is known to take place with the Lys-7 in bilayers. Furthermore, plots of π vs. time (t) for penetration of different ions showed differences that were found to be in accord with the data on bilayers with melittin, i.e., $CH_3COO^- < Cl^-$ (see Figs. 8.12-8.14).

8.2.3. VALINOMYCIN MONOLAYERS AT THE AIR–WATER INTERFACE

Valinomycin is one of a number of cyclic depsipeptides (Fig. 8.15) that have been shown to function by increasing membrane cation permeability by as much as a factor of 10^8. These kinds of ionophores have been found to increase the electrical conductivity of the lipid barrier in membranes of

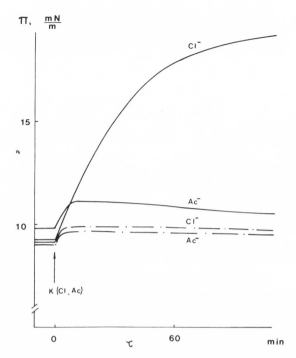

FIGURE 8.14. Plots of change in π of monolayers of melittin (——) and fragment 8-26 (—•—) on the addition (at arrows) of electrolytes (0.125 M). (Cl) KCl; (Ac) CH_3COOH. From Gevod and Birdi (1984).

bilayers because of their ion-complexing ability. Whether the conductivity increase is ion-selective or not has been found to be dependent on the complexing properties of the ionophore (Haydon and Hladky, 1972; Ovchinnikov, 1974). It has been reported that valinomycin acts in a variety

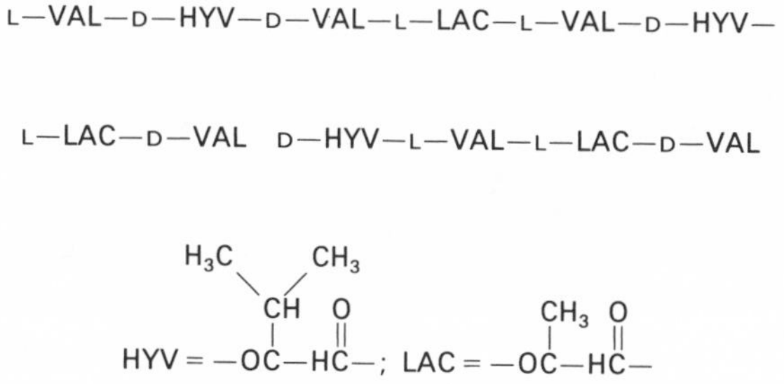

L—VAL—D—HYV—D—VAL—L—LAC—L—VAL—D—HYV—

L—LAC—D—VAL   D—HYV—L—VAL—L—LAC—D—VAL

$$HYV = \underset{\overset{|}{CH}}{\overset{H_3C \diagdown \diagup CH_3}{}} \underset{OC}{\overset{O}{\parallel}} HC—; \quad LAC = \underset{OC}{\overset{CH_3}{\underset{|}{}}} \underset{}{\overset{O}{\parallel}} HC—$$

FIGURE 8.15. Valinomycin structure.

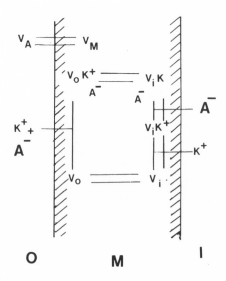

FIGURE 8.16. Mode of transport of K^+ ion by valinomycin (V) in membranes. (O) Outside; (M) membrane; (I) inside. From Ovchinnikov (1974).

of cell membranes as well as on liposomes and lipid bilayers (Haydon and Hladky, 1972; Ovchinnikov, 1974; Grell *et al.*, 1972; Eisenman *et al.*, 1973; Lauger, 1972; Graven *et al.*, 1966).

The current proposed model of valinomycin's mode of function is that the antibiotic forms a complex with the cation on one side of the membrane, transports the cation across the membrane, and releases it on the other side (Ovchinnikov, 1974) (see Fig. 8.16). Furthermore, it has been suggested that the valinomycin molecule shields the cation, which is normally highly insoluble in the "hydrocarbon" part of the membrane (bilayer), while the hydrophobic side chains solubilize the complex within the membrane (Pinkerton *et al.*, 1969; Ivanov *et al.*, 1969). Although the function of valinomycin in bilayers and membranes is much studied in the current literature, the thermodynamic analysis of valinomycin monolayers has not been as extensively investigated (Kemp and Wenner, 1972, 1973; Colacicco *et al.*, 1968; Capers *et al.*, 1981; Birdi, 1977; Birdi and Gevod, 1984).

As known from monolayer and bilayer studies, valinomycin binds K^+ ions very strongly in comparison, for example, to Na^+. This particular quality is the one that gives it antibiotic behavior. Surface pressure, π vs. area, A, and surface potential, $\Delta\psi$, vs. A isotherms of very dilute and condensed valinomycin monolayers have been reported as a function of $K^+(KCl)$ concentration in the subphase, as well as the rate of compression of monolayers (Birdi and Gevod, 1984). These results can be described as follows:

1. At low surface concentrations of peptide (Fig. 8.17a), the addition of K^+ to the subphase leads to a small increase in π, the effect becoming

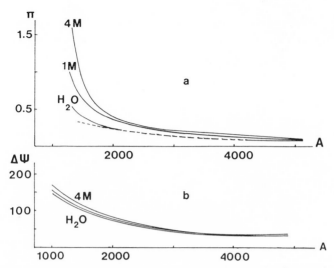

FIGURE 8.17. (a) π vs. A isotherms of valinomycin monolayers at the air–water interface as a function of KCl concentration in the subphase. (- - -) Curve of the equation $\pi(A - A_0) = kT$, where $A_0 = 170$ Å^2/molecule. (b) $\Delta\psi$ vs. A isotherms under the experimental conditions described for (a). From Birdi and Gevod (1984).

larger at higher surface concentrations. On the other hand, the measured values of $\Delta\psi$ are not so sensitive to K$^+$ concentration in the subphase (Fig. 8.17b). This observation suggests that the charging of the neutral peptide molecule gives rise to an increase in the repulsion forces (π_{el}, long-range) at the interface, but not to an appreciable effect on $\Delta\psi$. The magnitude of $\Delta\psi$ of monolayers has been found to be related to the dipole-charge distribution in the peptide sequence (Birdi and Gevod, 1984).

The shape of isotherms of the dilute peptide monolayer remains unaffected at different rates of compression (Fig. 8.18).

2. At high surface concentrations of peptide, the isotherms show much different behavior, depending on the surface composition and the rate of compression. In the case of pure water as subphase, the isotherms exhibit the shape depicted in Fig. 8.18a, in which only one collapse state is observed. As described in Chapter 4, the collapse of films takes place when the film molecules are highly compressed and a new phase is beginning. This compression can cause the molecules either to stack atop each other or to intermingle. This collapse was observed at $\pi_{col} \approx 28$ mN/m and $A_{col} = 180$ Å^2. These same values have been reported by various investigators (Kemp and Wenner, 1972, 1973; Birdi and Gevod, 1984). However, when valinomycin monolayers were investigated on concentrated KCl subphase solutions, the shapes of isotherms were found to be different, depending on the rate of monolayer compression (Fig. 8.18b). Such differences were

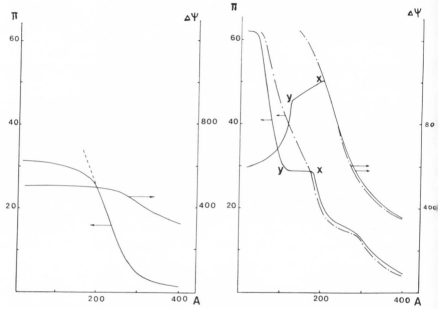

FIGURE 8.18. π and $\Delta\psi$ vs. A isotherms of valinomycin on water (a) and 4 M KCl aqueous solutions (b), as a function of rate of compression: (——) 40-min compression, 1 cm/min; [- - - (a), — · — (b)] 4-min compression, 10 cm/min. From Birdi and Gevod (1984).

especially observed when the close-packed state of the monolayer reached $A < 200$ Å^2. At slower rates of compression, isotherms exhibited three distinct collapse states:

First collapse: $A_{col,1} = 280$ Å^2

Second collapse: $A_{col,2} = 170$ Å^2

Third collapse: $A_{col,3} = 50$ Å^2

The transition from the second collapse state, $\pi_{col,2}$, to the third collapse state, $\pi_{col,3}$, is first-order, i.e., $d\pi/dA = 0$ (Birdi, 1981; Chattoraj and Birdi, 1984). The magnitude of $A_{col,2}$ was found to be about twice the value of the area where π starts to increase again before the $\pi_{col,3}$. A most interesting result as found from these investigations was the effect of compression rates on the $\Delta\psi$ vs. A isotherms. It was found that the value of ΔY increases until the second collapse state. At the transition plateau, $X - Y$, a linear decrease in ΔY was reported, followed by an abrupt decrease. This is the same region in which the π vs. A isotherms exhibited a transition from the second to the third collapse state (Figure 8.18b). This rather complicated shape of ΔY in comparison to the π vs. A isotherms provides evidence that three well-defined structural rearrangements in valinomycin monolayers could probably take place on concentrated KCl subphase solutions.

The data on π vs. A and $\Delta\psi$ vs. A were analyzed with the help of different equations of state. From the π vs. A isotherms of the monolayer (Fig. 8.18a), it can be seen that the data fit the following equation:

$$\pi(A - A_0) = kT \tag{4.29}$$

However, large deviations are observed when the subphase is KCl solution owing to the increased charging of the valinomycin monolayer. These data were further analyzed by using the equation of state for charged biopolymers [see equation (4.66)]. The calculated values for the number of charges, $Q = A/A_{el}$, are given in Fig. 8.19. It can be seen that $Q = 1$ on the 0.5 M KCl subphase, while the value of Q was reported to be much greater than 1

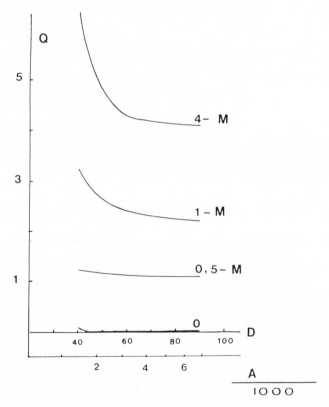

FIGURE 8.19. Calculated values for the number of charges, Q ($= A/A_e$), in valinomycin monolayers as a function of KCl concentration in the subphase [Davies equation (4.66)]. From Birdi and Gevod (1984).

when the KCl concentration in the subphase was greater than 0.5 M. Since only unit charge can exist when valinomycin interacts with K^+ ions, the deviation was accordingly analyzed in further detail.

Analogous to the corresponding data on melittin monolayers, the discrete-charge behavior was accounted for by the fact that the data could be satisfactorily analyzed by using the diffuse-charge model of Davies [equation (4.66)]. The valinomycin monolayer with one charge would give a value of $\pi_{el,gc} < 0.1$ mM/m when $A > 100$ Å^2 and $[KCl] = 1.0$ M. These findings suggest that only $\pi_{el,dis}$ and π_{kin} may act on the monolayer surface pressure. On the basis of this estimation, the π vs. A isotherms presented in Fig. 8.18B were analyzed with the help of the discrete-charge equation (8.3). The results are shown in Fig. 8.20, in comparison to the experimental data. It can be seen that when π_{kin} is taken into consideration, the calculated values of the π terms are found to be lower than the measured values. Only the calculated and experimental values of the $\pi_{col,2}$ state of the film were found to be of the same magnitude. On the other hand, when the term $\pi_{el,dis}$ is added to π_{kin}, the calculated π values are of the same magnitude as those measured.

It was concluded from these studies that the K^+ cation is able to penetrate the peptide molecule in the monolayer, thereby giving rise to

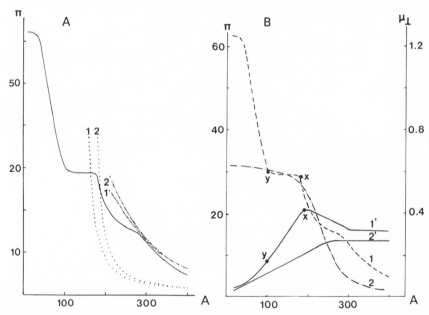

FIGURE 8.20. (A) Calculated values of $\pi_{kin} = \pi$ from equation (8.3) with $A_0 = 150$ Å^2 (1) or 170 Å^3 (2). (B) Calculated values of dipole moment, μ [from equation (3.18)]. From Birdi and Gevod (1984).

discrete-charge behavior. This conclusion is reasonable, considering that valinomycin is a neutral molecule. Furthermore, the stability of the discrete-charge peptide–cation complex was reported to be dependent on the magnitude of π. The compression monolayers in the region between $\pi_{col,1}$ and $\pi_{col,2}$ were shown to cause the disappearance of discrete-charge behavior. From this observation, it was concluded that the precipitation of the peptide hydrophilic core by K^+ ions occurs at $\pi > 15$ mN/m. The fact that the value of $\pi_{col,2}$ is almost twice the value of the monolayer π at which $\pi_{col,3}$ begins suggests that formation of a *dimeric* structure, i.e., a monolayer of double thickness, could be taking place. This collapse state could thus be arising from the compression of the duplex complex. When monolayer valinomycin is present on pure, i.e., neutral, water, the isotherms have been found to show a simple π vs. A isotherm. This observation suggested that formation of dimers (stacks) could not take place owing to the interfacial forces stabilizing these structures.

In accord with data presented earlier (Haydon and Hladky, 1972), the increase in polarity of the medium surrounding the valinomycin molecule leads to the breakup of the intermolecular hydrogen bonds. When the peptide is placed at the air–water interface, the hydrophilic side of the molecule interacts with the water (surface) to form hydrogen bonds, while the other side (hydrophobic) is oriented toward the air. This orientation of the valinomycin molecule suggests that almost all intermolecular bonds are broken. This conformation is very stable in the absence of complexing cations and prevents oligomerization of the peptide at the monolayer on compression. On the other hand, when K^+ is added to the subphase, the formation of complex takes place even at very low values of π. In the K^+ complexes, the ester carbonyls form six-ion bonds with the centrally situated K^+ ion. As a result, the conformation of the molecule becomes more compact. The K^+ ion complex is effectively screened from interaction with the aqueous phase by the despeptide skeleton and by pendant side chains. This screening is suggested to lead to the discrete (net)-charge behavior in the monolayer under conditions of low π and moderate packing, i.e., $\pi_{el} < \pi_{el,dis}$.

The basis for the cooperative uptake of cations by valinomycin is not at present completely understood. However, the following mechanisms could be present in such ionophore processes:

1. The cooperativity may be a consequence of the classic allosteric model, in which a conformation change induced by substrate binding alters the affinity for a substrate at a second binding site.

2. The cooperativity may result at least in part from an induced change in surface dielectric constant, ε, resulting from the valinomycin conformational change.

Thus, increased hydrophobicity of a molecule on undergoing conformational change might be propagated to and experienced by neighboring molecules as a change in interfacial ε, thereby increasing the ease with which neighboring molecules may form cation complexes.

Examples of cooperativity in membrane systems already exist. The peptide alamethicin, a channel-forming transport antibiotic, may form two-dimensional aggregates on the surface of a membrane in which adjacent channels interact to yield cooperativity between channels (Gordon and Haydon, 1972). Lipid (palmitic acid)-valinomycin and lipid-antamanide monolayers were investigated (Caspers *et al.*, 1981). Surface potential measurements were analyzed with the help of Gouy-Chapman theory, and the ionophore binding capacity was described from these data.

8.3. GRAMICIDIN MONOLAYERS

Gramicidin, isolated from *Bacillus brevis*, consists of a mixture of four related linear pentadecapeptides composed largely of hydrophobic and aromatic residues having formyl and ethanolamine end groups: HCO-L-Val-Gly - L-Ala - D-Leu - L-Ala - D-Val - L-Val - D-Val - L-Trp - D-Leu - L-Trp - D-Leu-L-Trp-D-Leu-L-Trp-NHC_2H_4-OH (Ramachandran, 1963; Sarges and Witkop, 1965). In lipid bilayers, it is thought to form a head-to-head dimer of helices, with an intramolecular channel permeable to monovalent alkali metal ions (Haydon and Hladky, 1972; Hladky and Haydon, 1972; Heitz and Gavach, 1983). This dimerization of gramicidin has been suggested to be responsible for the ion-conducting transmembrane channels in bilayers (Urry, 1971; Goodall, 1970). The solubility of gramacidin in water is very low (Kemp *et al.*, 1972).

The hydrophobic-hydrophilic character of gramicidin thus suggests that it can form stable monolayers at the air-water interface. π and ΔV vs. A isotherms have been investigated (Kemp and Wenner, 1972). An initial rise in π is observed at $A = 350 \text{ Å}^2$ and $\pi_{col} \approx 15 \text{ mN/m}$. The compression is reported to be reversible. The close-packed gramicidin film near its collapse state (which was not easily detected) was 145 Å^2. This value is consistent with the cross-sectional area, with the helices being oriented perpendicular to the surface. This orientation would allow dipole interactions between the exposed terminal acyl oxygens and the aqueous phase.

In contrast to valinomycin, gramicidin in the presence of high KCl in the subphase does not show the elevated ΔV or the new collapse state.

The penetration of gramicidin into lipid monolayers was investigated and found to proceed at a constant rate for at least an hour, even at $5 \times 10^{-7} \text{ M}$ initial gramicidin concentration in the bulk phase. It was suggested that dimers or larger aggregates could penetrate the lipid films. From spectroscopic data [e.g., infrared (IR), circular dichroism, nuclear magnetic resonance], it has been shown that Ca^{2+} interacts with gramicidin and that a head-to-head gramicidin dimer can have two Ca^{2+} binding sites located near the $-COOH$ termini. The transport of Cs^+ and K^+ is blocked by this Ca^{2+} binding.

8.4. TRINACTIN MONOLAYERS

π and ΔV vs. A isotherms of trinactin and nonactin have been investigated (Kemp and Wenner, 1973). Monolayers of trinactin spread on 3 M solutions of monovalent cations showed selectivity of binding in the sequence $NH_4^+ > K^+ > Rb^+ > Cs^+$, Na^+, and Li^+, which was found to be in agreement with partitioning studies (Eiseman et al., 1969). The π vs. A isotherms did not show appreciable differences. However, it is worth mentioning that the sensitivity in these studies was not high, and differences that might have been apparent at higher sensitivity may have been precluded.

The minimum packing area of the trinactin-K^+ complex, approximately 130 $Å^2$, was in satisfactory agreement with the value of 125 $Å^2$ obtained from the space-filling model based on the structure of the nonactin-K^+ complex obtained by X-ray crystallography. This prediction was based on the assumption that trinactin is oriented at the interface in a manner that allows loading and releasing of the cation. Such an orientation would allow anchoring of the molecule to the interface by hydrophilic interaction with the aqueous subphase. The monolayer method in such systems thus yields much useful information on the molecular dimensions, and thus the conformations of such relatively small hydrophobic peptides and macrotetrolide antibiotics can be elucidated.

Trinactin does not undergo a large conformational change on cation binding, as is indicated by the similarity in the π vs. A isotherms of the complexed and the uncomplexed molecules.

8.5. IONOPHORES AND MONOLAYERS

The behavior of melittin in bilayers is similar to that of another membrane protein, alamethicin. The bee venom melittin resembles alamethicin in amphiphatic nature, lytic activity, and secondary structure, but is completely different in amino acid composition.

Alamethicin and melittin both exhibit the same structural prerequisites: a lipophilic α-helix of adequate length and a hydrophilic C-terminal part. Both form multilevel pores. The peculiar amino acid α-aminoisobutyric acid in alamethicin is not essential for pore formation. Pore state distributions and pore state lifetimes of alamethicin and melittin are strongly dependent on ionic strength, which is consistent with the action of electrostatic forces (Boheim *et al.*, 1984).

8.5.1. LATERAL PROTON CONDUCTION AT THE. LIPID–WATER INTERFACE

Protons are now recognized as being the major driving force in the coupling between electron transport and ATP synthesis catalyzed by the membranes of mitochondria, chloroplasts, and bacteria. This coupling is described as being delocalized (Boyer *et al.*, 1977), semilocalized (Kell, 1979), or localized (Williams, 1978). In the delocalized theory, the membrane plays no role in the transmission of energy, which is assumed to occur via the two bulk aqueous phases, one on either side of the membrane. Although many experimental results support the delocalized theory, recent data from different laboratories appear to be in agreement with theoretical models in which proton movements are assumed to be localized within the membrane or at its surface (Bell, 1979; Haines, 1983; Lange *et al.*, 1975). In the latter case, the proton fluxes between the energy source and the ATP synthetase system occur laterally along the membrane in parallel with the aforementioned delocalized fluxes.

This kind of diffusion along a surface thus affords a good model system to be investigated by the monolayer method. Comparative studies of diffusion along a surface and in a volume have been reported (Hardt, 1979; Adam and Delbruck, 1968). One could conjecture that even if the diffusion coefficient is slightly smaller on the surface (two-dimensional space) than in a volume (three-dimensional space), the reduction of dimension will imply that the apparent diffusion is larger along the interface. Furthermore, it has been speculated that membranes and even tubular assemblies in the cytoplasm might be preferential pathways for the transfer of ions and metabolites in the cell (Skulachev, 1980; Berry, 1981; Sumper and Trauble, 1973).

Fast lateral proton conduction along the lipid–water interface has been experimentally demonstrated in a recent study (Tessie *et al.*, 1985). These experiments were carried out as follows (Prats *et al.*, 1985) (see Fig. 8.21): The trough was filled with the aqueous subphase. The fluorescence of the pH-sensitive chromophore covalently bound to the phospholipid was monitored (Tessie, 1979, 1981). A lipid film was spread and the fluorescence was monitored. The subphase pH was changed by injecting acid. It was concluded that very fast proton transfer can occur at the lipid–water interface. These studies provided direct evidence that such an interface can be considered an efficient proton conductor. The data were analyzed by assuming that the lipid polar head region is one of minimal energy for protons.

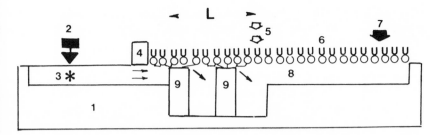

FIGURE 8.21. Schematic drawing of the monolayer trough. (1) Plexiglas trough; (2) injection of acid; (3) stirrer; (4) Teflon barrier; (5) fluorescence observation; (6) monolayer [PE with chromophore/phosphatidylethanolamine (2% mole/mole)]; (7) π measurement; (8) aqueous subphase; (9) glass barrier for proton diffusion. The arrows indicate the diffusion pathway of protons. L = 4.3 cm; monolayer area = 12.6 cm^2. Redrawn from Prats *et al.* (1985) with changes.

This thermodynamic conclusion would explain the fast lateral transfer of protons observed.

8.5.2. DIVERSE MEMBRANE PROTEIN MONOLAYERS

A hydrophobic lipid- and pigment-free polypeptide from the chromatophore membrane of *Rhodospirillum rubrum* was investigated as a spread monolayer on chloroform–methanol solutions at the air–water interfaces (Kopp *et al.*, 1979). From these investigations, an average molecular area of 12.9 nm^2/molecule was estimated at π = 20 dynes/cm. Multilayers built up on germanium plates at different surface pressures were investigated by total attenuated reflectance IR spectroscopy. In all cases, the amide I and II absorption bands were typical of α-helical and random conformation. Electron microscopy of transferred monolayers replicated by rotary platinum shadowing revealed domains of regular texture in specimens prepared at 10 and 30 dynes/cm. Light optical diffractometry of the ordered arrays yielded a smallest repetitive area of 13.5 nm^2, which agreed well with the molecular area obtained from the monolayer surface.

8.6. LIPID–MEMBRANE PROTEIN MONOLAYERS

From monolayer studies, it was shown that the surface behavior of glycosphingolipids depends on the type of oligosaccharide chain in the polar head group (Maggio *et al.*, 1981), and this can influence the interaction between lipids and Ca^{2+} (Maggio *et al.*, 1978a,b, 1980). With respect to the interactions with proteins, the penetration of the membrane protein melittin (and of other, nonmembrane proteins, e.g., albumin, myelin basic protein, glucophorin) into pure lipid monolayers of eight neutral or acidic glycosphingolipids was investigated (Fidelio et al., 1981, 1982). Since melittin and myelin basic protein contain sialosyl residues, the increase in π obtained for their penetration into gangliosides suggested that the interaction was dependent on the number of negative charges in the glycolipid.

Complex formation between ionophore A23187 and Ca^{2+} was investigated (Brasseur et al., 1982). Carboxylic ionophores such as A23187 and bromolasalocid are known to be able to facilitate Ca^{2+} ion transport across natural membranes (Scarpa and Baldassare, 1972; Caswell and Pressman, 1972; Case et al., 1974; Pressman, 1976) and artificial membranes (Hyono et al., 1975; Puskin and Gunter, 1975; Kafka and Holz, 1976; Wulf and Pohl, 1977). These ionophores are reported to form complexes in which each Ca^{2+} is bound to two molecules of the same ionophore or of different ionophores. The conformation of such complexes in nonpolar solvents and in the crystalline form has been extensively investigated. The conformational stability (kd) observed by a simulated monolayer method has been reported (Ferrieira et al., 1981). The conformational characteristics of Ca^{2+}-A23178 in monolayers were analyzed by the strategy described earlier for such polypeptides and other molecules (Ralston and de Coen, 1974a,b; Brasseur et al., 1981; Tanford, 1980). In the procedure used, the conformational energy of the interfacial complex was empirically calculated as the sum of all the contributions resulting from local interactions—e.g., van der Waals energy, the torsional potential, the electrostatic interactions, and the energy of transfer. The magnitude of electrostatic energy was estimated as a function of dielectric constant (ε). The magnitudes of ε chosen were 30 and 3 for hydrophobic and hydrophilic media, respectively. The magnitude of ε was further assumed to increase linearly over a distance of 4.5 Å along the z axis, i.e., perpendicular to the interface. This distance of 4.5 Å is the same as the distance between Ca^{2+} and the line passing between the N atoms of the two benzoxazole rings of the A23178 molecule in the all-trans configuration. The free energy of transfer for each part of the molecule was estimated as described in Chaper 5 for the unfolding energy of proteins.

The most probable conformation was characterized as a twofold axial symmetry that was maintained during transition to the hydrophobic bulk conformation. It was concluded that this method could be useful in predicting the behavior of extracellular molecules as they come into contact with and are eventually incorporated into the phospholipid domains of the plasma membrane.

The interaction of streptomycin and streptomycylamine derivatives with negatively charged bilipid membranes (BLMs) has been investigated (Brasseur et al., 1985). All streptomycin derivatives were found to extend parallel to fatty acid chains and across the hydrophobic–hydrophilic interface, whereas gentamicin is oriented largely parallel to this interface and above the plane of the phospho groups. These investigations allowed the conclusion that the inhibition of phospholipase activities is related to the position of an aminoglycoside in the BLM.

In a recent study (Fidelio et al., 1984), the miscibility and stability of mixed lipid-protein monolayers at the air-water (0.145 M NaCl) interface was investigated by using six glycosphingolipids (acidic and neutral), three different types of proteins (soluble, extrinsic, and highly amphipathic), and some phospholipids. The ΔG_{ex} for different melittin and lipids was measured. It was concluded that ΔG_{ex} becomes more negative as the complexity of the polar group of the lipid increases and the lipid interface is more liquidlike or when the protein surface concentration is lower. Mixed films of protein, vinculin, and phospholipids were investigated by using monolayers and compared with membranes (Fringeli et al., 1986). These interactions were also analyzed with the help of IR spectra. Mixed monolayers of indomethacin and phospholipids at the air-water interface were investigated (Cordoba and Sicre, 1985). The degree of miscibility in lipid-protein monolayers was reported recently (Cornell and Carroll, 1985).

Mixed monolayers of a carboxylic ionophore antibiotic, grisorixin, and egg lecithin were studied by π and ΔV measurements by varying the pH and ionic strength of the subphase as well as the composition of the film (Van Mau et al., 1980; Van Mau and Amblard, 1983). Grisorixin is an ionophore antibiotic of the carboxylic polyether family, of which the best known is nigericin. It is classed in the group of polyether ionophores, which specifically complex monovalent cations. ΔV isotherms of grisorixin monolayers showed that reactions of proton exchange and alkaline cation complexation take place between the subphase and the ionophore spread at the surface. The stability constant was estimated from these monolayer investigations. These studies suggested that neutral forms of grisorixin are located inside the nonpolar core of the bilayer, while the dissociated g-form of this antibiotic is located in a rather aqueous environment near the phospholipid head groups.

The interactions between melittin and ionic surfactants, both

$$\text{Cationic} = n\text{-}C_{20}H_{41}\text{-}N(CH_3)_3\text{-}Br(C_{20}NBr)$$

and

$$\text{Anionic} = n\text{-}C_{20}H_{41}\text{-}SO_4Na(C_{20}SO_4Na)$$

were studied by measuring the π and $\Delta\psi$ of these mixed monolayers (Birdi and Gevod, 1987). The change in π and $\Delta\psi$ of the surfactants (spread from $CHCl_3$ solutions) was monitored as a function of time (min) (Figs. 8.22 and 8.23). The injection of melittin under cationic $C_{20}NBr$ has no appreciable

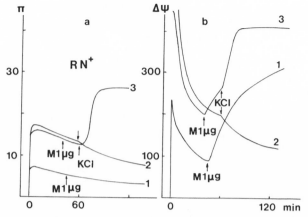

FIGURE 8.22. Rate of change of π vs. time (a) and of $\Delta\psi$ vs. time (b) of films of $C_{20}NBr$ on injection of melittin (1), KCl (0.1 M) (2), and melittin followed by KCl injection (3). The arrows indicate the time of addition to the subphase (25°C). From Birdi and Gevod (1987).

effect, while the film of anionic $C_{20}SO_4Na$ shows a large increase in π. The addition of KCl to both surfactants had no appreciable effect, as one would expect. However, the presence of melittin gave a large increase with both surfactant systems. These data clearly show that melittin interacts strongly with negatively charged lipids in monolayers, as it does in vesicles, BLMs, and membranes. The $\Delta\psi$ data provide further evidence for the selective

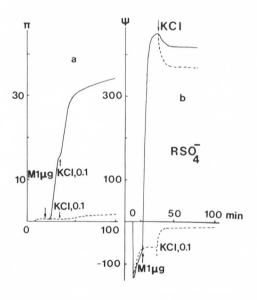

FIGURE 8.23. Rate of change of π (a) and $\Delta\psi$ (b) with time of films of $C_{20}SO_4Na$ on addition of melittin (M) or KCl (0.125 M). See the Figure 8.22 caption for details. From Birdi and Gevod (1987).

penetration of Cl⁻ ions into these monolayers, suggesting that melittin forms ionic channels of the same structure as in lipid bilayers.

8.6.1. DIVERSE LIPID MEMBRANE SYSTEMS

As a result of the observation that resistant species might grow with the use of insecticides, as well as of concern with environmental problems, there has been interest in finding different means of mosquito control (McMullen, 1973). The air–water interface is crucial to the different stages in the life cycle of the mosquito (and, for that matter, in the life cycles of other insects that live on water surfaces). Many insects deposit their eggs on the surface of water. The larvae and pupae of mosquitoes are known to spend appreciable time breathing and feeding at the water surface. Furthermore, the process of eclosion is known to take place through the interface.

The effects on surface tension of dissolved surface-active substances have been known for many years (Wigglesworth, 1972; Senior, 1943; Singh and Micks, 1957; Mulla and Chadhury, 1968; van Emdenkroon et al., 1974). However, it is obvious that surfactants will also be toxic to other, nontarget organisms. It was therefore considered of interest to investigate the effects of spread lipid monolayers on mosquitoes (Eckstein, 1939; McMullen et al., 1977; Reiter and McMullen, 1978; Reiter, 1978). It was found that an egg lecithin monolayer could kill mosquito larvae in the laboratory. Later, other lipids, such as lauryl alcohol, were also found to be effective. Natural lipids with maximum π_{eq} above 45 mN/m were reported to be lethal to the pupae of several species. These effects were valid in both laboratory and field tests. Monolayers of egg lecithin at maximum π_{eq} were tested against eight species of mosquitoes in the laboratory. The larvae of most species were prevented from surfacing by the lipid film (i.e., they were prevented from penetrating) and thus died in the water from lack of oxygen. It was suggested that a stable soap-bubble type of film formation is what prevents the insect from surfacing.

8.7. GENERATION OF MEMBRANE STRUCTURE DURING EVOLUTION

The structure of biological cell membranes requires that lipid and protein molecules be at some definite distance apart (4–5 Å) such that self-assemblies can be created. During prebiotic evolution, this assembly could have taken place through a variety of processes. However, the preceding description of the system could easily suggest that the liquid–air interface must have played a rather important role in the initial stages of prebiotic

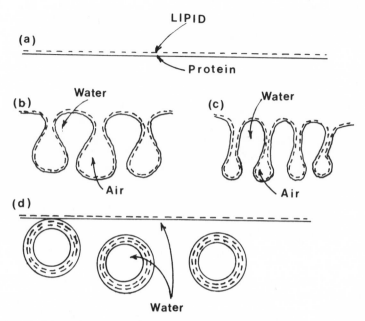

FIGURE 8.24. Schematic representation of vesicle formation from a lipid-protein monolayer spread at the air-water interface.

evolution. This does not mean that other interfaces, such as $liquid_1$-$liquid_2$ or solid-liquid, have not been of interest.

The prerequisite that lipid and protein molecules be close enough is easily satisfied by the lipid-protein monolayer (Fig. 8.24). When this structure is disturbed by a wind blowing along the surface or by air bubbles, the monolayer buckles and can form closed vesicles (Goldcare, 1958; Calvin, 1969). This process whereby vesicle formation from monolayer assemblies occurs at the air-water interface could thus have been one of the initial steps toward biological cell formation. This postulate is further supported by the fact that all lipids and proteins exhibit very characteristic surface adsorption properties. These properties must have been of some significance in the selection processes during prebiotic evolution.

9

ADSORPTION OF MONOLAYERS ON SOLIDS (LANGMUIR–BLODGETT FILMS)

Some decades ago (Langmuir, 1920), it was reported that when a clean glass plate was dipped into water covered by a monolayer of oleic acid, an area of the monolayer equal to the area of the plate dipped was deposited on the plate when the plate was withdrawn. Later (Blodgett, 1934, 1935), it was found that any number of layers could be deposited successively by repeated dippings; this later came to be known as the Langmuir–Blodgett (LB) method.

In another context, the electrical properties of thin films obtained by different procedures, e.g., thermal evaporation in vacuum, have been investigated in much detail. However, the films deposited by the LB technique have only recently been used in electrical applications. The thickness of LB films can be as little as one monomolecular layer (≈ 25 Å $= 25 \times 10^{-10}$ m), which is not possible by evaporation procedures.

9.1. LANGMUIR–BLODGETT FILMS

9.1.1. TRANSFER OF MONOLAYER FILMS ONTO SOLID SURFACES

Langmuir (1920) investigated the process of transferring a spread monolayer film to a solid surface by raising the solid surface through the interface. The process of transfer is depicted schematically in Fig. 9.1. It can be seen that if the monolayer is in a closely packed state, then it is transferred to the solid surface, most likely without any change in packing density. Detailed investigations have shown, however, that the process of transfer is not as simple as shown in Fig. 9.1. The monolayer of the solid surface may or may not be stable, and defects may be present.

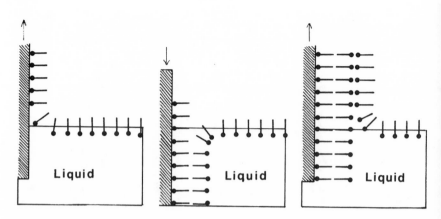

FIGURE 9.1. Schematic representation of the transfer of a monolayer to a solid substrate. The alkyl part of the monolayer molecules is represented by the tail of the "lollipop" figures and the polar group by the head; the solid substrate is cross-hatched.

Monomolecular layers (LB films) of lipids are of interest in a variety of applications, including the preparation of very thin controlled films for interfaces in solid-state electronic devices (Gaines, 1966; Vincent and Roberts, 1980). Scientists are currently using LB film assemblies as solutions to problems in diverse areas: microlithography, solid-state polymerization, mirror optics, electron tunneling, and photovoltaic effects (*Thin Solid Films*, 1980; Blodgett, 1939).

The simplest procedure generally used is one in which a clean, smooth solid surface (of suitable surface area) is dipped through the interface with the monolayer. Alternatively, the solid sample is immersed in the water before a monolayer is spread·and then drawn up through the interface to obtain the film transfer. It is obvious that such processes involving monomolecular film transfers will easily introduce defects arising from various sources. As will be shown below, these defects are in most cases easily detected.

Structural analysis of the molecular ordering within a single LB monolayer is important both to understand how the environment in the immediate vicinity of the surface (i.e., the solid) affects the structure of the molecular monolayer and to ascertain how the structure of one layer forms a template for subsequent layers in a multilayer formation. Studies of the order within surfactant monolayers have been reported for many decades. Multilayer assemblies have been studied by electron diffraction (Peterson and Russel, 1984; Bunnerot *et al.*, 1985) and X-ray diffraction (Prakash *et al.*, 1984), as well as infrared (IR) absorption (Bunnerot *et al.*, 1985; Chollet and Messier, 1983; Rabolt *et al.*, 1983).

Motivated by a proposed older model for the orientation of molecules (Langmuir, 1933; Epstein, 1950) and by recent theoretical calculations (Safran *et al.*, 1987), investigations of these two potential models for tilt disorder in the monolayer have been carried out (Garoff *et al.*, 1986). The principle of both models is that the monolayer structure tries to compensate

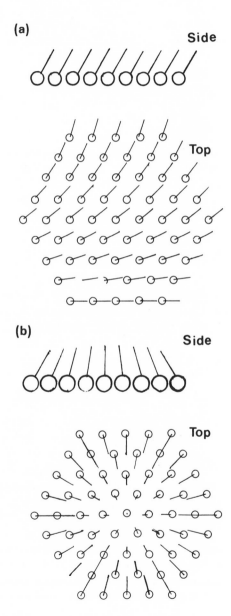

FIGURE 9.2. Schematic representations of alkyl tail directions in a cluster model with gradually varying tilt direction (a) and in a micellar cluster model (b). Redrawn from Garoff *et al.* (1986) with changes.

for the difference between the equilibrium of head–head and chain–chain distances that each piece of the molecule would attain if it were independent. In one model, the magnitude of the tilt is fixed, but the tilt direction varies gradually throughout the lattice (Fig. 9.2a). In the second model, the tilt always points to the center of the cluster, but the magnitude varies from straight up at the center of the cluster to some maximum tilt at the cluster edge. Such a structure displaying the distortion of the tail ordering is pictured in Fig. 9.2b and has been called a "micellar cluster."

Both models exhibit contributions to translational disorder beyond that existing in the head group lattice. The micellar cluster model preserves strong bond orientational correlations, while the first model degrades them to some extent. Simulations have shown that the micellar model can fit the general features of the diffraction pattern better than the first model. The precise nature of the positional disorder, liquidlike or microcrystalline, is difficult to determine for the strongly disordered structure of complex molecules found in the LB monolayer. It would be especially interesting to know whether the structure is similar to a stacked-hexatic phase (long-range bond orientation with exponentially decaying translational correlations) or microcrystalline (perfect translational correlations).

9.1.2. ELECTRICAL BEHAVIOR OF LANGMUIR–BLODGETT FILMS

Insulating thin films in the thickness range of 100–20,000 Å have been a subject of varied interest among the scientific community because of their potential application in devices for optical, magnetic, electronic, and other uses. Some of the unusual electrical properties possessed by thin LB films that are unlike those of bulk materials led to consideration of their technological applications; consequently, interest in thin-film studies grew rapidly. Earlier studies did not prove to be very inspiring because the LB films obtained always suffered from pinholes, stacking faults, and other defects, and hence the results were not reproducible. It is only in the past few decades that many sophisticated methods have become available for producing and examining thin films and exercising better quality control over these processes. Despite these advances, however, the unknown nature of the inherent defects and the wide variety of thin-film systems still complicate the interpretation of many experimental data and thus hinder their technological application.

The study of the breakdown of conduction in thin films, the major subject of investigation, has been based on films prepared by thermal evaporation under vacuum or similar techniques. LB films have remained less known among researchers in this field, but various interesting physical properties of LB films have been investigated in the current literature.

9.1.3. STRUCTURE AND ORIENTATION OF ORGANIZED LAYER ASSEMBLIES

The structure and orientation of the deposited amphiphile molecules have been found to be governed by the angle of contact between the monolayer and the solid surface. The deposited monolayers, in general, have been characterized as x-, y-, and z-types, the molecular arrangements of which are shown in Figure 9.3. In the case of x- and z-type films, the molecules are oriented in the same direction, and thus the surface is composed of carboxyl and methyl groups, respectively. In contrast, in a y-type film, the molecules in adjacent layers are oriented oppositely and the LB film surface is composed of methyl groups. Of these three types of LB films, the one that has been most studied is the y-type, in the production of which monolayer transfer takes place both ways, on each dipping and withdrawal of the solid through the interface. On the other hand, with an x-type film, transfer takes place only when the slide is dipped, and with a z-type film, it takes place only when the slide is withdrawn. The orientation is termed "exotropic" when the methyl groups touch the solid surface and the polar (carboxyl) groups remain away from it in the first layer. If the orientation of these two groups is reversed, the monolayer is termed "endotropic." Thus, an x-type film is made up of a series of exotropic layers, a y-type film is made up of a series of alternating exotropic and endotropic layers, and a z-type film is made up of a series of endotropic layers. The z-type films are rather uncommon.

There are two crucial factors in making satisfactory electrical measurements on monolayer films: uniform thickness and packing. These two requirements are generally fulfilled to a greater extent with the use of LB

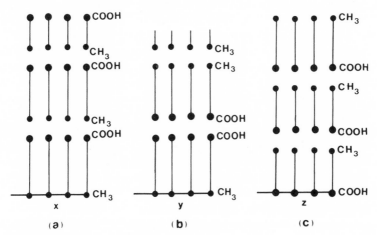

FIGURE 9.3. Molecular orientation of LB films: (a) x-type; (b) y-type; (c) z-type.

films. The thicknesses of several fatty acid monolayers have been measured using the best-known optical methods. Thickness measurements were made with the use of a protective colloid ion layer, which had high-quality fringes and, accordingly, high precision (Srivastava and Verma, 1962, 1966; Sirhatta and Scott, 1971). As regards the packing uniformity of LB films, it is known that the selection and purity of substrate must be monitored very carefully. It has been reported (Hold, 1967) that increasing the number of monolayers increases the degree of uniformity of the film, and this observation has been confirmed by other investigators. In fact, in achieving the required packing uniformity of LB films, the deposition of the first monolayer has been found to be very critical, since any voids or imperfections in the first layer generally lead to major disruption of the subsequent monolayers. It has therefore been suggested that the solid slide removed from the liquid surface must be completely dried before it is reimmersed. Generally, if a wet slide is reimmersed, no multilayers are adsorbed [a finding in the case of lecithin, fatty acids, and protein monolayers (Birdi, unpublished)]. Ba- or Cu-stearate films could be built up to as many as 3000 layers simply by repeated dipping and withdrawal (Blodgett, 1935; Blodgett and Langmuir, 1937). On the other hand, fatty acid films or films of other substances tended to crack and fog. It was suggested that the addition of Cu^{2+} ion (10^{-6} M) eliminated these difficulties (Blodgett and Langmuir, 1937).

9.1.4. STRUCTURE OF DEPOSITED FILMS

The most useful method for investigating the detailed structure of deposited films is the well-known electron diffraction technique. The molecular arrangements of mono- and multilayer films of fatty acids and their salts deposited by this technique have been reported (Germer and Storks, 1938). These analyses showed that the molecules were almost perpendicular to the solid surface in the first monolayer. It was also reported that Ba-stearate molecules have a more precise normal alignment than do the molecules in stearic acid monolayers. In some investigations (Hold, 1967), the thermal stability of these films has been found to be remarkably uniform up to 90°C. On the basis of the structural analyses obtained by the electron diffraction technique, these deposited films are known to be monocrystalline in nature and thus can be regarded as a special case of a layer–bilayer mechanical growth forming almost "two-dimensional" crystals. There is evidence, however, that Ba-behenate multilayers do in fact show an absence of crystallization, which has been demonstrated by electron micrographic studies (Turnit and Schidlovsky, 1961).

In many early reports, it was shown that deposited films obtained by the usual process of monolayer transfer invariably contained holes, cracks, or similar imperfections. These observations are not surprising, because at higher π, the more compact film will be removed. Nevertheless, it would be an oversimplification to regard films transferred at high π as perfectly uniform, coherent, and defect-free. Artifacts are indeed introduced if proper care has not been taken during the transfer process, or the subsequent thermal evaporation of metal electrodes over the film may disturb the film structure. In some cases, radioautographic investigations have indicated that these deposited layers are generally uniform with no apparent gross defects (Handy and Scala, 1966). Thus, recent modifications in the deposition technique have made it possible to obtain films largely free of gross defects or imperfections. Moreover, any defects or pinholes are easily detected, since the resistance around defects is very low (Birdi, unpublished).

In a recent study (Mizuno et al., 1983), an LB film of iron stearate was deposited from aqueous solution onto a substrate only during immersion. On removal, the substrate was withdrawn through a clean area. The condition of the aqueous surface was the same as for the preparation of a y-type layer. ^{55}Fe tracer techniques were used to examine the direction of molecules. The distance from the surface to the ^{55}Fe-labeled atoms could be determined with an error of $\pm 10\%$.

The unidirectional surface conductance of monolayers of stearic acid deposited on a glass support was investigated (Ksenzhek and Gevod, 1975). The influence of the two-dimensional pressure on the tangential permeability of the layers for various cations was examined. The current in the channel formed by the monomolecular film and the substrate is carried by hydrated cations. The permeability of this channel to various cations diminishes in the order $Ca^{2+} > Zn^{2+} > Ba^{2+} > Mg^{2+} > H^+ > Cs^+ > K^+ > Na^+ > Li^+$. The contact angles and adhesional energy changes during the transfer of monolayers from the air–water interface to solid (hydrophobic glass) supports has been analyzed (Gaines, 1977).

9.1.5. MOLECULAR ORIENTATION IN MIXED DYE MONOLAYERS ON POLYMER SURFACES

In the last few years, there has been increasing interest in the use of surface-active dyes to study properties of biological and artificial membranes and to construct monomolecular systems by self-organization (Waggoner et al., 1977; Mobius, 1978). When these dyes are incorporated into lamellar systems, it is found that the paraffin chains stand perpendicularly on the plane of the layer, while the chromophores lie flat near the hydrophilic interface. To develop new molecules as functional components of monolayer assemblies, a series of nine surface-active azo and stilbine compounds were synthesized (Heesemann, 1980). Their monolayer properties at the air–water interface were investigated by π and spectroscopic techniques. By means of molecular "Corey–Pauling–Kihava" models, several molecular arrangements that were found to be consistent with the experimental results were described.

The adsorption of mixed monolayers of n-octadecyltrichlorosilane [(OTS) $C_{18}H_{37}SiC_{13}$] and long-chain substituted cyanine dyes on stretched polyvinyl alcohol (PVA) films was investigated (Sagiv, 1979). These systems were selected for the following reasons:

1. The solid substrate (PVA) possesses a simple organization with uniaxial symmetry, which allows a straightforward correlation with the molecular orientation induced in the monolayer.
2. PVA is transparent and the adsorbed molecules contain elongated dye chromophores, so that the molecular orientation within the monolayer can be readily determined by means of polarized absorption spectroscopy (linear dichroism).
3. It is easy to produce the support in large quantities and adsorption is easy for oleophobic OTS monolayers.

Furthermore, PVA is not soluble and does not swell in the organic solvents used in monolayer studies. The orientational effects were estimated by linear dichroism.

An application of neutron activation analysis for the determination of inorganic ions in LB multilayers was reported (Kuleff and Petrov, 1979). A special technique for the removal of multilayers from the solid substrate was used. Multilayers of arachidic acid and octadecylamine were analyzed with respect to Cd^{2+} and HPO_4^{2-}.

9.1.6. DIELECTRIC PROPERTIES OF LANGMUIR–BLODGETT FILMS

Interest in the dielectric properties, e.g., capacitance, resistance, and dielectric constant, of deposited LB films of fatty acids and their metal salts was one of the main focuses of investigations in the early stages of research on LB films. In early investigations (Porter and Wyman, 1938), measurements on the impedance of films and on related phenomena were carried out on Cu-, Ba-, and Ca-stearate using both x- and y-type films. Initially, mercury droplets were used for small-area probe measurements and an AC bridge was used for impedance measurements. The resistance of the films was found to be very low ($<1 \Omega$) with signal voltage, whereas it was of the order of megohms with signals of 1 or 2 V. The results were not satisfactory, and later measurements were carried out by replacing the AC bridge with a radiofrequency bridge. Measurements at frequencies of 1 and 0.244 MHz on z films containing 7–41 deposited monolayers determined dielectric constant values ranging between 1.9 and 3.5, with an average value of 2.5.

In both types of films, the capacitance decreased with thickness, as would be expected from the following relationship:

$$C = A\varepsilon / 4\pi Nd \qquad (9.1)$$

in which C is the capacitance, ε is the dielectric constant, N is the number of layers, d is the layer thickness, and A is the area of contact between drop and film. On the other hand, the values of the resistance per layer showed a definite increase with increasing thickness of the film. The specific resistance of the films as determined in this way from the values of resistance per layer was approximately 10^{13} Ω.

Figure 9.4 shows the results of capacitance measurements on some 75 samples, in which $1/C$ is given as a function of predicted number of layers (N_p) (Handy and Scala, 1966). The capacitance measurements thus performed on stearate films (1–10 layers) yielded ε values between 2.1 and 4.2, with a bulk value of 2.5.

In Fig. 9.5, a typical plot of the variation of C as a function of frequency (upper curve) over the frequency range 0.1–20 kHz is shown. These investigations indicated some slight dependence on frequency. The lower curve in Fig. 9.5 shows the variation of the imaginary part of the complex dielectric constant (proportional to Cd, where d is the dissipation vector) with frequency. Evidently, there is a slight maximum near 700 Hz, which has

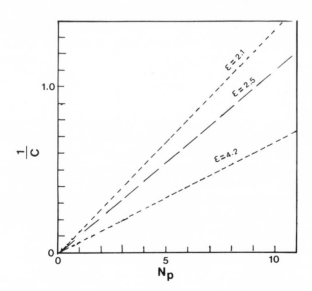

FIGURE 9.4. Plot of $1/C$ vs. number of layers, N_p. (——) Bulk phase $\varepsilon = 2.5$; (– – –) $\varepsilon = 2.1$ and 4.2. Redrawn from Handy and Scala (1966) with changes.

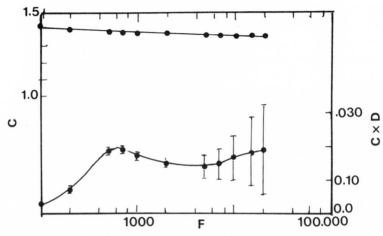

FIGURE 9.5. Plots of C vs. frequency (F) (upper curve) and CD (dissipation factor) vs. F (lower curve). Redrawn from Handy and Scala (1966) with changes.

been proposed to indicate the presence of a weak polar adsorption mechanism with a characteristic relaxation time ($\frac{1}{2} \pi f_n$) of 0.23 msec.

In many of the measurements reported in the literature, the organic film was sandwiched between evaporated-aluminum electrodes. It was realized that an oxide layer grows on the base of an aluminum electrode, but its effect on the capacitance values of the device was neglected because its resistivity is small compared with the resistivity of the organic LB layers. The presence of such a thin oxide film between metal electrodes and fatty acids can be analyzed by the following equation (Mann and Kuhn, 1971):

$$1/C = (\varepsilon_a A)^{-1}[(d_{ox}/\varepsilon_{ox}) + (Nd/\varepsilon)] \qquad (9.2)$$

where d_{ox} and ε_{ox} are the thickness and the dielectric constant, respectively, of the oxide layer, ε_a is the dielectric constant of air, and all other symbols have their usual meanings. Capacitance has been reported to be a linear function of $1/C$ with respect to the number of transferred monolayers (Fig. 9.6) (Leger *et al.*, 1971).

LB films of Ba salts of fatty acids deposited at $\pi = 50$ mN/m gave the following relationships between $1/C$ and N (Birdi, unpublished):

Ba-stearate:

$$1/C = 15.9N + 1.13 \qquad (10^6 F^{-1}) \qquad (9.3)$$

Ba-behenate:

$$1/C = 17.2N + 8 \qquad (10^6 F^{-1}) \qquad (9.4)$$

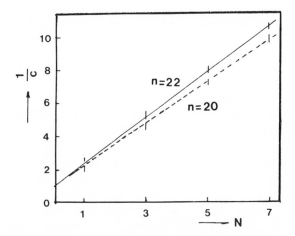

FIGURE 9.6. Plots of reciprocal capacitance, $1/C$, vs. number of layers, N, of Cd-stearate ($n = 20$) and Cd-behenate ($n = 22$). Redrawn from Mann and Kuhn (1971) with changes.

To explain the significant difference in ε values of monolayers and multilayers obtained theoretically as well as experimentally, systematic investigations on the thickness dependence of ε from 1 to 80 layers have been reported (Khanna, 1973). It was observed that ε increased slowly with thickness initially, but saturated at a particular thickness (≈ 1000 Å) for all types of LB films.

The dielectric anisotropy of long-chain fatty acid monolayers was analyzed (Agarwal, 1979). These fatty acids were considered as being oriented in a cylinder cavity with length $L \gg$ diameter (D). With each bond in these molecules considered as a polarization ellipsoid with axial symmetry about the $-C-C-$ bonds, the mean polarizability of the bonds was calculated. The axial ratio $f(L/D)$ affects the geometric factor (G) and hence the dielectric constant, ε. With the use of arbitrary f values of 0.5–10.0, G and ε were estimated for several fatty acids. These parameters varied with respect to f. For palmitic acid, ε was less dependent on the axial ratio of $f > 4$, and the actual L/D ratio yielded ε values quite close to 1, corresponding to $L/D > 4$. By comparing ε values obtained from two different f values, the assumption that the molecules are best represented by a cylinder cavity ($L \gg D$) was confirmed. Theroetical values of ε were suggested to be an indirect method to verify ellipsometric data on the index of refraction of biological membranes.

The LB films of fatty acid salts are known to retain breakdown strength of approximately 10^6 V/cm for long periods. In the earliest study, breakdown of the films under high DC voltages was measured by using a galvanometer to read the current (Porter and Wyman, 1938). With the use of a mercury drop as the upper electrode, no current was detected until the magnitude of the applied voltage across the specimen reached a critical value in the case of a 30-layer LB film, whereas in thinner films, appreciable currents were present almost from the start. The breakdown per layer was found to rise sharply at a thickness of approximately 20 layers, when its order of magnitude was approximately 10^6 V/cm. The specific resistance was found to be approximately 10^{13} Ω, below the breakdown voltages. In later studies, a water drop with electrolyte was used instead of mercury, and no difference was found between the two methods.

The destructive breakdown in LB films has been found to comprise two events, i.e., the destructive breakdown voltage and the maximum breakdown voltage. These events have been determined in various fatty acid soaps in the thickness range corresponding to 16–80 layers (Thiessen *et al.*, 1940). A typical plot of applied voltage, V, vs. current density, J, in the nondestructive phase of 40-layer Ba-stearate, has an area of approximately 0.5 cm^2. Point A in Fig. 9.7 corresponds to the initial abrupt rise of current (onset of breakdown voltage) and point B to the destructive breakdown voltage, V, at which the "visible" destruction of the film commences. The destructive breakdown voltages have been reported to be independent of the material and thickness of the film, but the maximum breakdown voltages have been reported to vary slightly with increasing film thickness. These

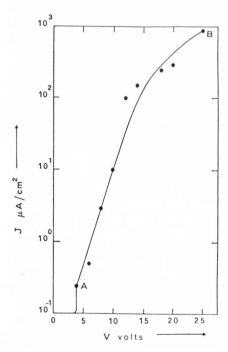

FIGURE 9.7. Plot of V vs. J for a 40-layer Ba-stearate sandwich with an area of approximately 0.5 cm^2. Redrawn from Agarwal and Srivastava (to be published) with changes.

breakdown voltage systems have been reviewed by various investigators (Klein, 1969; Klein and Gafni, 1966; Budenstein and Hayes, 1967; Budenstein et al., 1969; Smith and Budenstein, 1969). The breakdown mechanism in LB films has been described in terms of a statistical model (Klein, 1972; Agarwal, 1974a, b). The AC breakdowns in films are not yet well understood.

The study of current-voltage characteristics in metal-oxide-semiconductor (MOS) structures has been found to produce much useful information about the mechanism of a certain type of device. LB films of Ca-behenate have been investigated with o-phenanthroline with three stearate chains (Tanguy, 1972). In these studies, p-type silicon surfaces (metal–insulator–semiconductor), after being etched in CP_4, cleaned, and dried, were used to deposit LB films and were covered with aluminum in the form of two circles connected by a channel. From the linear relationship $1/C$ vs. N, the thickness was estimated to be 15 Å. A small variation in bias voltage gave rise to a large modulation of C, which has been ascribed to the small thickness of the insulator and to the high resistivity ($\approx 600 \, \Omega \, cm^{-1}$) of the p-type silicon. In other studies (Snow et al., 1965; Bui et al., 1972) on MOS structures, the two types of hysteresis, normal and abnormal, were suggested to arise from ion displacement in the insulator and from trapping at the interface states.

The inclusion of sites of radiation-induced polymerization has been used to provide increased film stability and has been described as an application of high-resolution electron beam lithography to the fabrication of microcircuitry (Barrand et al., 1979a,b), 1980; Nagarajan and Shah, 1981). So far, only a few types of polymerizable sites have been described, and monomolecular layers involving radiation-cleavable sites do not appear to have been reported. One approach to producing such a site is to include in the chain a quaternary center or a heavy element or both to enhance electron-beam and X-ray cross sections. Essentially nothing appears to be known about the film-forming properties of lipid acids containing a heavy element such as Si, Sn, Ge, or Pb in the chain.

The transfer rates of dipalmitoylphosphatidylcholine (DPPC) lecithin monolayers from an aqueous phase to a glass surface were investigated (Sato and Kishimoto, 1982). The change in contact angle of the DPPC monolayer-glass interface, when DPPC was kept to a fixed area/molecule of 134 Å^2, was reported to be sigmoidally increased. These rates were analyzed as first-order kinetics, and the transfer rate was found to be 2.59×10^{-3} liters sec^{-1}.

Synthesis and monolayer studies (Brown et al., 1983) on a geminally branched acid—α,α-dimethylarachidic acid—to assess the feasibility of forming more stable films were reported. It was hoped the assemblies would possess enhanced radiation sensitivity and provide a model for films of acids containing the $M(CH_3)_3$ subunit. Surprisingly, the quaternary center did not markedly disturb formation of monomolecular films, and films formed and transferred to a solid surface were studied by optical and X-ray photoelectron spectroscopic measurements.

The interactions between insoluble monolayers of ionic surfactants and ions dissolved in the subsolution have been an object of continuing scientific interest since the pioneering investigations of Adam and Langmuir. This interest rests on the desire to model important natural and technological processes such as enzymatic activity, permeability of cellular membranes, ion-exchange processes in soils, adsorption in ion-exchange resins, ion flotation, and chromatography.

A recent report describes the ion adsorption of radiolabeled chromium (III) on stearic acid LB films (Simovic and Dobrilovic, 1978). The adsorption of chromium(III) on a stearic acid monolayer on the surface of $CrCl_3$ was described. Stearic acid monomolecular films on 10^{-4} M $CaCl_2$ subsolutions were deposited on paraffin-coated microscope glass slides by the LB technique (pH range 2–9) (Neuman, 1978). A combination of the LB technique and the neutron activation method of analysis was used to determine the stoichiometry of the interaction between the fatty acid (arachidic acid) and the metal ions dissolved in the subphase. The experimental data were used to estimate the stability constants of arachidic acid and bivalent metal ions (Cd^{2+} and Ba^{2+}) (Petrov et al., 1982, 1983). The method used was that described by Matsubara et al. (1965). The data were explained as an interaction between metal ions and monolayer as an adsorption process:

$$2RCOO^- + Me^{2+} \Leftrightarrow (RCOO)_2Me \qquad (9.5)$$

The equilibrium constant is given by the mass action law:

$$\beta = [X_{R_2Me}]/[XR^-]^2[M_{Me}^{2+}] \qquad (9.6)$$

where $R^- = RCOO^-$. Furthermore, the following equation can be derived from these:

$$\beta = [X_{R_2Me}X_{H^+}^2]/[X_{RH}^2 X_{Me^{2+}}](K_d^2) \qquad (9.7)$$

where $K_d = [X_{R^-}X_{H^+}]/[X_{RH}]$. Equation (9.7) represents the Langmuir adsorption isotherms, in which the Me^{2+} ion occupies two adsorption sites, the number of the lattice sites being dependent on the subphase pH. The interactions in the monolayer and the subphase are neglected. These can be taken into account by introducing the appropriate activities (A, i).

In the case that very dilute concentrations are present in the subphase, one can safely assume that $A_{H^+} = X_{H^+}$ and $A_{Me^{2+}} = X_{Me^{2+}}$. If necessary, the bulk activity coefficients can be estimated with the help of the Debye-Hückel theory. To estimate the activity coefficients in the monolayer, Matsubara *et al.* (1965) considered the latter as a two-dimensional regular solution. By application of the Bragg-Williams approximation to the case of a square two-dimensional lattice, the following relationship was obtained:

$$A_{R_2Me} = X_{R_2Me} \exp[2\Omega(1 - X_{R_2Me})^2 / RT] \tag{9.8}$$

$$A_{RH} = (1 - X_{R_2Me}) \exp(\Omega X_{R_2Me} / RT) \tag{9.9}$$

where Ω is the energy of mixing of the monolayer components. The magnitude of $[X_{R_2Me}]$ can be estimated from the experimentally known ratio Me/RCOOH

$$[X_{R_2Me}] = 2(Me/RCOOH)$$

$$= \phi \tag{9.10}$$

If the magnitudes of $[X_{H^+}]$ and $[X_{Me^{2+}}]$ are estimated from the measured pH values and $C_{Me^{2+}}$, then one can write

$$\log[\phi/(1 - \phi)^2] - 2(\text{pH}) \log C_{Me^{2+}} - 1.74$$

$$= 1.74\Omega\phi / RT - 0.87\Omega / RT + (\log \beta)K_d^2 \tag{9.11}$$

The quantity on the left-hand side is linearly dependent on ϕ. A plot of the quantity on the left-hand side vs. ϕ allows one to estimate the value of Ω. Setting $\phi = 0.5$, where the first two terms on the right-hand side become zero, gives the value of (βK_d^2). The data for various binding constants are

TABLE 9.1. Stability Constants, β, of Bivalent Fatty Acids and Energies of Mixing, Ω, as Calculated from Equation (9.7)[a]

System	Ω (kcal/mole)	$(\log \beta)K_d^2$	β (dimensionless)	Experimental conditions
Cd^{2+}-arachidic acid	−0.07	−9.15	4×10^3	T constant pH constant
Cd^{2+}-arachidic acid	−0.13	−8.75	1×10^4	T constant C_{CdCl_2} constant
Ba^{2+}-arachidic acid	−1.60	−10.80	89.3	T constant C_{BaCl_2} constant

[a] From Petrov *et al.* (1982).

given in Table 9.1 (Petrov *et al.*, 1982). These data were reported to be in good agreement with those of other literature studies.

9.2. PHYSICAL PROPERTIES OF LANGMUIR–BLODGETT FILMS

Fourier transform IR (FTIR)-attenuated total reflection (ATR) spectra of LB films of stearic acid deposited on a germanium plate with 1, 2, 3, 5, and 9 monolayers have been measured (Kimura *et al.*, 1986). These spectra are given in Fig. 9.8. In this study, special attention was given to obtaining accurate and unequivocal examinations of changes in structure, subcell packing, and molecular orientation as a function of number of monolayers deposited. The C=O stretching band at $1702 \, \text{cm}^{-1}$ was missing for monolayer 1. The intensity increased linearly for the multilayer samples. A CH_2 scissoring band at $1468 \, \text{cm}^{-1}$ appeared as a singlet in the case of monolayer 1. A doublet at 1473 and $1465 \, \text{cm}^{-1}$ was observed for films containing more than three monolayers. Band progression due to $-CH_2-$ wagging vibration of the *trans*-zigzag hydrocarbon chains is known to

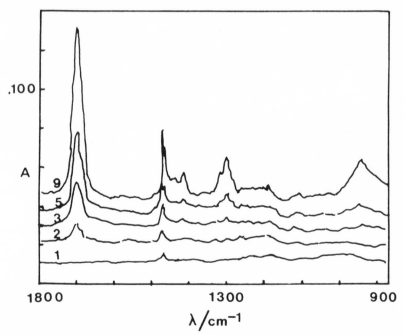

FIGURE 9.8. FTIR–ATR spectra of LB films of stearic acid on germanium. (1, 2, 3, 5, 9) Number of layers. From Kimura *et al.* (1986).

appear between 1400 and 1180 cm^{-1}. The intensities increased in this region with the number of layers.

Monolayers of a surface-active dye (1-methyl-1'-octyadecyl-2,2'-cyanine perchlorate, S-120) incorporated into inert matrix material (e.g., methyl arachidate + arachidic acid) were transferred onto structurally defined silver films by the LB method (Cotton et al., 1986). The dye-containing monolayers were spaced from the surface by accurately known increments by deposition of inert spacer monolayers. Surface-enhanced resonance Raman spectra were observed from dye molecules spaced as distant as six spacer increments ($\approx$16 nm = 1.6 Å) from the silver surface. These studies suggested that an electromagnetic mechanism is operative in this assembly, in contradistinction to a chemical mechanism, which would require direct contact between the Raman-active species and the metal surface. These studies are of relevance in the study of chromophoric species in biological membranes (e.g., enzymes, redox proteins, and chlorophylls).

To compare the monolayer properties at the air–water interface and the spectral characteristics in the visible range of purple, membrane films were transferred to glass slides (Hwang et al., 1977). Purple membrane monolayers transferred to the glass support only on withdrawal across the interface. The slides emerged wet and needed to be dried thoroughly before being dipped again (after about 20 min). If the wet slide is dipped again without being dried, the adsorbed film will float off the slide as it is being dipped. This behavior is general and has been observed for different proteins (e.g., hemoglobin, bovine serum albumin ovalbumin) and lipids (e.g., lecithin, cholesterol) (Birdi, unpublished). These LB films were prepared with up to 160 layers. The visible spectrum of 80 layers of bacterio-rhodopsin/soya–phosphatidylcholine showed a single absorption band at λ_{max} at 570 nm and is identical to the spectrum of an aqueous suspension of purple membrane. The absorption at 570 nm was found to be linearly proportional to the number of LB layers. These analyses also indicated that all bacteriorhodopsin molecules at the interface remain spectroscopically intact.

In a recent study (Claesson et al., 1986), the LB monolayers of dimethyl-dioctadecylammonium ions on molecularly smooth muscovite mica surfaces were investigated. Direct measurements of the interaction between such surfaces were carried out by using the surface force apparatus. A long-range attractive force considerably stronger than the expected van der Waals force was measured. Studies on the electrolyte dependence of this force indicate that it does not have an electrostatic origin, but that the water molecules were involved in the force.

The electrical potential across an LB film of dioleoyl lecithin deposited onto a fine-pore membrane, imposed between equimolar aqueous solutions of NaCl and KCl, was reported to exhibit rhythmic and sustained pulsing or oscillations of electrical potential between the two solutions (Ishi *et al.*, 1986). These oscillations were attributed to the change of permeability of Na^+ and K^+ ions across the membrane, which originated from the phase transition of the lecithin.

LB monolayers of cyanine dye have been studied by various investigators (Khun *et al.*, 1972). A new method was proposed (Hada *et al.*, 1985) for preparing and removing the *J* aggregates of some cyanine dyes that do not have long alkyl chains. The cyanine dye and arachidic acid were mixed in a solution, and the film was deposited on a quartz plate by the LB method. The dye molecules were found to be contained in the film, and some films showed a remarkable *J* band and more resonance fluorescence in the longer wavelength region than in the corresponding monomer band.

Catalysis and LB. Catalysis by electrodeposited platinum in an LB film on a glassy carbon electrode was investigated (Fujihara and Poosittisak, 1986).

9.3. APPLICATION OF LANGMUIR–BLODGETT FILMS IN INDUSTRY

Thin films of phthalocyanine compounds, in general, and those prepared by the LB method, in particular, display novel electrical properties (Baker, 1985). The LB technique for depositing mono- and multilayer coatings with well-controlled thickness and morphology offers excellent compatability with microelectronic technology. Such films have recently been reviewed for their potential applications (Roberts, 1983). The combination of LB supramolecular films with small dimensionally comparable microelectronic substrates affords new opportunities for generation of fundamental chemical property information and evaluation of new organic thin-film semiconductors as microelectronic components and devices (Snow, 1986).

10

DIVERSE APPLICATIONS OF MONOMOLECULAR FILMS

10.1. INTRODUCTION

In Chapters 1-9, experimental procedures and theoretical analyses of the monomolecular films of amphiphiles and biopolymers have been delineated. In this final chapter, the application of such studies to industrial and biological systems is outlined. The data on such applied examples in the current literature are so vast that it would be impossible to cover the whole area in the space available here. Therefore, this chapter discusses only some typical but important examples, which should allow any future investigator to devise further related monolayer studies as needed.

The most significant property of such films is that they are made up of an *organized, one-molecule*-thick layer of molecules. Various organized structures and processes involve monolayers:

- Foams (soap bubbles).

- Vesicles.

- Bilayers [bilipid membranes (BLMs)].

- Adsorbed amphiphiles on solids (Langmuir–Blodgett films) catalysis, enzyme activity on solids, synthetic polymers, and metal implants (thrombosis and blood compatibility).

- Emulsions and microemulsions (oil–water).

This means that the information one obtains by studying such films is directly related to the molecular size, shape, and orientation in the interfaces of these various systems. Some of these investigations can thus provide more understanding of related systems, such as monomolecular films on solids and catalysis. The monolayer technique is the only procedure that can provide direct information about the interfacial forces.

10.2. FOAMS AND EMULSIONS

10.2.1. FOAMS

As is known, if one blows air bubbles in pure water, no foam is formed. On the other hand, if a detergent or protein (amphiphile) is present in the system, adsorbed surfactant molecules at the interface give rise to foam or soap bubble formation. Foam can be characterized as a coarse dispersion of a gas in a liquid, in which gas is the major phase volume. The foam, or the lamina of liquid, will tend to contract because of its surface tension, and a low surface tension would thus be expected to be a necessary requirement for a good foam-forming property. Furthermore, to be stable, the lamina should be able to maintain slight differences of tension in its different regions. It is therefore also clear that a pure liquid, which has uniform surface tension, cannot meet this requirement. The stability of such foams or bubbles has been shown to be related to the structure and stability of the monomolecular film. For instance, the stability of a foam has been shown to be related to its surface elasticity or surface viscosity, η_s, in addition to other interfacial forces. Foam destabilization has also been found to be related to the packing and orientation of molecules in mixed films, which can be determined from monolayer studies.

It is also worth mentioning that foam formation from monolayers of amphiphiles constitutes the most fundamental process in everyday life. Other assemblies, such as vesicles and BLMs, are somewhat more complicated systems, which are also found to be in equilibrium with monolayers (see Fig. 1.3).

Although the surface potential, $\Delta\psi$, i.e., the electrical potential due to the charge on a monolayer, will clearly affect the actual pressure required to thin the lamella to any given thickness, let us assume for the purpose of a simple illustration that $1/k$, the mean Debye–Hückel thickness of the ionic double layer, will influence the ultimate thickness when the liquid film is under a relatively low pressure. Let us also assume that each ionic atmosphere extends a distance of only $3/k$ into the liquid when the film is under a relatively low excess pressure from the gas in the bubbles; this value corresponds to a repulsion potential of only a few millivolts. Thus, at about 1 atm pressure

$$h_\infty = 6/k + 2(\text{monolayer thickness}) \tag{10.1}$$

For charged monolayers adsorbed from 10^{-3} N sodium oleate, the final total thickness, h_∞, of the aqueous layer should thus be of the order of 600 Å (i.e., $6/k$ or $18/\sqrt{c}$ Å). To this value, one needs to add 60 Å ($=60 \times 10^{-10}$ m) for the two films of oriented soap molecules, giving a total of

660 Å. The experimental value is 700 Å. The thickness decreases on the addition of electrolytes, as also suggested by equation (10.1). For instance, the value of h_∞ is 120 Å in the case of 0.1 M NaCl. Addition of a small amount of any of certain nonionic surface-active agents (e.g., n-lauryl alcohol, n-decyl glycerol ether, laurylethanolamide, laurylsufanoylamide) to anionic detergent solutions has been found to stabilize the foam (Schick and Fowkes, 1967). It was suggested that the mode of packing is analogous to the palisade layers of micelles and the surface layers of foam lamellae.

Measurements have been carried out on the excess tensions, equilibrium thicknesses, and compositions of aqueous foam films that were stabilized by either n-decyl methyl sulfoxide or n-decyl trimethyl ammonium-decyl sulfate and contained inorganic electrolytes (Ingram, 1972).

It was recognized very early (Marangoni, 1871; Gibbs, 1828) that the stability of a liquid film must be greatest when the surface pressure strongly resists deforming forces. In the case that the area of the film is extended by a shock, the change in surface pressure, π, is given as

$$\pi = -(d\pi/dA)(A_2 - A_1) \qquad (10.2)$$

where A_1 and A_2 are the available areas per molecule of foam-stabilizing agent in, respectively, the original and the extended parts of the surface. This can be written as

$$\pi = -A_1(d\pi/dA)(A_2/A_1 - 1) \qquad (10.3)$$

$$= -A_1(d\pi/dA)(j - 1) \qquad (10.4)$$

$$= C_s^{-1}(j - 1) \qquad (10.5)$$

where j is the area extension factor. The term $-A_1(d\pi/dA)$ is the surface compressibility modulus of the monolayer [see equation (2.45)]. For a large restoring pressure $\Delta\pi$, this modulus should be large. The local reduction of the surface pressure in the extended region to $\pi - \Delta\pi$ results in a spreading of molecules from the adjacent parts of the monolayer to the extended region.

It has been shown (Friberg, 1976) that there exists a correlation between foam stability and the elasticity (E) of the film, i.e., the monolayer. For E to be large, surface excess must be large. Maximum foam stability has been reported in systems with fatty acid and alcohol concentrations that are well below the minimum in γ. Similar conclusions have been observed with n-$C_{12}H_{25}SO_4Na$ (sodium dodecyl sulfate (SDS)]-n-$C_{12}H_{25}OH$ systems, which give a minimum in γ vs. concentration with maximum foam at the minimum point (Chattoraj and Birdi, 1984) [as a result of mixed monolayer formation (see Chapter 4) and liquid crystalline structures].

Foam drainage, surface viscosity, and bubble size distributions have been reported for different systems consisting of detergents and proteins (Jashnani and Lenlich, 1974). Foam drainage was investigated by using an incident light interference microscope technique (Rao *et al.*, 1982). The foaming of protein solutions is of theoretical interest and also has wide application in the food industry (Friberg, 1976) and in fire fighting practice. Furthermore, in the fermentation industry, where foaming is undesirable, the foam is generally caused by proteins. Since mechanical defoaming is expensive owing to the high power required, antifoam agents are generally used, although antifoam agents are not desirable in some of these systems, e.g., in food products. Further, antifoam agents deteriorate gas dispersion owing to increased coalescence of the bubbles. It has long been known that foams are stabilized by proteins and that they are dependent on pH and electrolyte (Bikerman, 1973; Peter and Bell, 1930; Thuman *et al.*, 1949; Cumper, 1953).

High foaming capacity is explained by the stability of the gas–liquid interface due to the denaturation of proteins, especially due to their strong adsorption at the interface, which gives rise to the stable monomolecular films at the interface. Foam stability is provided by film cohesion and elasticity. In a recent study (Kalischewski and Schogrerl, 1979), the degree of foaming of bovine serum albumin aqueous solution was investigated, as were the effects of electrolytes and alcohol. A good correlation was found between adsorption kinetics (Chapter 6) and foaming properties.

The effects of partial denaturation on the surface properties of ovalbumin and lysozyme have been reported (Kato *et al.*, 1981). Most protein molecules exhibit increased hydrophobicity at the interface as denaturation proceeds, owing to exposure of their buried hydrophobic residues (in the native state) to the outer surface. The hydrophobicity of proteins (as described in Chapter 5) has been found to give fairly good correlation with emulsion stability in food proteins. The surface tension of these proteins decreased greatly as denaturation proceeded. The emulsifying and foaming properties of proteins were remarkably improved by heat denaturation without coagulation. The emulsifying properties increased and were found to exhibit correlation with surface hydrophobicity. The protein foaming properties increased with denaturation. The foaming power and foam stability of SDS–ovalbumin complexes did not improve as much as with heat-denatured protein. The surface hydrophobicity showed an increase. It is thus safe to conclude that heat- and detergent-denatured proteins are unfolded by different mechanisms. These studies are in accord with studies in progress on comparing urea or SDS unfolding by fluorescence (Birdi, unpublished).

10.2.2. SURFACE VISCOSITY AND FOAM FORMATION OF BEER

Surface and bulk viscosities not only reduce the draining rate of the lamella, but also help in restoration from mechanical, thermal, or chemical shocks. The highest foam stability is associated with appreciable surface viscosity, η_s, and yield value. The overfoaming characteristic of beer (gushing) has been the subject of many investigations (Beattie, 1951, 1958; Brenner, 1957; Carrington et al., 1972; Gjertson, 1967; Thorne, 1964; Gardner, 1972, 1973). The relationship between η_s and gushing was reported by several investigators (Gardner, 1972, 1973). Various factors in the gushing process that could lead to protein denaturation were described: pH, temperature, and metal ions. The stability of a gas (i.e., CO_2) bubble in a solution depends on its dimensions. A bubble with a radius greater than a certain critical magnitude will continue to expand indefinitely, and degassing of the solution will take place. Bubbles with a radius equal to the critical value would be in equilibrium, while bubbles with a radius less than the critical value would be able to redissolve in the bulk liquid. The magnitude of the critical radius, R_c, varies with the degree of saturation of the liquid; i.e., the higher the level of supersaturation, the smaller the R_c.

The work required for the formation of a bubble of radius R_c is given by (La Mer, 1952)

$$W = (16 \times 3.14\gamma^3)/[3(p' - p'')^2] \tag{10.6}$$

where γ is the surface tension of beer, p' is the pressure inside the bubble, and p'' is the pressure in the bulk liquid. It has been suggested that there is nothing unusual in the stability of beer, and although carbon dioxide is far from an ideal gas, empirical work supports this conclusion.

A possible connection between η_s and gushing has been reported. Nickel ion, a potent inducer of gushing, has been reported to give rise to a large increase in the η_s of beer. Other additives besides Ni that cause gushing, such as Fe or humulinic acid, have also been reported to give a large increase in η_s. On the other hand, additives that are reported to inhibit gushing, such as EDTA, have been reported to decrease the η_s of beer. This relationship between η_s and gushing suggested that an efficient gushing inhibitor should be very surface-active (to be competitive with gushing promoters), but incapable of forming rigid surface layers (i.e., layers with high η_s). Unsaturated fatty acids, such as linoleic acid, are potent gushing inhibitors, since they destabilize the surface films. Surface viscosity, η_s (g/sec), was investigated by the oscillating-disk method (Gardner, 1972). It was found that beer surfaces with low η_s (0.03–0.08 g/sec) gave nongushing behavior. Beers with high η_s (2.3–9.0 g/sec) gave gushing.

10.2.3. EMULSIONS

Analogous to colloidal systems, emulsions can be formed by methods of either dispersion or condensation. The principle is that shear energy is added to a mixture of oil and aqueous phases, breaking the two phases into tiny droplets. To reduce the amount of mechanical work required for emulsification, the oil–water interfacial tension, γ_{ow}, must be lowered by adding surface-active agents.

The rheological properties of emulsions, such as high viscosity, give unique behavior. It is known that yield stress and apparent viscosity increase with increasing volume fraction of the dispersed phase (Bikerman, 1973; Lissant, 1966; Nixon and Beerbower, 1969). Flow behavior has been found to be dependent on the solid–liquid interface (Mannheimer, 1972). It is also well established that both foams and emulsions can be destroyed under very high shear.

Emulsifying agents have been classified numerically on the hydrophile-lipophile balance (HLB) scale. Table 10.1 lists the HLB values found experimentally for different oil–water emulsion systems. HLB values were initially reported as empirical values assigned to different emulsifiers. Later, it was found that HLB values could be estimated from the chemical formula, using empirically determined group numbers (Davies, 1957), by the following relationship:

$$\text{HLB} = \sum (\text{hydrophilic group number})$$

$$- m(\text{group number per } -CH_2- \text{ group}) + 7 \qquad (10.7)$$

$$= 7 + 0.36 \ln(C_w/C_o) \qquad (10.8)$$

HLB was further shown to be related to the distribution constant of the amphiphile ($K_d = C_w/C_o$, where C_w and C_o are the concentrations of the

TABLE 10.1. Classification of Emulsifiers according to HLB values[a]

Range of HLB values	Applications
3.5–6	Water–oil emulsifiers
7–9	Wetting agents
8–18	Oil–water emulsifiers
13–15	Detergents
15–18	Solubilizing agents

[a] From Davies and Rideal (1963).

TABLE 10.2. HLB Group Numbers

Group	Group number
Hydrophilic groups	
$-SO_4Na$	38.7
$-COOH$	21.1
$-COONa$	19.1
Sulfonate	≈ 11
Ester	6.8–2.4
$-COOH$	2.1
Hyroxyl (free)	1.9
$-O-$	1.3
Hydroxyl (ring)	0.5
Lipophilic groups	
$-CH-$	0.475
$-CH_2-$	—
$-CH_3$	—
$=CH-$	—
Derived group	
$-(CH_2-CH_2-O)-$	0.33

emulsifier in the two phases, water and oil, respectively) (Davies and Rideal, 1963).

The HLB system is not concerned with the stability of the emulsion once it is formed, but is merely a correlation term. The contributions of various chemical groups are given in Table 10.2. It can be easily seen that since HLB is related to the number of $-CH_2-$ groups, it will be related to any other quantity that is also determined by the $-CH_2-$ groups. Thus, one finds that the distribution of alcohol between the oil and water phases is determined by the free energy of transfer from water to oil. The free

TABLE 10.3. HLB Values of Some Typical Food Emulsifiers[a]

Emulsifier	HLB
Sorbitan tristearate	2.1
Sorbitan monostearate	5.7
Sorbitan monopalmitate	6.6
Sorbitan monolaurate	8.5
Glycerol monostearate	3.7
Diglycerol monostearate	5.5
Tetraglycerol monostearate	9.1
Sodium oleate	18.1

[a] From Friberg (1976).

energy changes linearly with $-CH_2-$ groups by 800 cal/mole per $-CH_2-$ (Chapter 5). Hence, one finds a correlation between free energy and HLB. Table 10.3 lists some typical values for emulsifiers used in the food industry (Friberg, 1976).

10.2.4. FOAMS AND IMPROVED OIL RECOVERY

The application of surfactants or polymers or both in improved oil recovery has been reported by various investigators. The coalescence behavior of crude oil drops at an oil-water interface and the interdroplet coalescence in crude oil-water emulsions containing petroleum sulfonates and cosurfactant as surfactant systems with other additives were described in terms of interfacial surface viscosity, η_s, interfacial tension, interfacial charge, ΔV, and thickness of the films (Wasan et al., 1978; Shah et al., 1978). A qualitative correlation was reported between interfacial surface viscosity, η_s, and the coalescence rates. No direct correlation with interfacial tension was found.

The flow of foam through porous media in improved oil recovery processes has been investigated (Rancell and Ziritt, 1983; Rosmalen et al., 1985; Ziritt et al., 1985; Robin, 1985).

10.3. POLYMERIZATION IN MONOLAYERS

It was suggested half a century ago (Carothers, 1936) that the biological synthesis of proteins from amino acids during evolution might have required the orientation of the polypeptide in some interfacial systems in order to prevent the formation of cyclic compounds, which are known to be the usual result of polymerization in homogeneous solutions. This evolutionary aspect of amino acid polymerization needs to be further studied at this stage. The oxidized form of maleic anhydride-β-eliostearin adduct was found to polymerize rapidly in a monolayer under given conditions (Gee, 1935).

In another study, monolayers of stearyl aldehyde have been shown to react with certain polyfunctional amines in the subphase, and measurements of surface viscosity, η_s, suggested that polymers were formed (Bresler et al., 1941). With tetraaminobiphenyl additions, the monolayers became brittle, which indicated cross-linking in the two-dimensional polymer.

Vinyl isobutylether, which forms an unstable monolayer by itself, was found to be polymerized on the introduction of BF_3 gas or on cospreading with benzoyl peroxide (Scheibe and Schuller, 1955). The polymerization of vinyl stearate monolayers following the injection of $NaHSO_3$ into the surface by ultraviolet radiation has also been investigated (Friedlander, 1962).

Polymerization studies have recently been carried out in conjunction with formation of stable Langmuir–Blodgett films (Chapter 9).

10.4. DIVERSE OTHER SYSTEMS

10.4.1. ADSORPTION OF PROTEINS AND POLYELECTROLYTES ON SOLIDS

Protein molecules adsorb readily to solid surfaces from the bulk phase, analogously to other macromolecules, although it can be seen that not all the molecules can adsorb to the solid. The fraction bound varies from 0.2 to 0.4 (Theies, 1966; Peyser et al., 1967; Killman and Eckart, 1971; Herd et al., 1971; Fleer et al., 1972; Fox et al., 1974). The adsorption mechanisms of proteins on solids have been described by various investigators (James and Augenstein, 1966; Brash and Lyman, 1971; Miller and Bach, 1973; Norde, 1976).

Extensive studies have been conducted during the past decade on the concept of rendering biomaterials nonthrombogenic (Salzman, 1971; Forbes and Prentice, 1978). It has been suggested that relationships exist between the blood compatability of foreign substrates and substrate characteristics such as surface charge, hydrophilicity, surface free energy, and surface protein interactions (Lyman et al., 1968). The most important adsorption on solids that remains to be described at the molecular level is that encountered in blood-contacting devices, which are not as biologically compatible with blood as blood is with the endothelial lining of the vessels. As is well known, when blood contacts medical devices or foreign surfaces (e.g., glass, plastics, metal), thrombosis immediately follows. An in vitro study has shown several important aspects relating to localized heparin delivery at a foreign surface–blood interface (Winterton et al., 1986).

The adsorption of albumin, γ-globulins, and fibrinogen onto the poly(L-lactide) on the surface of microcapsules and the zeta-potential of protein-adsorbed microcapsules were measured as a function of protein concentration at various ionic strengths (Makino et al., 1987). From these investigations, it was concluded that the driving force for the adsorption of proteins onto poly(L-lactide) microcapsules is probably not of an electrostatic nature, but may be due mostly to hydrophobic and/or hydrogen bonding. A simple model was proposed in an attempt to explain the dependence of the zeta-potential on the protein concentration and ionic strength. It was also suggested that the data indicate that adsorbed fibrinogen molecules form a charge layer on the microcapsule in such a way that the outer portion of the surface layer (closer to the surrounding solution) is rich in positively charged groups, while the inner one is rich in negatively charged groups.

BLOOD PROTEIN-SOLID ADSORPTION

Blood protein adsorption at solid–solution interfaces is known to be a factor in determining the thromboresistance of synthetic materials introduced into the cardiovascular system. It has been established that the first event that takes place on such implantation is the adsorption of a layer of proteinaceous material at the solid–blood interface, probably because of van der Waals forces (Vroman, 1969; Baier and Dutton, 1969). Thrombogenesis takes place subsequent to the formation of this initially adsorbed protein layer. Some materials, such as isotropic pyrolytic carbon, have exhibited thromboresistance, while other materials high in the electromotive series, such as metals (i.e., gold or platinum), are generally unacceptable. The thromboresistance of a material is difficult to relate to its physical-chemical properties. It has been shown that there is instantaneous protein film formation on any surface regardless of charge; thus, it is possible that a completely independent mechanism could be responsible for such spontaneously adsorbed films (Baier and Dutton, 1969). It is not clear what the role of the electrochemical properties of the solid–solution interface is, but either electrostatic attraction or an electron-transfer reaction involving the surface and protein film is involved in the thromboresistance of a synthetic biomaterial. The role of the adsorbed film will not be clear until there is fuller understanding of the mechanisms by which the structure and activity of the protein in the film are affected by surface electrolytic processes.

The effects of serum protein preadsorption on short-term thrombosis on polymeric surfaces were investigated using an *ex vivo* shunt model in concert with radiolabeling, immunolabeling, and scanning and transmission electron microscopy (Barber *et al.*, 1979). These investigations showed that the complex response of blood flowing to polymeric surfaces may be influenced by the ability of serum proteins to modify leukocyte as well as platelet activity in thrombosis. The results suggested that modulators of leukocyte function be applied in regulating thrombotic reactions on polymeric surfaces.

In a recent study using infrared (IR)-transparent electrodes consisting of approximately 250 Å of carbon on germanium, and internal-reflection spectroelectrochemistry, transitory coadsorption of an unexpected form of fibrinogen was observed along with adsorption of the native material at potentials more positive than the point of zero charge (Mattson and Jones, 1976). It was shown that two different forms of fibrinogen appeared to adsorb at potentials anodic to (more positive than) the onset potential, at which only one form was expected. Furthermore, the unexpected form was found to disappear when the potential was adjusted to be more negative than the onset potential.

A number of investigations have been made on protein adsorption at solid–liquid interfaces by estimating the free energy of protein adsorption and determining its adsorption mode (Lee and Kim, 1974; Shirahama and Shuzawa, 1985; Schmidt *et al.*, 1983; Dillonman and Miller, 1973; Nyilas *et al.*, 1974; Brash *et al.*, 1974; Uzgiris and Fromageot, 1976; Bagchi and Birnbaum, 1981; Soberquist and Walton, 1980; Chan and Brash, 1981; Chiu *et al.*, 1976; Lyman *et al.*, 1968; Norde and Lyklema, 1978; Dulm *et al.*, 1981; Dulm and Norde, 1983).

THERMODYNAMICS OF ADSORPTION OF PROTEINS ON SOLID SURFACES

The Gibbs free energy of adsorption, ΔG_{ads}, at a given temperature, T, is given as

$$\Delta G_{ads} = \Delta H_{ads} - T\Delta S_{ads} \qquad (10.9)$$

where ΔH_{ads} and ΔS_{ads} are the enthalpy and entropy of adsorption, respectively. When ΔH_{ads} is negative and ΔS_{ads} is positive, adsorption takes place, since ΔG_{ads} has a negative value. Calorimetric measurements of adsorption of proteins have been reported (Norde, 1976; Birdi, unpublished).

10.4.2. IMMOBILIZED BIOPOLYMERS (ENZYMES) ON SOLIDS

There is an extensive literature available on the immobilization of biopolymers (mainly enzymes) on solids. An immobilized enzyme may be defined as any biopolymer molecule that is not freely soluble. At present, it is agreed that there is no universal immobilization technique, nor is there any universal solid support. There are a number of reviews on this subject in the literature (Silman and Katchalski, 1966; Weetall, 1971; Gryszkiewicz, 1971; Zaborsky, 1973; Lartigue and Yaverbaum, 1976).

The immobilization technique has been used because enzymes can retain their biological activity after this process. The versatility of biologically active molecules thus becomes unlimited in that they can be immobilized on organic or inorganic (e.g., glass) supports, the surfaces of vessels, columns, a variety of membranes, and particles. These supports can be used in static or dynamic processes. Since the activity is fixed on a solid support it can easily be separated from the system after the desired interval. This flexibility therefore allows the automation and re-use of the immobilized enzyme system. Some of the major uses of immobilization have been for proteolytic enzymes in the beer industry and the dairy industry, for making cheese and milk, and in the hydrolysis of cornstarch.

Although immobilization can be effected by ordinary adsorption forces, this practice will eventually lead to loss in activity after use. Various procedures have been reported whereby the biopolymer can be covalently coupled to a solid (glass) (Donald and Yaverbaum, 1976). In most cases, controlled-pore glass (CPG) has been used as the solid support. CPG is derived from borosilicate glass, the class of glasses that also includes the glass material used in laboratory equipment. This glass consists of 96% SiO_2 and 4% B_2O_3. The pore size can range from 30 to 3000 Å, with the surface area of the pores ranging from 1 to 100 m^2/g. The CPG surface has two types of acidic sites; a pK_a of 7 is found for most silica surfaces, a pK_a of 5.1 for the more strongly acid sites. IR techniques have been used to determine these binding sites.

10.4.3. ADSORPTION OF POLYMERS ON METAL SURFACES

The adsorption of polymers on metal surfaces has special importance in a variety of industrial uses (e.g., in the food industry in processing milk) and biological uses (e.g., in implants). The interaction mechanisms on metals are therefore of interest. Adsorption processes of interest in the food industry have been reviewed (Sandu and Lundi, 1979).

Estimates of changes in molecular conformation or orientation on adsorption can be made from surface-concentration data and the known hydrodynamic dimension and weight of the polymer. The mass distribution of the adsorbed polymer has been estimated from ellipsometry studies (Mosbach, 1976). The influence of surface potential on the extension and refractive index of fibrinogen layers was investigated (Morrisey et al., 1976). The thickness and refractive index of protein layers adsorbed on chromium surfaces having different energy were measured as a function of time by using dynamic ellipsometry (Cuypers et al., 1977). In the technique of ellipsometry used for these studies, an ellipsometer measures the change in the degree of polarization of light on reflection from a metal surface, which allows measurements of the optical constants of the solid surface to be made (McCrackin, 1969; McCrackin et al., 1963).

The presence of a metal surface causes redistribution of electrical charges due to the electrical field and thus gives rise to an electrochemical double layer at the solid–solution interface (Bockris et al., 1980; Chattoraj and Birdi, 1984). This situation is analogous to the charge distribution near the air–water or the $liquid_1$–$liquid_2$ interface. The differential capacitance of this double layer as a function of applied potential before and after protein adsorption on a solid surface has been investigated (Stoner and Srinivasan, 1970). These studies indicated that a more flattened protein was adsorbed on platinum when the electrostatic attraction between the net negatively charged protein, fibrinogen, and platinum was favored.

In a recent study (Ivarsson et al., 1985), the adsorption of lysozyme and ovalbumin was investigated by ellipsometric and capacitance measurements. Both proteins reached a plateau in the surface concentration, Γ, much faster on platinum than on other metal surfaces. Lysozyme continued

TABLE 10.4. Ellipsometric and Capacitance Data after Final Adsorbing of Lysozyme (LY) and Ovalbumin (OV)[a]

| | Solid surface | | | | | |
| | Platinum | | Titanium dioxide | | Zirconium oxide | |
Property	LY	OV	LY	OV	LY	OV
Γ (mg/m^2)	4	2	4	1	5	1
n	2	2	1.6	1.5	1.5	1.3
d (Å)	11	8	33	13	62	225
$\Delta C/C_0$	-29	-19	-12	0	-4	0
ΔE (mV)	-7	-48	$+108$	-20	-71	-33

[a] From Ivarsson *et al.* (1985).

to adsorb without reaching saturation Γ even after 60 min on titanium and zirconium. Ovalbumin demonstrated the most dependence on the characteristics of the metal surface. The degree of desorption on rinsing was greater for lysozyme than for ovalbumin. This difference can be ascribed to the more polar nature of lysozyme in comparison with ovalbumin (Chapter 5). Various data from these ellipsometric and electrical measurements are given in Table 10.4. The relative decrease in capacitance, $-\Delta C/C_0$, and the change in potential, ΔE, were measured simultaneously with the change in optical parameters. The change in refractive index, n, was also measured. The thickness of the adsorbed protein layer, d (in Å), was also estimated. The differences between the two proteins were related to the difference in charge at the pH studied (at which lysozyme is positive and ovalbumin negative).

10.5. SYNTHETIC POLYMER MONOLAYERS ON ORGANIC LIQUIDS

Monomolecular Film of Poly[methyl-(n-alkyl)siloxanes] on Organic Liquids. Because of the low surface free energies associated with the poly(dimethyl-siloxane) group, and the low solubility of these polymers in many liquids, poly(dimethyl-siloxanes) and their derivatives have been effective as surface-active agents in aqueous as well as nonaqueous systems (Bass, 1959; Schwartz and Reid, 1964; Kanner *et al.*, 1967; Bascom *et al.*, 1969; Gaines, 1969). Poly(dimethyl-siloxanes) of sufficiently high molecular weight have also been reported to give stable, insoluble monomolecular films on water and on several organic liquid substrates (Fox *et al.*, 1947; 1950; Banks, 1954, 1957; Ellison and Zisman, 1956; Trapeznikov *et al.*, 1965; Noll, 1966; Noll *et al.*, 1963, 1965, 1970; Jarvis, 1966, 1969).

Experiments with these polymers on a water surface indicate that at large A values, the adsorbed polymer molecules are fully extended and each Si–O–Si group is in contact with water. It has been reported that in monolayers of poly(dimethylsiloxane), the molecules assume on compression an α-helical configuration with the polar Si–O–Si directed primarily toward the axis of the α-helix (Fox *et al.*, 1947, 1950; Noll *et al.*, 1963, 1965, 1970; Noll, 1966). The proposed helical configuration of poly(dimethylsiloxane) polymer may not be unique to the interface, for several investigators have suggested that these polymer molecules in bulk are also in a helical coil configuration.

REFERENCES

Abraham, B. M., and Ketterson, J. B. (1985), *Langmuir*, **1**, 708.

Abraham, B. M., Miyano, K., Xu, S. Q., and Ketterson, J. B. (1982), *Phys. Rev. Lett.*, **49**, 1643.

Abraham, B. M., Miyano, K., Xu, S. Q., and Ketterson, J. B. (1983a), *Rev. Sci. Instrum.*, **54**, 213.

Abraham, B. M., Miyano, K., Ketterson, J. B., and Xu, S. Q. (1983b), *Phys. Rev. Lett.*, **51**, 1975.

Abramson, M. B., Katzman, R., Wilson, C. E., and Gregor, H. P. (1964), *J. Biol. Chem.*, **239**, 4066.

Adam, G. (1970), in *Physiological Principles of Biological Membranes* (Snell, F., ed.), Gordon and Breach, New York.

Adam, G., and Delbruck, M. (1968), in *Structural Chemistry and Molecular Biology* (Rich, A., and Davidson, N., eds.), Freeman, San Francisco.

Adam, N. K. (1923), *Proc. R. Soc. London Ser. A*, **103**, 676.

Adam, N. K. (1928), *Proc. R. Soc. London Ser. A*, **119**, 628.

Adam, N. K. (1933), *Trans. Faraday Soc.*, **29**, 90.

Adam, N. K. (1941), *The Physics and Chemistry of Surfaces*, 3rd ed., Oxford University Press, London.

Adam, N. K. (1968), *The Physics and Chemistry of Surfaces*, Dover, New York.

Adam, N. K., and Harding, J. B. (1932), *Proc. R. Soc. London Ser. A*, **138**, 411.

Adam, N. K., Askew, F. A., and Danielli, J. F. (1935), *Biochem. J.*, **29**, 1786.

Adams, D. J., Evans, M. T. A., Mitchell, J. R., Phillips, M. C., and Rees, P. M. (1971), *J. Poly. Sci.*, **C34**, 167.

Adams, F. H., Desilets, D. T., and Towers, B. (1967), *Respir. Physiol.*, **2**, 302.

Adamson, A. W. (1982), *Physical Chemistry of Surfaces*, 4th ed., John Wiley, New York.

Aellison, A. H. (1962), *J. Phys. Chem.*, **66**, 1867.

Agarwal, V. K. (1974a), *Thin Solid Films*, **22**, 367.

Agarwal, V. K. (1974b), International Atomic Energy Agency, Internal Report, IC/74/33, Trieste, Italy.

Agarwal, V. K. (1979), Annu. Rep. Conf. Electr. Insul. Dielectr. Phenom., Philadelphia.

Agnihotri, V. G., and Giles, C. H. (1972), *J. Chem. Soc., Perkin Vol. II*, 2241.

Ahkong, Q. F., Fisher, D., Tampion, W., and Lucy, J. A. (1973), *Biochem. J.*, **136**, 147.

Ahmid, J., and Hansen, R. S. (1972), *J. Colloid Interface Sci.*, **38**, 601.

Albon, N. (1983), *J. Chem. Phys.*, **78**, 4676.

Albrecht, O., and Sackmann, E. (1980), *J. Phys. Sci. Instrum.*, **13**, 512.

Alexander, A. E. (1949), *Suppl. Surface Chem.*, **56**, 123.

Alexander, A. E. (1963), *J. Colloid Sci.*, **18**, 458.

Alexander, A., and Goodrich, F. C. (1964), *J. Colloid Interface Sci.*, **19**, 468.

Alexander, A. E., and Schulman, J. H. (1937), *Proc. R. Soc. London Ser. A*, **161**, 115.

Alexander A. E., and Teorell, T. (1939), *Trans. Faraday Soc.*, **35**, 727.

Allan, A. J., and Alexander, A. E. (1954), *Trans. Faraday Soc.*, **50**, 863.

293

Amidon, G. L., Yalkowsky, S. H., and Leung, S. (1974), *J. Pharm. Sci.*, **63**, 1858.

Amidon, G. L., Yalkowsky, S. H., Anik, S. T., and Valvani, S. C. (1975), *J. Phys. Chem.*, **79**, 2239.

Andersen, O. S., and Fuchs, M. (1975), *Biophys. J.*, **15**, 795.

Anderson, P. A., and Evett, A. A. (1952), *Rev. Sci. Instrum.*, **23**, 485.

Andrade, J. D., ed. (1985), *Biomedical Polymers*, Vol. 2, Plenum Press, New York.

Anik, S. T. (1978), *J. Phys. Chem.*, **80**, 2220.

Antonow, G. (1907), *J. Chim. Phys.*, **5**, 372.

Antonow, G. (1932), *Kolloid-Z.*, **59**, 7.

Archer, R. J., and La Mer, V. K. (1954), *Ann. NY Acad. Sci.*, **58**, 807.

Archer, R. J., and La Mer, V. K. (1955), *J. Phys. Chem.*, **59**, 200.

Archer, R. J., and Shank, D. (1967), *J. Colloid Interface Sci.*, **25**, 53.

Asakura, T., Ohnishi, T., Friedman, S., and Schwartz, E. (1974), *Proc. Natl. Acad. Sci. U.S.A.*, **71**, 1594.

Askew, F. A., and Danielli, J. F. (1940), *Trans. Faraday Soc.*, **36**, 785.

Avery, M. E., and Mead, J. (1959), *Am. J. Dis. Child.*, **97**, 517.

Bagchi, P., and Birnbaum, S. M. (1981), *J. Colloid Interface Sci.*, **83**, 460.

Baglioni, P., Dei, L., Ferroni, E., and Gabrielli, G. (1986), *J. Colloid Interface Sci.*, **109**, 109.

Baier, R. E., and Dutton, R. C. (1969), *J. Biomed. Mater. Res.*, **3**, 191.

Baines, G. T., Quickenden, T. I., and Seylor, J. E. (1970), *J. Colloid Interface Sci.*, **33**, 236.

Baker, S. (1983), in *Proc. Int. Symp. Future Electronic Devices, Bioelectronic and Molecular Electronic Devices*, Tokyo, November 1985: see Proceedings of the First International Conference on Langmuir–Blodgett Films, in *Thin Solid Films*, **99** (1983).

Bakker, G. (1914), *Z. Phys. Chem.*, **86**, 129.

Bakker, G. (1928), *Handbuch der Experimental Physik*, Vol. VI, Akademie Verlagsgesellschaft, Leipzig.

Bangham, A. D., and Dawson, R. M. C. (1962), *Biochim. Biophys. Acta*, **59**, 103.

Bangham, A. D., Dingle, J. T., and Lucy, J. A. (1964), *Biochem. J.*, **90**, 133.

Bangham, A., Morley, C., and Phillips, M. (1979), *Biochim. Biophys. Acta*, **573**, 552.

Banks, W. H. (1954), *Nature*, **174**, 365.

Banks, W. H. (1957), in *Proceedings of the Second International Congress on Surface Active Agents*, Vol. 1, Academic Press, New York.

Barber, T. A., Lambrecht, L. K., Mosher, D. L., and Cooper, S. L. (1979), *Scanning Electron Microsc.*, III, 881.

Baret, J. F. (1969), *J. Colloid Interface Sci.*, **30**, 1.

Baret, J. F., Bois, G., Casalta, L., Dupin, J. J., Firpo, J. L., Gonella, J., Melinon, J. P., and Rodeau, J. L. (1975), *J. Colloid Interface Sci.*, **53**, 50.

Barnes, G. T., Elliot, A. J., and Grigg, E. C. M. (1968), *J. Colloid Interface Sci.*, **26**, 230.

Barrand, A., Rosilio, C., and Ruandel-Teixier, A. (1979a), *Solid State Technol.*, **22**, 120.

Barrand, A., Rosilio, C., and Ruandel-Teixier, A. (1979b), *J. Colloid Interface Sci.*, **62**, 307.

Barrand, A., Rosilio, C., and Ruandel-Teixier, A. (1980), *Thin Solid Films*, **68**, 99.

Bascom, W. D., Halper, L. A., and Jarvis, N. L. (1969), *Ind. Eng. Chem. Prod. Res. Dev.*, **8**, 118.

Bass, R. L. (1959), *Chem. Ind.*, 912.

Beattie, G. B. (1951), *Wallerstein Labs. Commun.*, **14**, 81.

Beattie, G. B. (1958), *Bottling*, **135**, 3.

Behr, B., Bialowolska, M., and Chodkowski, J. (1973), *J. Electroanal. Chem.*, **46**, 223.

Bell, G. M., Mingins, J., and Taylor, J. A. G. (1978), *Trans. Faraday Soc.*, **74**, 223.

Bell, K. H. (1979), *Biophys. Biochem. Res. Commun.*, **66**, 199.

Bellamy, L. J. (1975), *The Infrared Spectra of Complex Molecules*, Chapman and Hall, London.

Benard, H. (1900), *Rev. Gen. Sci. Pures Appl. Bull. Assoc. France, Avan. Sci.*, **11**, 1261.

Benjamins, J., De Feijter, J. A., Evans, M. T. A., Graham, D. E., and Phillips, M. C. (1975), *Discuss. Faraday Soc.*, **59**, 218.

Ben-Naim, A. (1980), *Hydrophobic Interactions*, Plenum Press, New York.
Benson, G. C., and McIntosh, R. L. (1948), *J. Colloid Sci.*, **3**, 323.
Berg, H., Granath, K., and Nygard, B. (1972), *Electroanal. Chem.*, **36**, 167.
Berry, M. N. (1981), *FEBS Lett.*, **134**, 134.
Berry, R. S., Rice, S. A., and Ross, J. (1980), *Physical Chemistry*, John Wiley, New York.
Bertrand, G. L. (1982), *J. Colloid Interface Sci.*, **88**, 512.
Betso, S., Kalpper, M., and Anderson, L. (1972), *J. Am. Chem. Soc.*, **94**, 8197.
Bewig, K. (1958), U.S. Naval Res. Lab. Report, 5096.
Bewig, K. W. (1964), *Rev. Sci. Instrum.*, **35**, 1160.
Bienkowski, R., and Skolnick, M. (1972), *J. Colloid Interface Sci.*, **39**, 323.
Bikerman, J. J. (1973), *Foams*, Springer Verlag, Berlin.
Birdi, K. S. (1972), *Kolloid Z. Z. Polym.*, **250**, 222.
Birdi, K. S. (1973a), *Proc. Vth Int. Congr. Surface Active Agents*, Zurich, 1972, Carl Hanser Verlag, Munich, p. 257.
Birdi, K. S. (1973b), *J. Colloid Interface Sci.*, **43**, 545.
Birdi, K. S. (1976a), *J. Colloid Interface Sci.*, **57**, 228.
Birdi, K. S. (1976b), in *Micellization, Solubilization, and Microemulsions* (Mittal, K. L., ed.), Plenum Press, New York.
Birdi, K. S. (1977), in *Proc. IVth Winter School on Biophysics of Membrane Transport*, Wisla, Poland.
Birdi, K. S. (1981a), *J. Theor. Biol.*, **93**, 1.
Birdi, K. S. (1981b), in *Proc. VIIth Scand. Symp. Surface Chemistry* (Birdi, K. S., ed.), Holte, Denmark.
Birdi, K. S. (1987), *Langmuir*, **3**, 132.
Birdi, K. S. (1988), in *Aqueous Size-Exclusion Chromatography* (Dubin, P., ed.), Elsevier, Amsterdam.
Birdi, K. S., and Fasman, G. D. (1972), *J. Polym. Sci.*, *A-1*, **10**, 2483.
Birdi, K. S., and Fasman, G. D. (1973), *J. Polym. Sci.*, **42**, 1099.
Birdi, K. S., and Gevod, V. S. (1984), in *VIIth School on Biophysics of Membrane Transport*, School Proceedings, Zakopane, Poland.
Birdi, K. S., and Gevod, V. S. (1987), *Colloid Polym. Sci.*, **265**, 1.
Birdi, K. S., and Nikolov, A. (1979), *J. Phys. Chem.*, **83**, 365.
Birdi, K. S., and Sander, B. (1981), in *Proc. VIIth Scand. Symp. Surface Chemistry* (Birdi, K. S., ed.), Holte, Denmark.
Birdi, K. S., and Schack, P. (1973), *Makromol. Chem.*, **166**, 319.
Birdi, K. S., and Sørensen, K. (1979), *Colloid Polym. Sci.*, **257**, 942.
Birdi, K. S., and Steinhardt, J. (1978), *Biochim. Biophys. Acta*, **534**, 219.
Birdi, K. S., Gabrielli, G., and Puggelli, M. (1972), *Kolloid Z. Z. Polym.*, **250**, 591.
Birdi, K. S., Gevod, V. S., Ksenzhek, O. S., Stenby, E. H., and Rasmussen, K. L. (1983), *Colloid Polym. Sci.*, **261**, 767.
Birdi, K. S., Sanchez, R., and Fredenslund, Aa., to be published.
Blank, M. (1962), in *Retardation of Evaporation by Monolayers* (La Mer, V. K., ed.), Academic Press, New York.
Blank, M. (1964), *J. Phys. Chem.*, **68**, 2793.
Blank, M. (1970), *J. Colloid Interface Sci.*, **30**, 233.
Blank, M., and Britten, J. (1965), *J. Colloid Sci.*, **20**, 780.
Blank, M., Lee, B. B. and Britten, J. S. (1973), *J. Colloid Interface Sci.*, **43**, 539.
Blank, M., and Roughton, F. J. W. (1960), *Trans. Faraday Soc.*, **56**, 1832.
Blodgett, K. B. (1934), *J. Am. Chem. Soc.*, **495**, 1007.
Blodgett, K. B. (1935), *J. Am. Chem. Soc.*, **57**, 1007.
Blodgett, K. B. (1937), *J. Phys. Chem.*, **41**, 975.

Blodgett, K. B. (1939), *Excursions in Science* (Reynolds, N., and Manning, E. L., eds.), McGraw-Hill, New York.

Blodgett, K. B., and Langmuir, I. (1937), *Phys. Rev.*, **51**, 964.

Blume, A. (1979), *Biochim. Biophys. Acta*, **557**, 32.

Bockris, J. O. M., Conway, B. E., and Yeager, E., eds. (1980), *Comprehensive Treatise of Electrochemistry*, Vol. 1: *The Double Layer*, Plenum Press, New York.

Boheim, G., Hanke, W., and Jung, G. (1984), *Chem. Peptides Proteins*, **2**, 281.

Bouhet, C. (1931), *Ann. Phys. (Leipzig)*, **55**, 7.

Boyd, G. E. (1958), *J. Phys. Chem.*, **62**, 537.

Boyd, G. E., and Harkins, W. D. (1939), *J. Am. Chem. Soc.*, **61**, 1188.

Boyd, G. E., and Schubert, J. (1957), *J. Phys. Chem.*, **61**, 1271.

Boyer, P. D., Chance, B., Ernster, L., Mitchell, P., Racker, E., and Slater, E. C. (1977), *Annu. Rev. Biochem.*, **46**, 955.

Brady, A. P. (1949), *J. Colloid Sci.*, **4**, 417.

Bragadin, E. (1972), *Rev. Sci. Instrum.*, **43**, 468.

Brash, J. L., and Lyman, D. J. (1971), in *The Chemistry of Biosurfaces* (Hair, M. L., ed.), Marcel Dekker, New York.

Brash, J. L., Uniyal, S., and Samak, Q. (1974), *Trans. Am. Soc. Artif. Inter. Organs*, **20**, 69.

Brasseur, R., Goormaghtigh, E., and Ruysschaert, J. M. (1981), *Biochem. Biophys. Res. Commun.*, **103**, 301.

Brasseur, R., Deleers, M., Malaisse, W. J., and Ruysschaert, J. M. (1982), *Proc. Natl. Acad. Sci. U.S.A.*, **79**, 2895.

Brasseur, R., Carlier, M. B., Laurent, G., Claes, P. J., van der Haeghe, H. J., Tulkens, P. M., and Ruysschaert, J. M. (1985), *Biochem. Pharm.*, **34**, 1035.

Brdicka, R. (1933), *Collect. Czech. Chem. Commun.*, **5**, 122.

Brenner, M. W. (1957), *Proc. Am. Soc. Brew. Chem.*, **39**, 5.

Bresler, S., Judin, M., and Talmud, D. (1941), *Acta Physicochem., URSS*, **14**, 71.

Brockman, H. L., Kezdy, F. J., and Law, J. H. (1975), *J. Lipid Res.*, **16**, 67.

Brockman, H. L., Law, J. H., and Kezdy, F. J. (1973), *J. Biol. Chem.*, **248**, 4965.

Brody, S. S. (1971), *Z. Naturforsch.*, **26b**, 922.

Brooks, J. H., and Alexander, A. E. (1960), in *Proc. 3rd Int. Congr. Surface Active Agents*, Vol. II, University of Mainz Press, Cologne.

Brooks, J. H., and Alexander, A. E. (1962a), *J. Phys. Chem.*, **66**, 1851.

Brooks, J. H., and Alexander, A. E. (1960), in *Proc. 3rd Int. Congr. Surface Active Agents*, Vol. II, University of Mainz Press, Cologne.

Brooks, J. H., and Alexander, A. E. (1962b), in *Retardation of Evaporation by Monolayers* (La Mer, V., ed.), Academic Press, New York.

Brooks, J. H., and Pethica, B. A. (1964), *Trans. Faraday Soc.*, **60**, 208.

Brown, C. A., Burns, F. C., Knoll, W., and Swalen, J. D. (1983), *J. Phys. Chem.*, **87**, 3616–3619.

Brown, H. D., and Hasselberger, F. X. (1971), in *Chemistry of the Cell Interface* (Brown, H. D., ed.), Academic Press, New York.

Brown, P. K., and Wald, G. (1956), *J. Biol. Chem.*, **222**, 865.

Bruner, L. J. (1975), *J. Membrane Biol.*, **22**, 125.

Budenstein, P. P., and Hayes, P. J. (1967), *J. Appl. Phys.*, **38**, 2837.

Budenstein, P. P., Hayes, P. J., Smith, J. L., and Smith, W. B. (1969), *J. Vac. Sci. Technol.*, **6**, 289.

Buff, F. P. (1960), *Discuss. Faraday Soc.*, **30**, 52.

Buff, F. P., Goel, N. S., and Clay, J. R. (1972), *J. Chem. Phys.*, **56**, 2405.

Buhaenko, M. R., Goodwin, J. W., and Richardson, R. M. (1985), *Thin Solid Films*, **134**, 217.

Bui, A., Carchano, H., and Sanchez, D. (1972), *Thin Solid Films*, **13**, 207.

Bülbring, E., and Tomita, T. (1970), *J. Physiol., London*, **210**, 217.

Bulkin, B. J., and Krishnamachari, N. I. (1970), *Biochim. Biophys. Acta*, **211**, 592.

Bull, H. B. (1945), *J. Am. Chem. Soc.*, **67**, 4, 8.

Bull, H. B. (1947a), *J. Biol. Chem.*, **3**, 95.

Bull, H. B. (1947b), *Advan. Protein Chem.*, **3**, 95.

Bull, H. B. (1972), *J. Colloid Interface Sci.*, **41**, 310.

Bunnerot, A., Chollet, P. A., Frisby, H., and Hoclet, A. (1985), *Chem. Phys.*, **97**, 365.

Burachik, M., Craig, L. C., and Chang, J. (1970), *Biochemistry*, **9**, 3293.

Burger, M. M. (1986), *Proc. Natl. Acad. Sci. U.S.A.*, **83**, 1315.

Burgess, A. W., Pnnuswamy, P. K., and Scheraga, H. A. (1974), *Israel J. Chem.*, **12**, 239.

Burke, L. I., Paril, G. S., Panganamal, R. V., Geer, J. C., and Cornwell, D. G. (1973), *J. Lipid Res.*, **14**, 9.

Byren, D., and Burnshaw, J. C. (1979), *J. Phys. D*, **12**, 1145.

Cadenhead, D. A. (1971), in *Chemistry and Physics of Interfaces* (Ross, S., ed.), American Chemical Society, Washington, D.C.

Cadenhead, D. A., and Mueller-Landau, F. (1980), in *Ordering in Two Dimensions* (Sinha, R., ed.), Elsevier, North-Holland, Amsterdam.

Caille, A., Pink, D., de Vertuil, F., and Zuckermann, M. J. (1980), *Can. J. Phys.*, **58**, 581.

Calvin, M. (1969), *Chemical Evolution*, Oxford Press, London.

Cameron, A., Giles, C. H., and MacEwan, T. H. (1957), *J. Chem. Soc. London*, **62**, 4304.

Cameron, A., Giles, C. H., and MacEwan, T. H. (1958), *J. Chem. Soc. London*, **63**, 1224.

Capaldi, R. A. (1982), *Trends Biochem. Sci.*, **7**, 292.

Carothers, W. H. (1936), *Trans. Faraday Soc.*, **32**, 39.

Carr, S. W. (1953), *Arch. Biochem. Biophys.*, **46**, 424.

Carrington, R., Collett, R. C., Dunkin, R. C., and Halek, G. (1972), *J. Inst. Brew.*, **78**, 243.

Cary, A., and Rideal, E. (1925), *Proc. R. Soc. London Ser. A*, **109**, 318.

Case, G. C., Vanderkooi, J. M., and Scarpa, A. (1974), *Arch. Biochem. Biophys.*, **162**, 174.

Caspers, J., L.-Cauferiez, M., Ferrieira, J., Goorghtigh, E., and Ruysschaert, J. M. (1981), *J. Colloid Interface Sci.*, **81**, 410.

Caswell, A. H., and Pressman, B. C. (1972), *Biochem. Biophys. Res. Commun.*, **49**, 292.

Chan, B. E. C., and Brash, J. L. (1981), *J. Colloid Interface Sci.*, **82**, 217.

Chandler, W. K., Hodgkin, A. L., and Meves, H. (1965), *J. Physiol.*, **180**, 821.

Chapman, D. (1967a), *Trans. Faraday Soc.*, **86**, 3559.

Chapman, D. (1967b), *Biological Membranes*, Academic Press, New York.

Chapman, D., Williams, R. M., and Ladbrooke, B. D. (1967), *Chem. Phys. Lipids*, **1**, 445.

Chatelain, P., Berliner, C., Ruysschaert, J. M., and Jaffe, J. (1975), *J. Colloid Interface Sci.*, **51**, 239.

Chattoraj, D. K., and Birdi, K. S. (1984), *Adsorption and the Gibbs Surface Excess*, Plenum Press, New York.

Chen, C.-H. (1980), *J. Phys. Chem.*, **84**, 2050.

Chen, C.-H. (1981), *J. Phys. Chem.*, **85**, 603.

Chen, C.-H. (1982), *J. Phys. Chem.*, **86**, 3559.

Chen, C.-H., Berns, D. S., and Berns, A. S. (1981), *Biophys. J.*, **36**, 359.

Chen, S. (1976), Ph.D. Thesis, City University of New York.

Chen, S. S., and Rosano, H. L. (1977), *J. Colloid Interface Sci.*, **61**, 207.

Chiu, T-H., Nyilas, E., and Lederman, D. M. (1976), *Trans. Am. Soc. Artif. Inter. Organs*, **22**, 498.

Chollet, P. A., and Messier, J. (1983), *Thin Solid Films*, **99**, 197.

Chou, P., and Fasman, G. D. (1978), *Adv. Enzymol. Relat. Subj. Biochem.*, **47**, 45.

Christensen, H. N. (1975), *Biological Transport*, 2nd ed., W. A. Benjamin, London.

Chu, B. (1967), *Molecular Forces*, Interscience Publishers, New York.

Cini, R., Loglio, G., and Ficalbi, A. (1972a), *J. Colloid Interface Sci.*, **41**, 287.

Cini, R., Loglio, G., and Ficalbi, A. (1972b), *J. Colloid Interface Sci.*, **27**, 287.

Claesson, P. M., Blom, C. E., Herder, P. C., and Ninham, B. W. (1986), *J. Colloid Interface Sci.*, **114**, 234.

Clark, S. G., and Holt, P. F. (1957), *Trans. Faraday Soc.*, **53**, 1509.

Clark, S. G., Holt, P. F., and Went, C. W. (1957), *Trans. Faraday Soc.*, **53**, 1500.

Clements, J. A. (1956), *Am. J. Physiol.*, **187**, 592.

Clements, J. A. (1957), *Proc. Soc. Exptl. Biol. Med.*, **95**, 170.

Cohen, H., Shen, B. W., Snyder, W. R., Law, J. H., and Kezdy, F. J. (1976), *J. Colloid Interface Sci.*, **56**, 240.

Colacicco, G., and Rapport, M. M. (1965), *Fed. Proc.*, **24**, 295.

Colacicco, G., Gordon, E. E., and Berchenco, G. (1968), *Biophys. J.*, **8**, 22a.

Colin, M., and Bangham, A. (1981), *Prog. Respir. Res.*, **15**, 188.

Cordoba, J., and Sicre, P. (1985), *An. R. Acad. Farm.*, **51**, 731.

Cornell, D. G. (1982), *J. Colloid Interface Sci.*, **88**, 36.

Cornell, D. G., and Carroll, R. J. (1985), *J. Colloid Interface Sci.*, **108**, 226.

Corpe, W. A. (1970), in *Adhesion in Biological Systems* (Manly, R. S., ed.), Academic Press, New York.

Cotterill, R. M. J. (1976), *Biochim. Biophys. Acta*, **433**, 264.

Cotton, T. M., Uphaus, R. A., and Mobius, D. (1986), *J. Phys. Chem.*, **90**, 6071.

Crawford, G., and Earnshaw, J. C. (1986), *Biophys. J.*, **49**, 869.

Crisp, D. J. (1946), *Trans. Faraday Soc.*, **42**, 619.

Crisp, D. J. (1949), *Surface Chemistry*, Butterworths, London.

Crone, A. H. M., Snik, A. F. M., Poulis, J. A., and Kruger, A. J. (1980), *J. Colloid Interface Sci.*, **74**, 1.

Cumper, C. W. (1953), *Trans. Faraday Soc.*, **49**, 1360.

Cuypers, P. A., Hermens, W. T., and Hemker, H. C. (1977), *Ann. NY Acad. Sci.*, **283**, 77.

Cuypers, P. A., Hermens, W. T., and Hemker, H. C. (1978), *J. Colloid Interface Sci.*, **65**, 483.

Dalton, J., and Gilberts, D. B. (1803), *Ann. Phys.*, **15**, 121.

Dandliker, W. B., and de Saussure, V. A. (1971), in *The Chemistry of Biosurfaces* (Hair, M. L., ed.), Vol. 1, Marcel Dekker, New York.

Danielli, J. F., and Davson, H. (1935), *J. Cell Physiol.*, **5**, 495.

Davies, J. (1951a), *Proc. R. Soc. London*, **A208**, 224.

Davies, J. T. (1951b), *Z. Electrochem.*, **55**, 559.

Davies, J. T. (1952), *Trans. Faraday Soc.*, **48**, 1052.

Davies, J. T. (1954), *J. Colloid Sci., Suppl.*, **1**, 9.

Davies, J. T. (1956), *J. Colloid Interface Sci.*, **11**, 377.

Davies, J. T. (1957), in *Proc. 2nd Int. Congr. Surface Active Agents*, Vol. 1, Butterworths, London, p. 426.

Davies, J. T., and Llopis, J. (1955), *Proc. R. Soc. London*, **A227**, 537.

Davies, J. T., and Rideal, E. K. (1963), *Interfacial Phenomena*, Academic Press, New York.

Davis, J. (1956), *Catalysis—Investigation of Heterogenous Processes* (Russian translation), Moscow.

Dawson, R. M. C. (1968), in *Biological Membranes* (Chapman, D., ed.), Academic Press, New York.

Dawson, R. M. C., and Hauser, H. (1970), in *Calcium and Cellular Function* (Cuthbert, A. W., ed.), Macmillan, London.

Defay, R. (1934), *Etude thermodynamique de la superficielle*, D.U.N.O.D., Paris.

Defay, R., Prigogine, I., Bellemans, A., and Everett, D. H. (1966), *Surface Tension and Adsorption*, Longmans, Green and Co., London.

Degrado, W. F., Kezdy, F. J., and Kaiser, E. T. (1981), *J. Am. Chem. Soc.*, **103**, 679.

Delahay, P., and Fike, C. T. (1958), *J. Am. Chem. Soc.*, **80**, 2628.

Delahay, P., and Trachtenberg, I. (1957), *J. Am. Chem. Soc.*, **79**, 2355.

Deleage, G., Tinland, B., and Roux, B. (1987), *Anal. Biochem.*, **163**, 292.

Demel, R. A., Geurts van Kessel, W. S. M., Zwaal, R. F. A., Roelofsen, B., and van Deenen, L. L. M. (1975), *Biochim. Biophys. Acta*, **406**, 97.

Deo, A. V., Kulkarni, S. B., Gharpurey, M. K., and Biswas, A. B. (1961), *J. Colloid Interface Sci.*, **19**, 820.

Deo, A. V., Kulkarni, S. B., Gharpurey, M. K., and Biswas, A. B. (1962), *J. Phys. Chem.*, **66**, 1361.

Dervichian, D. G., and Joly, M. (1939), *J. Phys. Radium*, **10**, 375.

Desnuelle, P. (1961), *Advan. Enzym. Related Subj. Biochem.*, **23**, 7.

Deuel, H. J. (1951), *The Lipids*, Vol. 1, Interscience, New York.

Devaux, P., and McConnell, H. M. (1972), *J. Am. Chem. Soc.*, **94**, 4475.

Dickerson, R. E., and Gies, I. (1969), *Structure and Action of Proteins*, Harper and Row, New York.

Dickinson, E. (1978), *J. Colloid Interface Sci.*, **60**, 161.

Dillonman, W. J., Jr., and Miller, I. F. (1973), *J. Colloid Interface Sci.*, **44**, 221.

Dluhy, R. A., and Cornell, D. G. (1985), *J. Phys. Chem.*, **89**, 3195.

Dmaskin, B., Petry, O., and Batrakov, V. (1968), in *Adsorption of Organic Substances at Electrodes*, Nauka, Moscow.

Dörfler, H. D., and Rettig, W. (1980), *Kong. Poverkhn. Akt.*, Veschestvam, 7th, **2**, 570.

Donald, W., and Yaverbaum, C. (1976), *Arch. Biochem. Biophys.*, **182**, 200.

Doniach, S. (1978), *J. Chem. Phys.*, **68**, 4912.

Doty, P., and Schulman, J. H. (1949), *Discuss. Faraday Soc.*, **6**, 21.

Dreher, K. D., and Wilson, J. E. (1970), *J. Colloid Interface Sci.*, **32**, 248.

Drost-Hansen, W. (1971), in *Chemistry of the Cell Interface* (Brown, H. D., ed.), Part B, Academic Press, New York.

Dulm, P. V., and Norde, W. (1983), *J. Colloid Interface Sci.*, **91**, 248.

Dulm, P. V., Norde, W., and Lyklema, J. (1981), *J. Colloid Interface Sci.*, **82**, 77.

Dzyaloshinskii, I. E., Lifshitz, E. M., and Pitaevkii, L. P. (1961), *Adv. Phys.*, **10**, 165.

Eckstein, F. (1939), *Z. Hyg. Zool. Schaed.*, **31**, 237.

Edidin, M. (1974), *Ann. Rev. Biophys. Bioeng.*, **3**, 179.

Ehrenfest, P. (1933), *Proc. K. Ned. Akad. Wet.*, Amsterdam, **36**, 153.

Eisenberg, D., Terwillinger, T. C., and Tsui, F. (1980), *Biophys. J.*, **32**, 252.

Eisenman, G., Ciani, S., and Szabo, G. (1969), *J. Membrane Biol.*, **1**, 294.

Eisenman, G., Szabo, G., Ciani, S., McLaughlin, S., and Krasne, S. (1973), *Prog. Surf. Membrane Sci.*, **6**, 139.

Eley, D. D. (1939), *Trans. Faraday Soc.*, **35**, 1281, 1421.

Eley, D. D., and Hedge, D. G. (1956), *J. Colloid Sci.*, **11**, 445.

Eley, D. D., and Hedge, D. G. (1957), *J. Colloid Sci.*, **12**, 419.

Ellis, S. C., and Pankhurst, K. G. A. (1954), *Discuss. Faraday Soc.*, **16**, 170.

Ellison, A. H., and Zisman, W. A. (1956), *J. Phys. Chem.*, **60**, 416.

Elworthy, P. H., and Mysels, K. Y. (1966), *J. Colloid Interface Sci.*, **21**, 331.

Emanuel, N., ed. (1965), *The Oxidation of Hydrocarbons in the Liquid Phase*, Pergamon Press, New York.

Emerson, M. F., and Holtzer, A. (1967), *J. Phys. Chem.*, **71**, 3321.

Enever, R. P., and Pilpel, N. (1967a), *Res. Tech. Instrum.*, **1**, 7.

Enever, R. P., and Pilpel, N. (1967b), *Trans. Faraday Soc.*, **63**, 781.

Enever, R. P., and Pilpel, N. (1967c), *Trans. Faraday Soc.*, **63**, 1559.

Epstein, C. J., Goldberger, R. F., and Anfinson, C. B. (1963), *Cold Spring Harb. Symp. Quant. Biol.*, **28**, 439.

Epstein, H. T. (1950), *J. Phys. Colloid Chem.*, **54**, 1053.

Eriksson, J. C. (1972), *J. Colloid Interface Sci.*, **41**, 287.

Esposito, S., Semeriva, M., and Desnuelle, P. (1973), *Biochim. Biophys. Acta*, **302**, 293.

Evans, M. T. A., Mitchell, J., Musselwhite, P. R., and Irons, L. (1970), in *Surface Chemistry of Biological Systems* (Blank, M., ed.), Vol. 7, Plenum Press, New York.

Evans, R. W., Williams, M. A., and Tinoco, J. (1980), *Lipids*, **15**, 7.

Fasman, G. D., ed. (1967), *Poly-α-Amino Acids*, Vol. 1, Marcel Dekker, New York.

Felmeister, A., and Schaubman, R. (1969), *J. Pharm. Sci.*, **58**, 64.

Ferrieira, J., Caspers, J., Brasseur, R., and Ruysschaert, J. M. (1981), *J. Colloid Interface Sci.*, **81**, 158.

Fidelio, G. D., Maggio, B., Cumar, F. A., and Caputto, R. (1981), *Biochem. J.*, **193**, 643.

Fidelio, G. D., Maggio, B., and Cumar, F. A. (1982), *Biochem. J.*, **203**, 717.

Fidelio, G. D., Maggio, B., and Cumar, F. A. (1984), *Chem. Phys. Lipids*, **35**, 231.

Finkelstein, A., and Holzer, R. (1973), in *Membranes* (Eisenman, G., ed.), Marcel Dekker, New York.

Finkelstein, A., and Pititsyn, D. B. (1971), *J. Mol. Biol.*, **62**, 613.

Fishman, P. H., and Brady, R. O. (1976), *Science*, **194**, 906.

Flaim, T., Friberg, S. E., and Plummer, P. P. Lim (1981), *J. Biol. Phys.*, **9**, 201.

Fleer, G. J., Lyklema, J., and Koopal, L. K. (1972), *Kolloid. Z. Z. Polym.*, **250**, 689.

Flory, P. J. (1942), *J. Chem. Phys.*, **10**, 51.

Forbes, C. D., and Prentice, C. R. (1978), *Brit. Med. Bull.*, **34**, 201.

Fordham, S. (1954), *Trans. Faraday Soc.*, **50**, 593.

Fourt, L., and Harkins, W. D. (1938), *J. Phys. Chem.*, **42**, 897.

Fowkes, F. M. (1962), *J. Phys. Chem.*, **66**, 1863.

Fowler, R. H., and Guggenheim, E. A. (1949), *Statistical Thermodynamics*, Cambridge University Press, Cambridge.

Fox, H. W., Taylor, P. W., and Zisman, W. A. (1947), *Ind. Eng. Chem.*, **39**, 1401.

Fox, H. W., Solomon, E. M., and Zisman, W. A. (1950), *J. Phys. Coll. Chem.*, **54**, 723.

Fox, U. K., Robb, I. D., and Smith, R. (1974), *Chem. Soc. Faraday Trans. I*, **70**, 1186.

Franklin, B. (1976), in *The Complete Works of Benjamin Franklin* (Bigelow, J., ed.), Putnam and Sons, New York.

Freundlich, H. (1930), *Kapillarchemie*, 3rd ed.

Friberg, S. (1976), *Food Emulsions*, Marcel Dekker, New York.

Friedlander, H. Z. (1962), U.S. Patent 3031721, May, 1; *Chem. Abstr.* 5714008.

Fringeli, U. P., Leutert, P., Thurnhofer, H., Fringeli, M., and Fromherz, P. (1975), *Rev. Sci. Instrum.*, **46**, 1380.

Frisch, H. L., and Simha, R. (1956), *J. Chem. Phys.*, **24**, 652.

Frisch, H. L., and Simha, R. (1957), *J. Chem. Phys.*, **27**, 702.

Fromherz, P. (1971), *Biochim. Biophys. Acta*, **225**, 382.

Frumkin, A. N. (1925), *Z. Phys. Chem.* (*Leipzig*), **116**, 485.

Fujihira, M., and Poosittisak, S. (1986), *J. Electroanal. Chem. Interfacial Electrochem.*, **199**, 481.

Fukuda, K., Nakahara, K., and Kato, T. (1976), *J. Colloid Interface Sci.*, **54**, 430.

Furlong, D. N., and Hartland, S. (1979), *J. Colloid Interface Sci.*, **71**, 301.

Gabrielli, G. (1975), *J. Colloid Interface Sci.*, **53**, 148.

Gabrielli, G., and Puggelli, M. (1972), *J. Appl. Polym. Sci.*, **16**, 2427.

Gabrielli, G., Puggelli, M., and Ferroni, E. (1970), *J. Colloid Interface Sci.*, **32**, 242.

Gabrielli, G., Puggelli, M., and Ferroni, E. (1971), *J. Colloid Interface Sci.*, **36**, 401.

Gabrielli, G., Baglioni, P., and Fabbrini, A. (1981), *Colloid Surf.*, **3**, 147.

Gaines, G. L., Jr. (1961), *J. Phys. Chem.*, **63**, 382.

Gaines, G. L., Jr. (1966), *Insoluble Monolayers at Liquid–Gas Interfaces*, Interscience Publishers, New York.

Gaines, G. L., Jr. (1969), *J. Phys. Chem.*, **73**, 3143.

Gaines, G. L., Jr. (1977), *J. Colloid Interface Sci.*, **59**, 438.

Galla, H. J., and Sackmann, E. (1974), *Biochim. Biophys. Acta*, **339**, 103.
Galler, D. F. J., McConnell, H. M., and Noy, B. T. (1986), *J. Phys. Chem.*, **90**, 2311.
Gardner, R. J. (1972), *J. Inst. Brew.*, **78**, 391.
Gardner, R. J. (1973), *J. Inst. Brew.*, **79**, 275.
Garnier, A., Tosi, L., and Herve, M. (1978), *Biochem. Biophys. Res. Commun.*, **74**, 1280.
Garoff, S., Deck, H. W., Dunsmuir, J. H., Alvarez, M. S., and Bloch, J. M. (1986), *J. Phys. (Paris)*, **47**, 701.
Gee, G. (1935), *Proc. R. Soc. London Ser. A*, **153**, 129.
Germer, L.-H., and Storks, K. H. (1938), *J. Chem. Phys.*, **6**, 280.
Gershfeld, N. L. (1964), Proceed. IVth Intern. Congr. Surface Active Agents, Brussels.
Gershfeld, N. L. (1976), *Ann. Rev. Phys. Chem.*, **27**, 349.
Gershfeld, N. L., and Tajima, K. (1979), *Nature*, **279**, 708.
Gevod, V. S., and Birdi, K. S. (1984), *Biophys. J.*, **45**, 1079.
Ghosh, S. B., and Bull, H. B. (1963), *Biochim. Biophys. Acta*, **66**, 150.
Gibbs, J. W. (1928), *Collected Works*, 2nd ed., Vol. I, Longmans, New York.
Giles, C. H. (1957), in *Proc. 2nd Int. Congr. Surface Active Agents*, London, Vol. 1, p. 92.
Giles, C. H., and Agnihotri, V. G. (1967), *Chem. Ind.*, 1874.
Giles, C. H., and Agnihotri, V. G. (1969), *Chem. Ind.*, 754.
Giles, C. H., and MacEwan, T. H. (1959), *J. Chem. Soc. London*, 1791.
Giles, C. H., and McIver, N. (1974a), *Textile Res. J.*, **44**, 587.
Giles, C. H., and McIver, N. (1974b), *J. Soc. Dyers Colourists*, **90**, 93.
Giles, C. H., and McIver, N. (1975), *J. Colloid Interface Sci.*, **53**, 155.
Giles, C. H., and Neustadter, E. L. (1952), *J. Chem. Soc.*, **918**, 3806.
Giles, C. H., Agnihotri, V. G., and Trivedi, A. S. (1970), *J. Soc. Dyers Colourists*, **86**, 451.
Giles, C. H., MacEwan, T. H., and McIver, N. (1974), *Text. Res. J.*, **44**, 580.
Gjertson, P. (1967), *Brew. Digest*, **42**, 80.
Gluck, L. (1979), *Am. J. Obstet. Gynecol.*, **135**, 57.
Gluck, L., Kulovich, M. V., Borer, R. C., Brenner, P. H., Anderson, G. G., and Spellacy, W. N. (1971), *Am. J. Obstet. Gynecol.*, **109**, 440.
Goddard, E. D. (1975), in *Monolayers* (Goddard, E. D., ed.), *Advances in Chemistry Series*, Vol. 144, American Chemical Society, Washington, D.C.
Goddard, E. D., Kao, O., and Kung, H. C. (1967a), *J. Colloid Interface Sci.*, **24**, 297.
Goddard, E. D., Smith, S. R., and Lander, L. H. (1967b), in *Proc. Int. Congr. Surf. Active Agents*, Brussels, Vol. II, p. 198.
Goddard, E. D., Kao, O., and Kung, H. C. (1968), *J. Colloid Interface Sci.*, **27**, 616.
Goldcare, R. J. (1958), *Surface Phenomena in Chemistry and Biology*, Pergamon Press, London.
Golian, C., Hales, R. S., Hawke, J. G., and Gebicki, J. M. (1978), *J. Phys. E: Sci. Instrum.*, Vol. II, 787.
Gonzales, G., and MacRitchie, F. (1970), *J. Colloid Interface Sci.*, **32**, 55.
Goodall, M. (1970), *Biochim. Biophys. Acta*, **219**, 28.
Goodrich, F. C. (1957), in *Proc. 2nd Int. Congr. Surface Activity*, London, Vol. I, p. 85.
Goodrich, F. C. (1961), *Proc. R. Soc. London Ser. A*, **260**, 990.
Goodrich, F. C. (1962), *J. Phys. Chem.*, **66**, 1858.
Goodrich, F. C. (1973), in *Progress in Surface Membrane Science* (Danielli, J. F., Rosenberg, M. D., and Cadenhead, D. A., eds.), Vol. 7, Academic Press, New York.
Goodrich, F. C. (1974), in *Surface and Colloid Science* (Matijevic, E., ed.), Vols. 1 and 2, Wiley, New York.
Goodrich, F. C., Allen, L. H., and Poskanzer, A. (1975), *J. Colloid Interface Sci.*, **52**, 201.
Gordon, L. G. M., and Haydon, D. A. (1972), *Biochim. Biophys. Acta*, **255**, 1014.
Gorter, E., and Grendel, F. (1925), *J. Exp. Med.*, **41**, 439.
Graven, S.-N., Lardy, H. A., Johnson, D., and Rutter, A. (1966), *Biochemistry*, **5**, 1729.

Green, D. E., Fry, M., and Blondin, G. A. (1980), *Proc. Natl. Acad. Sci. U.S.A.*, **77**, 257.

Grisner, C., and Langevin, D. (1978), in *Proc. Conf. Physicochimie des Amphiphiles*, Bordeaux, France, CNRS, Paris.

Gryszkiewicz, J. (1971), *Folia Biol. Warsawa*, **19**, 119.

Guastala, J. (1938), *C. R. Acad. Sci. Paris*, **206**, 993.

Guastala, J. (1939), *Compt. Rend. Acad. Sci.*, **208**, 1978.

Guastala, J. (1942), *Cah. Phys.*, **10**, 30.

Guggenheim, E. A. (1945), *J. Chem. Phys.*, **13**, 253.

Guggenheim, E. A. (1951), *Thermodynamics*, Interscience, Wiley & Sons, New York.

Gurnay, R. W. (1953), *Ionic Processes in Solutions*, Dover, New York.

Guyot, J. (1924), *Ann. Phys. (Paris)*, **2**, 506.

Haberman, V. E. (1972), *Science*, **177**, 314.

Haberman, V. E., and Jentsch, J. (1976), *Hoppe-Seylers Z. Physiol. Chem.*, **38**, 37.

Hada, H., Hanawa, R., Haraguchi, A., and Yonezawa, Y. (1985), *J. Phys. Chem.*, **89**, 560.

Haest, C. W. M., de Gier, J., and van Deenen, L. L. M. (1969), *Chem. Phys. Lipids*, **3**, 413.

Haines, T. H. (1983), *Proc. Natl. Acad. Sci. U.S.A.*, **80**, 160.

Hamaker, H. C. (1937), *Physica*, **4**, 1058.

Hamel, B. B., and Zimmerman, I. (1970), *J. Lipid Res.*, **10**, 1029.

Hammermeister, D., and Barnett, G. (1974), *Biochim. Biophys. Acta*, **332**, 125.

Hanai, T., Haydon, D. A., and Taylor, J. (1964), *Proc. R. Soc. London Ser. A*, **281**, 377.

Handy, R. M., and Scala, L. C. (1966), *J. Electrochem. Soc.*, **113**, 109.

Hansen, R. S. (1960), *J. Phys. Chem.*, **64**, 637.

Hansen, R. S. (1961), *J. Colloid Sci.*, **16**, 549.

Hard, S., and Lofgren, H. (1977), *J. Colloid Interface Sci.*, **60**, 529.

Hard, S., and Neuman, R. D. (1981), *J. Colloid Interface Sci.*, **83**, 315.

Harding, J. B., and Adam, N. K. (1933), *Trans. Faraday Soc.*, **29**, 837.

Hardt, S. L. (1979), *Biophys. Chem.*, **10**, 239.

Harkins, W. D. (1952), *The Physical Chemistry of Surface Films*, Reinhold, New York.

Harkins, W. D., and Anderson, P. A. (1937), *J. Am. Chem. Soc.*, **59**, 2189.

Harkins, W. D., and Copeland, L. E. (1942), *J. Chem. Phys.*, **10**, 272.

Harkins, W. D., and Fischer, E. K. (1933), *J. Chem. Phys.*, **1**, 852.

Harkins, W. D., and Kirkwood, J. G. (1938), *J. Chem. Phys.*, **653**, 298.

Harkins, W. D., and Nutting, G. (1939), *J. Am. Chem. Soc.*, **61**, 1702.

Harkins, W. D., Young, T. E., and Boyd, E. (1940), *J. Chem. Phys.*, **8**, 954.

Harwood, J. L., Desai, R., Hext, P., Tetley, T., and Richards, R. (1975), *Biochem. J.*, **151**, 707.

Hassager, O., and Westborg, H. (1987), *J. Colloid Interface Sci.*, **2**, 119.

Hatch, F. T. (1965), *Nature*, **206**, 777.

Hatefi, A., and Hanstein, D. E. (1969), *Biochim. Biophys. Acta*, **183**, 320.

Hawke, J. G., and Alexander, A. E. (1960), in *Proc. 3rd Int. Conf. of Surface Active Agents*, University of Mainz Press, Cologne.

Hawke, J. G., and Alexander, A. E. (1962), in *Retardation of Evaporation by Monolayers* (La Mer, V. K., ed.), Academic Press, New York.

Hawke, J. G., and Parts, A. G. (1964), *J. Colloid Sci.*, **19**, 448.

Hawkins, G. A., and Benedek, G. B. (1974), *Phys. Rev. Lett.*, **32**, 524.

Hayami, Y., Yano, T., Motomura, K., and Matuura, R. (1980), *Bull. Chem. Soc., Jpn.*, **53**, 3414.

Hayashi, M., Muramatsu, T., and Hara, I. (1972), *Biochim. Biophys. Acta*, **255**, 98.

Hayashi, M., Muramatsu, T., and Hara, I. (1973), *Biochim. Biophys. Acta*, **291**, 335.

Hayashi, M., Muramatsu, T., Hara, I., and Seimiya, T. (1975), *Chem. Phys. Lipids*, **15**, 209.

Hayashi, M., Kobayashi, T., Seimiya, T., Muramatsu, T., and Hara, I. (1980), *Chem. Phys. Lipids*, **27**, 1.

Haydon, D. A., and Hladky, S. B. (1972), *Q. Rev. Biophys.*, **5**, 197.

Hedestrand, G. (1924), *J. Phys. Chem.*, **28**, 1245.

Heesemann, J. (1980), *J. Am. Chem. Soc.*, **102**, 2167.

Heikkila, R. E., Deamer, D. W., and Cornwell, D. G. (1950), *J. Lipid Res.*, **11**, 195.

Heikkila, R. E., Kwong, C. N., and Cornwell, D. G. (1970), *J. Lipid Res.*, **11**, 190.

Heitz, F., and Gavach, C. (1983), *Biophys. Chem.*, **18**, 153.

Heller, S. (1954), *Colloid-Z.*, **136**, 120.

Herd, J. M., Hopkins, A. J., and Howard, G. J. (1971), *J. Polym. Sci.*, **C34**, 211.

Hermann, R. B. (1972), *J. Phys. Chem.*, **76**, 2754.

Heymann, L. (1931), *Kolloid-Z.*, **57**, 81.

Hildebrand, J. H., Prasnitz, J. M., and Scott, R. L. (1970), *Regular and Related Solutions*, Van Nostrand Reinhold, New York.

Hill, T. L. (1962), *Introduction to Statistical Thermodynamics*, Addison-Wesley, New York.

Hill, T. L. (1963), *Thermodynamics of Small Systems*, Vol. I, W. A. Benjamin, New York.

Hille, B. (1968), *J. Gen. Physiol.*, **51**, 199.

Hilton, B. D., and O'Brien, R. D. (1973), *Pesticide Biochem. Physiol.*, **3**, 206.

Hladky, S. B., and Haydon, D. A. (1973), *Biochim. Biophys. Acta*, **318**, 464.

Hodgkin, A. L., Huxley, A. F., and Katz, B. (1952), *J. Physiol. London*, **116**, 424.

Hold, L. (1967), *Nature*, **214**, 1105.

Holly, F. J. (1974), *J. Colloid Interface Sci.*, **49**, 221.

Hookes, D. E. (1971), Ph.D. thesis, University of London.

Hopfinger, A. J. (1973), *Conformational Properties of Macromolecules*, Academic Press, New York.

Hsu, J.-H. (1987), *J. Theor. Biol.*, **124**, 495.

Hsu, J.-P., and Wang, H.-H. (1987), *J. Theor. Biol.*, **124**, 405.

Hubbard, R., Gregermann, R. I., and Wald, G. (1952–1953), *J. Fen. Physiol.*, **36**, 415.

Huggins, M. L. (1942), *J. Phys. Chem.*, **46**, 151.

Huggins, M. L. (1965), *Makromol. Chem.*, **87**, 119.

Hühnerfuss, H. (1985), *J. Colloid Interface Sci.*, **107**, 84.

Hühnerfuss, H., Lange, P., and Walter, W. (1984), *J. Mater. Res.*, **42**, 737.

Hwang, S. B., Korenbrot, J. I., and Stoeckenius, W. (1977), *J. Membrane Biol.*, **36**, 115.

Hyono, A., Hendriks, T. H., Daeman, F. J., and Bonting, S. L. (1975), *Biochim. Biophys. Acta*, **389**, 34.

Ingram, B. T. (1972), *Faraday Trans.*, **68i**, 2230.

Ishi, T., Kuroda, Y., Omochi, T., and Yoshokawa, K. (1986), *Langmuir*, **2**, 319.

Israelachvili, J. N., and Tabor, D. (1976), in *Progress in Surface and Membrane Science* (Danielli, J. F., Rosenberg, M. D., and Cadenhead, D. A., eds.), Academic Press, New York.

Israelachvili, J. N., Mitchell, D. J., and Ninham, B. W. (1977), *Biochim. Biophys. Acta*, **470**, 185.

Ivanov, A., Laine, I. A., Abdulaev, N. D., Senyavina, L. B., Popov, E. M., Ovchinnikov, Yu. A., and Shemaylicin, M. M. (1969), *Biochim. Biophys. Res. Commun.*, **34**, 803.

Ivarsson, B. A., Hegg, P. O., Lundstrom, K. I., and Jonsson, U. (1985), *Colloid Surf.*, **13**, 169.

Jackson, C. M., and Yui, P. Y. J. (1975), in *Monolayers* (Goddard, E. D., ed.), *Advances in Chemistry Series*, No. 144, American Chemical Society, Washington, D.C.

Jacobs, E. E., Holt, A. S., and Rabinovitch, E. (1954), *J. Chem. Phys.*, **22**, 142.

Jaffe, J. (1954), *J. Chem. Phys.*, **51**, 243.

Jaffe, J., and Ruysschaert, J.-M. (1964), in *Proc. IVth Int. Congr. Surface Active Agents*, Brussels.

Jaffe, J., Ruysschaert, J. M., and Hecq, W. (1970), *Biochim. Biophys. Acta*, **207**, 11.

Jahnig, F. (1979), *J. Chem. Phys.*, **70**, 3279.

Jaing, T.-S., Chaen, J.-D., and Slattery, J. C. (1983), *J. Colloid Interface Sci.*, **96**, 7.

Jakobsson, E. (1971), *J. Theor. Biol.*, **33**, 77.

Jalal, I. M. (1978), Ph.D. thesis, University of Wisconsin.

Jalal, I., and Geografi, G. (1979), *J. Colloid Interface Sci.*, **68**, 196.

James, L. K., and Augenstein, L. G. (1966), *Adv. Enzymol.*, **28**, 1.

Jarvis, N. L. (1962), Surface Viscosity of Monomolecular Films on Water, U.S. Naval Res. Lab., Washington, D.C., N.R.L. Report 5743.

Jarvis, N. L. (1966), *J. Phys. Chem.*, **70**, 3027.

Jarvis, N. L. (1969), *J. Colloid Interface Sci.*, **29**, 647.

Jashani, I. L., and Lenlich, R. (1974), *J. Colloid Interface Sci.*, **46**, 13.

Jehring, H. (1969), *J. Electroanal. Chem.*, **21**, 77.

Jehring, H., and Horn, E. (1968), *Monatsber. Deut. Akad. Wiss. Berlin*, **10**, 295.

Joly, M. (1952), *Kolloid-Z.*, **126**, 35.

Joly, M. (1964), in *Recent Progress in Surface Science* (Danielli, J. F., Pankhurst, K. G. A., and Riddiford, A. C., eds.), Vol. 1, Academic Press, New York.

Jones, D. D. (1975), *J. Mol. Biol.*, **9**, 605.

Jones, R. O., Taylor, J., and Owens, F. N. (1969), *Colloid Surfaces* **2**, 201.

Joos, P. (1968), *Proc. Vth Int. Congr. Surface Active Agents*, Vol. 2, p. 513.

Joos, P. (1969), *Bull. Soc. Chim. Belg.*, **78**, 207.

Joos, P. (1975), *Biochim. Biophys. Acta*, **375**, 1.

Kafka, M. S., and Holz, R. W. (1976), *Biochim. Biophys. Acta*, **426**, 31.

Kahlenberg, A., Walker, C., and Rohrlick, R. (1974), *Can. J. Biochem.*, **52**, 803.

Kalischewski, K., and Schogrer, K. (1979), *Colloid Polym. Sci.*, **257**, 1099.

Kalousek, N. R. (1949), *J. Chem. Soc.*, 894.

Kan, S., Teruko, S., and Igaku, T. O. (1977), *J. Colloid Interface Sci.*, **93**, 243.

Kaneda, T. (1977), *Microbiol. Rev.*, **41**, 391.

Kannenberg, E., Bluime, A., McElhaney, R., and Poralla, K. (1983), *Biochim. Biophys. Acta*, **733**, 111.

Kanner, B., Reid, W. G., and Peterson, I. H. (1967), *Industr. Eng. Chem. Process Design Dev.*, **6**, 88.

Katachalski, E., Silman, I., and Goldman, R. (1971), *Adv. Enzymol.*, **34**, 445.

Kato, A., Tsutsui, N., Matsudomi, N., Kobayashi, K., and Nakai, S. (1981), *Agr. Biol. Chem.*, **45**, 2755.

Kato, T., Seki, K., and Kanako, R. (1986), *J. Colloid Interface Sci.*, **109**, 77.

Katti, S. S., and Sansare, S. D. (1970), *J. Colloid Interface Sci.*, **32**, 361.

Katz, R., and Samwel, K. (1929), *Annalen*, **472**, 24; **474**, 296.

Kaufer, H., and Schibe, G. (1955), *Z.-Elektrochem.*, **59**, 584.

Kell, D. S. (1979), *Biochim. Biophys. Acta*, **549**, 55.

Kelvin, W. T. Lord (1898), *Philos. Mag.*, **46**, 91.

Kemp, G., and Wenner, C. (1973), *Biochim. Biophys. Acta*, **323**, 161.

Kemp, G., Jacobson, K. A., and Wenner, C. E. (1972), *Biochim. Biophys. Acta*, **255**, 493.

Keshavaraz, E., and Nakai, S. (1979), *Biochim. Biophys. Acta*, **576**, 269.

Ketterer, B., Neumcke, B., and Lauger, P. J. (1971), *Membrane Biol.*, **5**, 225.

Keulegan, G. H. (1951), *J. Res. Natl. Bur. Std., USA*, **46**, 358.

Khaiat, A., Ketevi, P., Saraga, T. M., Cittanova, N., and Jayale, M. F. (1975), *Biochim. Biophys. Acta*, **401**, 1.

Khanna, U. (1973), Ph.D. thesis, University of Roorke, India.

Killman, E., and Eckart, R. (1971), *Makromolek. Chem.*, **144**, 45.

Kim, H. W., and Cannell, D. S. (1975), *Phys. Rev. Lett.*, **35**, 889.

Kimura, F., Umemura, J., and Takenaka, T. (1986), *Langmuir*, **2**, 96.

Kinloch, C. D., and McMullen, A. I. (1959), *J. Sci. Instrum.*, **36**, 347.

Kinsella, J. E. (1976), in *Critical Reviews in Food Science and Nutrition*, Vol. 7, Chemical Rubber, Cleveland, Ohio.

Kjelleberg, S., and Stenström, T. A. (1980), *J. Gen. Microbiol.*, **116**, 417.

Kjelleberg, S., Norkrans, B., Lofrgren, H., and Larsson, K. (1976), *Appl. Environ. Microbiol.*, **31**, 609.

Kjelleberg, S., Stenström, T. A., and Odham, G. (1979), *Marine Biol.*, **53**, 21.

Klein, N. (1969), *Adv. Electron. Electron Phys.*, **26**, 309.

Klein, N. (1972), *Adv. Phys.*, **21**, 605.

Klein, N., and Gafni, H. (1966), *IEEE Trans. Electron Devices*, **ED13**, 281.

Kontecky, J., and Brdicka, R. (1947), *Collection Czech. Chem. Commun.*, **12**, 337.

Kopp, F., Cuendet, P. A., Muhlethaler, K., and Zuber, H. (1979), *Biochim. Biophys. Acta*, **553**, 438.

Korenbrot, J. I. (1977), *Annu. Rev. Physiol.*, **39**, 19.

Korenbrot, J. I., and Jonrs, O. (1979), *J. Membrane Biol.*, **46**, 239.

Korenbrot, J. L., and Pramik, M. J. (1977), *J. Membrane Biol.*, **37**, 235.

Koryta, J. (1953), *Collect. Czech. Chem. Commun.*, **18**, 206.

Kotyk, A., and Janacek, K. (1975), *Cell Membrane Transport*, 2nd ed., Plenum Press, New York.

Kramer, L. (1971), *J. Chem. Phys.*, **55**, 2097.

Kretzschmar, G. (1969), *Kolloid Z. Z. Polym.*, **234**, 1030.

Krieg, R. D., Son, J. E., and Flumerfelt, R. W. (1981), *J. Colloid Interface Sci.*, **79**, 14.

Krigbaum, W. R., and Knuutton, S. P. (1973), *Proc. Natl. Acad. Sci., U.S.A.*, **70**, 2809.

Ksenzhek, O. S., and Gevod, V. S. (1975a), *Zh. Fiz. Khim.*, **49**, 2159.

Ksenzhek, O. S., and Gevod, V. S. (1975b), *Zh. Fiz. Khim.*, **49**, 2158.

Kubicki, J., Ohlenbusch, H. D., Schoeder, E., and Wollmer, A. (1976), *Biochemistry*, **15**, 5698.

Kuhn, H., Mobius, D., and Bucher, H. (1972), in *Physical Methods of Chemistry* (Weissberger, A., and Rossister, B., eds.), Wiley, New York.

Kuleff, I., and Petrov, J. G. (1979), *J. Radioanal. Chem.*, **49**, 239.

Kulovich, M. V., and Gluck, L. (1979), *Am. J. Obstet. Gynecol.*, **135**, 64.

Kulovich, M. V., Hallman, M., Kuramoto, N., Sekita, K., Motomurta, K., and Matuura, R. (1972), *Mem. Fac. Fac. Sci., Kyushu Univ.*, **C8**, 67.

Kuznetsov, B. (1971), *Soviet Phys. Semicond.*, **7**, 692.

Lagaly, G., Stuke, E., and Weiss, A. (1976), *Progr. Colloid Polym. Sci.*, **60**, 102.

Lagocki, J. W., Boyd, N. D., Law, J. H., and Kezdy, F. J. (1970), *J. Am. Chem. Soc.*, **92**, 2923.

Laidler, K. J., and Sundaram, P. V. (1971), in *Chemistry of the Cell Interface* (Brown, H. D., ed.), Academic Press, New York.

La Mer, V. K. (1952), *Indus. Eng. Chem.*, **44**, 1270.

La Mer, V. K., ed. (1962), *Retardation of Evaporation by Monolayers*, Academic Press, New York.

La Mer, V. K., Healey, T. W., and Aylmore, L. A. G. (1964), *J. Colloid Sci.*, **19**, 673.

Lange, Y., Ralph, E. K., and Redfield, A. G. (1975), *Biophys. Biochem. Res. Commun.*, **62**, 891.

Langevin, D. (1981), *J. Colloid Interface Sci.*, **80**, 412.

Langevin, D., and Bouchiat, M. A. (1971), *C.R. Acad. Sci. Ser. B*, **272**, 1422.

Langmuir, I. (1917), *J. Am. Chem. Soc.*, **39**, 1848.

Langmuir, I. (1920), *Trans. Faraday Soc.*, **15**, 68.

Langmuir, I. (1925), *Third Colloid Symposium Monograph*, Chemical Catalog Co., New York.

Langmuir, I., and Langmuir, B. B. (1927), *J. Phys. Chem.*, **31**, 1719.

Langmuir, I., and Schaefer, V. J. (1936), *J. Am. Chem. Soc.*, **58**, 284.

Langmuir, I., and Schaefer, V. J. (1937), *J. Am. Chem. Soc.*, **59**, 2400.

Langmuir, I., and Schaefer, V. J. (1943), *J. Franklin Inst.*, **235**, 119.

Langmuir, I., and Waugh, D. F. (1940), *J. Am. Chem. Soc.*, **62**, 2771.

Lanham, A. F., and Pankhurst, K. G. A. (1956), *Trans. Faraday Soc.*, **52**, 521.

Larsson, K., Lundquist, M., Stallhagen-Stenhagen, S., and Stenhagen, E. (1969), *J. Colloid Interface Sci.*, **29**, 268.

Larsson, K. (1973), in *Surface and Colloid Science* (Matijevic, E., ed.), John Wiley and Sons, New York.

Larsson, K. (1976), in *Food Emulsions* (Friberg, S., ed.), Marcel Dekker, New York, Chap. 2.

Lartigue, D. J., and Yaverbaum, S. (1976), in *Progress in Surface and Membrane Science* (Cadenhead, D. A., and Danielli, J. F., eds.), Vol. 10, Academic Press, New York.

Laueger, P. (1972), *Science*, **178**, 24.

Laueger, P., Lesslauer, W., Marti, E., and Richter, J. (1967), *Biochim. Biophys. Acta*, **135**, 20.

Lawrence, A. S. C. (1969), *Mol. Cryst. Liq. Cryst.*, **7**, 1.

Lee, R. G., and Kim, S. R. (1974), *J. Biol. Mater. Res.*, **8**, 251.

Leger, A., Klerin, J., Belin, M., and Defourneau, D. (1971), *Thin Solid Films*, **8**, 51.

Lewis, G. N., and Randall, M. (1923), *Thermodynamics*, McGraw-Hill, New York.

Lifshitz, E. M. (1956), *Sov. Phys.*, **2**, 73.

Lim, V. I. (1974), *J. Mol. Biol.*, **88**, 857, 873.

Lindblom, G., Wennerstrom, H., Arvidson, G., and Lindman, B. (1976), *Biophys. J.*, **16**, 1287.

Linden, C. D., Wright, K. L., McConnell, H. M., and Fox, C. F. (1973), *Proc. Natl. Acad. Sci. U.S.A.*, **70**, 2271.

Lippert, J. L., and Peticolas, W. L. (1971), *Proc. Natl. Acad. Sci. U.S.A.*, **68**, 1572.

Lissant, K. J. (1966), *J. Colloid Interface Sci.*, **22**, 462.

Loeb, G. I. (1969), *J. Colloid Interface Sci.*, **31**, 572.

Loeb, G. I. (1971), *J. Polymer Sci.*, **34**, 63. .

London, F. (1930), *Z. Phys.*, **63**, 245.

Longton, R. W., Cole, J. S., and Quinn, P. F. (1975), *Arch. Oral Biol.*, **20**, 103.

Low, M. G., Ferguson, A. J., Futerman, H. F., and Silman, I. (1986), *TIBS*, **11**, 212.

Lucassen, E. H., and Lucassen, J. (1969), *Adv. Colloid Interface Sci.*, **2**, 347.

Lucassen, J. (1968), *Trans. Faraday Soc.*, **64**, 2221.

Lucassen, J., and Hausen, R. S. (1966), *J. Colloid Interface Sci.*, **22**, 32.

Lusccan, C., and Fancon, J. F. (1971), *FEBS Lett.*, **19**, 186.

Lusted, D. (1973a), *Biochim. Biophys. Acta*, **307**, 270.

Lusted, D. (1973b), *J. Colloid Interface Sci.*, **44**, 72.

Luzzati, V. (1968), in *Biological Membranes* (Chapman, D., ed.), Academic Press, New York.

Luzzati, V., and Husson, F. (1962), *J. Cell Biol.*, **12**, 207.

Lyman, D. J., Brash, J. L., Chaikin, S. W., Klein, K. G., and Carini, M. (1968), *Trans. Am. Soc. Artif. Intern. Organs*, **14**, 250.

MacArthur, B. W. (1977), Ph.D. thesis, University of Washington.

MacGrath, A. E., Morgan, C. G., and Radda, G. K. (1976), *Biochim. Biophys. Acta*, **426**, 173.

MacIntyre, E. (1974), *Sci. Am.*, **20**, 62.

MacRitchie, F. (1977), *J. Colloid Interface Sci.*, **61**, 223.

MacRitchie, F. (1981), *J. Colloid Interface Sci.*, **79**, 461.

MacRitchie, F. (1985), *J. Colloid Interface Sci.*, **107**, 276.

MacRitchie, F. (1985), *J. Colloid Interface Sci.*, **1**, 276.

MacRitchie, F., and Alexander, A. E. (1963), *J. Colloid Sci.*, **18**, 453.

Maeda, Y., and Isemura, T. (1967), *Nature*, **215**, 765.

Maggio, B., and Lucy, J. A. (1974), *Biochem. J.*, **149**, 597.

Maggio, B., Cumar, F. A., and Caputto, R. (1978a), *Biochem. J.*, **175**, 1113.

Maggio, B., Cumar, F. A., and Caputto, R. (1978b), *Biochem. J.*, **171**, 559.

Maggio, B., Cumar, F. A., and Caputto, R. (1978c), *FEBS Lett.*, **90**, 149.

Maggio, B., Cumar, F. A., and Caputto, R. (1980), *Biochem. J.*, **189**, 435.

Maggio, B., Cumar, F. A., and Caputto, R. (1981), *Biochim. Biophys. Acta*, **650**, 69.

Makino, K., Ohshima, H., and Kondo, T. (1987), *J. Colloid Interface Sci.*, **115**, 65.

Malcolm, B. R. (1966), *Polymer*, **7**, 595.

Malcolm, B. R. (1968a), *Biochem. J.*, **110**, 733.

Malcolm, B. R. (1968b), *Proc. R. Soc. London Ser. A*, **305**, 363.

Malcolm, B. R. (1970a), *Biopolymers*, **9**, 911.

Malcolm, B. R. (1970b), *Nature*, **227**, 1358.

Malcolm, B. R. (1971), *J. Polym. Sci.*, **A34**, 87.

Malcolm, B. R. (1976), *Biochem. J.*, **120**, 82.

Malcolm, B. R., and Davies, S. R. (1965), *J. Sci. Instrum.*, **42**, 359.

Manery, G. (1966), *Fed. Proc.*, **25**, 1804.

Mann, B., and Kuhn, H. (1971), *J. Appl. Phys.*, **42**, 4398.

Mann, J. A., and Hansen, R. S. (1963), *Rev. Sci. Instrum.*, **34**, 702.

Mannheimer, R. J. (1972), *J. Colloid Interface Sci.*, **40**, 370.

Mannheimer, R. J., and Burton, R. R. (1970), *J. Colloid Interface Sci.*, **32**, 73.

Mansfield, W. W. (1959), *Aust. J.-Chem.*, **12**, 382.

Mansfield, W. W. (1974), in *Proc. 5th Natl. Convention on Evaporation Control*, Raci, Canberra.

Marangoni, C. (1871), *Nuovo Cimento*, [2], **16**, 239.

Marcelin, K. (1929), *C.R. Acad. Sci.*, **189**, 241.

Marcelja, S. (1974), *Biochim. Biophys. Acta*, **367**, 165.

Markel, D. T. F., and Vanderslice, T. A. (1960), *J. Phys. Chem.*, **64**, 1231.

Marra, J. (1986a), *J. Phys. Chem.*, **90**, 2145.

Marra, J. (1986b), *J. Colloid Interface Sci.*, **109**, 11.

Marshall, K. C. (1971), in *Soil Biochemistry* (McCllaren, A. D., and Skujins, J. J., eds.), Vol. II, Marcel Dekker, New York.

Marshall, K. C. (1975), *Annu. Rev. Phytopathol.*, **13**, 357.

Marshall, K. C. (1976), *Interfaces in Microbial Ecology*, Harvard University Press, Cambridge, Massachusetts.

Marshall, K. C., and Cruickshank, R. H. (1973), *Arch. Mikrobiol.*, **91**, 29.

Marsten, J., and Rideal, E. K. (1938), *J. Chem. Soc.*, **1163**, 51.

Matalon, R., and Schulman, J. (1949), *Discuss. Faraday Soc.*, **6**, 27.

Mateeva, R. L., Panayotov, I., Ivanova, M., Loshtilova, E. M., Prokopov, V. J., and Georgiev, G. A. (1975), *C.R. Acad. Bulg. Sci.*, **28**, 1223.

Matsubara, A., Matsumura, R., and Kimizuka, H. (1965), *Bull. Chem. Soc. Jpn.*, **38**, 369.

Matsumoto, M., Montandon, C., and Hartland, S. (1977), *Colloid Polym. Sci.*, **255**, 261.

Mattson, J. S., and Jones, T. T. (1976), *Anal. Chem.*, **48**, 2164.

Matuo, H., Hiromoto, K., Motomura, K., and Matuura, R. (1978), *Bull. Chem. Soc. Jpn.*, **51**, 690.

Matuo, H., Motomura, K., and Matuura, R. (1979a), *Bull. Chem. Soc. Jpn.*, **52**, 673.

Matuo, H., Motomura, K., and Matuura, R. (1979b), *J. Colloid Interface Sci.*, **69**, 192.

Matuo, H., Motomura, K., and Matuura, R. (1981), *Chem. Phys. Lipids*, **28**, 281, 385.

Matuo, H., Motomura, K., and Matuura, R. (1987), *Bull. Chem. Soc. Jpn.*, **61**, 529.

Mauer, F. A. (1954), *Rev. Sci. Instrum.*, **25**, 598.

Maun, J. A., and Hansen, R. S. (1960), *Rev. Sci. Instrum.*, **31**, 961.

Maun, J. A., and Hansen, R. S. (1963), *Rev. Sci. Instrum.*, **34**, 702.

McCrackin, F. L. (1969), *Natl. Bur. Stand., Tech. Note*, **479**, 1.

McCrackin, F. L., Passaglia, E., Stromberg, R. R., and Steinberg, J. (1963), *Res. Natl. Bur. Stand., U.S.A., Sect. A*, **67**, 363.

McElhaney, V. K. (1976), *Biochim. Biophys. Acta*, **363**, 59.

McLaughlin, S. G. A., Szabo, S., Eisenman, G., and Ciani, S. M. (1970), *Proc. Natl. Acad. Sci. U.S.A.*, **67**, 1268.

McLeod, D. B. (1923), *Trans. Faraday Soc.*, **19**, 38.

McMullen, A. I., Reiter, P., and Phillips, M. C. (1977), *Nature*, **267**, 244.

Melnik, E., Latorre, R., Hall, J., and Tosteson, D. C. (1977), *J. Gen. Physiol.*, **69**, 243.

Mendenhall, R. M., and Mendenhall, A. L. (1963), *Rev. Sci. Instrum.*, **34**, 1350.

Meyers, R. J., and Harkins, W. D. (1937), *J. Chem. Phys.*, **5**, 601.

Miller, I. R. (1961), *J. Mol. Biol.*, **3**, 229.

Miller, I. R. (1971), *Prog. Membrane Surface Sci.*, **4**, 299.

Miller, I. R., and Bach, D. (1973), in *Surface and Colloid Science* (Matijevic, E., ed.), Wiley, New York.

Miller, I. R., and Ruysschaert, J. M. (1971), *J. Colloid Interface Sci.*, **35**, 340.

Mingins, J., Owens, N. F., and Iles, D. H. (1969), *J. Phys. Chem.*, **73**, 2118.

Mingins, J., Owens, N. F., Taylor, J. A. G., Brooks, H., and Pethica, B. A. (1975), in *Monolayers* (Goddard, E. D., ed.), *Advances in Chemistry Series*, No. 144, p. 14, American Chemical Society, Washington, D.C.

Mitchell, J. R., Adams, D. J., Evans, M. T. A., Phillips, M. C., and Rees, P. M. (1971), *J. Polymer Sci.*, C**34**, 167.

Mitchell, A., Rideal, E. K., and Schulman, J. H. (1937), *Nature*, **139**, 625.

Miyano, K., Abraham, B. M., Ting, L., and Wasan, D. T. (1983), *J. Colloid Interface Sci.*, **92**, 297.

Mizuno, M., Mori, C., Noguchi, H., and Watanabe, T. (1983), *Jpn. J. Appl. Phys.*, **22**, 808.

Mobius, D. (1978), *Ber. Bunsenges. Phys. Chem.*, **82**, 848.

Moelwyn-Hughes, E. A. (1974), *The Kinetics of Reactions in Solutions*, 2nd ed., Oxford University Press, Oxford.

Momura, H., and Verrall, R. E. (1981), *J. Phys. Chem.*, **85**, 1042.

Montal, M., and Muller, P. (1972), *Proc. Natl. Acad. Sci. U.S.A.*, **69**, 3561.

Morrisey, B. W., Smith, L. E., Stromberg, R. R., and Fenstermaker, P. K. (1976), *J. Colloid Interface Sci.*, **56**, 557.

Morse, P. D., and Deamer, D. W. (1973), *Biochim. Biophys. Acta*, **298**, 769.

Mosbach, K. (1976), in *Methods in Enzymology* (Colwick, S. P., and Kaplan, N. O., eds.), Vol. 44, Academic Press, New York.

McMullen, A.-I. (1973), Mosquito Control: Some Perspectives for Developing Countries, National Academy of Science, Washington, D.C.

Motomura, K. (1974), *J. Colloid Interface Sci.*, **48**, 307.

Motomura, K. (1980), *Adv. Colloid Interface Sci.*, **12**, 1.

Motomura, K., and Matuura, R. (1963), *J. Colloid Sci.*, **18**, 52.

Motomura, K., Sakita, K., and Matuura, R. (1974), *J. Colloid Interface Sci.*, **48**, 319.

Motomura, K., Terazono, T., Matuo, H., and Matuura, R. (1976), *J. Colloid Interface Sci.*, **57**, 52.

Motomura, K., Yano, T., Ikematsu, M., Matuo, H., and Matuura, R. (1979), *J. Colloid Interface Sci.*, **69**, 209.

Motomura, K., Ikematsu, M., Hayami, Y., Matuo, H., and Matuura, R. (1980), *Bull. Chem. Soc. Jpn.*, **53**, 2217.

Mouritsen, O. G. (1984), *Computer Studies of Phase Transition and Critical Phenomena*, Springer Verlag, Heidelberg, West Germany.

Mouritsen, O. G., and Zuckermann, M. J. (1985), *Eur. Biophys. J.*, **12**, 75.

Mueller-Landau, F., Cadenhead, D. A., and Kellener, B. M. J. (1980), *J. Colloid Interface Sci.*, **73**, 264.

Mulla, M. S., and Chadhury, M. F. B. (1968), *Mosquito News*, **28**, 187.

Müller, O. (1963), in *Methods of Biochemical Analysis* (Glich, D., ed.), Vol. 11, Wiley, New York.

Muramatsu, M. (1959), *Bull. Chem. Soc. Jpn.*, **32**, 114.

Muramatsu, M., and Ohno, T. (1971), *J. Colloid Interface Sci.*, **35**, 469.

Mutafchieva, R., Panayotov, I., and Dimitrov, D. S. (1984), *Z. Naturforsch.*, **39**, 965.

Myers, K., and Harkins, W. D. (1937), *Nature*, **139**, 367.

Nagano, L. (1977), *Biochemistry*, **16**, 3484.

Nagarajan, M. K., and Shah, J. P. (1981), *J. Colloid Interface Sci.*, **80**, 7.

Nagle, J. F. (1980), *Ann. Rev. Phys. Chem.*, **31**, 157.

Nagle, J. F., and Scott, H. L. (1978), *Biochim. Biophys. Acta*, **513**, 236.

Nakahara, H., Fukuda, K., Akatsu, H., and Kyogoku, Y. (1978), *J. Colloid Interface Sci.*, **65**, 517.

Nakanaga, T., and Takenaka, T. (1977), *J. Phys. Chem.*, **81**, 645.

Narahashi, T., and Haas, H. G. (1968), *J. Gen. Physiol.*, **51**, 177.

Neuman, R. D. (1978), *J. Colloid Interface Sci.*, **63**, 106.

Neumann, A. W., Moscarello, M. A., and Epand, R. M. (1973), *Biopolymers*, **12**, 1945.

Neumcke, B. (1970), *Biophysik*, **6**, 231.

Nicholls, P., Mochan, E., and Kimelberg, H. K. (1967), *FEBS Lett.*, **3**, 242.

Nixon, J., and Beerbower, A. (1969), *Am. Chem. Soc. Div. Pet. Chem. Prepr.*, **14**, 49.

Noll, W. (1966a), *Pure Appl. Chem.*, **13**, 101.

Noll, W. (1966b), *Kolloid-Z.*, **211**, 98.

Noll, W., Steinbach, H., and Sucker, C. (1963), *Ber. Bunsenges. Phys. Chem.*, **67**, 407.

Noll, W., Steinbach, H., and Sucker, C. (1965), *Kolloid-Z.*, **204**, 94.

Noll, W., Steinbach, H., and Sucker, C. (1970), *Kolloid-Z.*, **236**, 1.

Noll, W., Buchner, W., Steinbach, H., and Sucker, C. (1972), *Kolloid-Z.*, **250**, 9.

Norde, W. (1976), Ph.D. thesis, Agr. Univ., Wageningen, Holland, 76-6.

Norde, W., and Lyklema, J. (1978a), *J. Colloid Interface Sci.*, **66**, 257.

Norde, W., and Lyklema, J. (1978b), *J. Colloid Interface Sci.*, **66**, 266.

Norde, W., and Lyklema, J. (1978c), *J. Colloid Interface Sci.*, **66**, 277.

Norde, W., and Lyklema, J. (1978d), *J. Colloid Interface Sci.*, **66**, 285.

Norde, W., and Lyklema, J. (1978e), *J. Colloid Interface Sci.*, **66**, 295.

Norkrans, B. (1979), in *Proc. Workshop: Microbial Degrad. Pollution in Marine Environment* (Bourquin, A. W., and Pritchard, P. H., eds.), Gulf-Breeze, Florida, Environ. Res. Lab.

Norkrans, B., and Sorensson, F. (1977), *Botanica Marina*, **20**, 473.

Nutting, G. C., and Harkins, W. D. (1939), *J. Am. Chem. Soc.*, **61**, 1180.

Nyilas, E., Chiu, T.-H., and Herzlinger, G. A. (1974), *Trans. Am. Soc. Artif. Inter. Organs*, **20**, 480.

O'Brien, R. N., Feher, A. I., and Leja, J. (1975), *J. Colloid Interface Sci.*, **51**, 366.

O'Brien, R. N., Feher, A. I., Li, K. L., and Tan, W. C. (1976a), *Can. J. Chem.*, **54**, 2739.

O'Brien, R. N., Feher, A. I., and Leja, J. (1976b), *J. Colloid Interface Sci.*, **56**, 469, 474.

Ohki, S. (1971), *J. Colloid Interface Sci.*, **37**, 318.

Ohki, S. (1973), *J. Theor. Biol.*, **42**, 593.

Ohki, S., and Ohki, C. (1976), *J. Theor. Biol.*, **62**, 389.

Ovchinnikov, Yu. A. (1974), *FEBS Lett.*, **44**, 1.

Paddy, J. F. (1951), in *Proc. 2nd Int. Congr. Surface Active Agents*, Vol. I, p. 1.

Pagano, R. E., and Gershfeld, N. L. (1972), *J. Colloid Interface Sci.*, **41**, 311.

Palau, J., and Puigdomenech, P. (1974), *J. Mol. Biol.*, **88**, 457.

Palacek, E., and Pechan, Z. (1971), *Anal. Biochem.*, **42**, 59.

Pankhurst, K. G. A. (1958), in *Surface Phenomena in Chemistry and Biology* (Danielli, J. F., Pankhurst, K. G. A., and Riddeiford, A. C., eds.), Pergamon, Oxford.

Papahadjopolous, D., Jacobson, K., Nu, S., and Isac, T. (1973), *Biochim. Biophys. Acta*, **311**, 330.

Pavlovic, O., and Miller, J. (1971), *J. Polym. Sci. C*, **34**, 181.

Patil, G. S., Matthews, H., and Cornwell, D. G. (1973), *J. Lipid Res.*, **14**, 26.

Patil, G. S., Dorman, N. J., and Cornwell, D. G. (1979), *J. Lipid Res.*, **20**, 663.

Pattle, R. E. (1955), *Nature*, **175**, 1125.

Payens, T. A. (1955), *Philips Res. Rep.*, **10**, 425.

Payens, T. A. (1960), *J. Biochim. Biophys. Acta*, **38**, 539.

Peter, H., and Bell, D. (1930), *J. Physical Chem.*, **34**, 1399.

Petermann, J. (1977), *Ber. Bunsenges. Phys. Chem.*, **81**, 649.

Petersen, L. C., and Birdi, K. S. (1983), *Scand. J. Clin. Lab. Invest.*, **43**, 41.

Peterson, I. R., and Russell, G. J. (1984), *Philos. Mag.*, **49A**, 463.

Pethica, B. A. (1955), *Trans. Faraday Soc.*, **51**, 1402.

Pethica, B. A., Glasser, M. L., and Mingins, J. (1981), *J. Colloid Interface Sci.*, **81**, 41.

Petlak, C. S., and Gershfeld, N. L. (1967), *J. Colloid Interface Sci.*, **25**, 503.

Petrov, J. G., Kuleff, I., and Platikanov, D. (1982), *J. Colloid Interface Sci.*, **88**, 29.

Petrov, I., Kulev, I., and Platikanov, D. (1983), *God. Sofil. Univ., Khim. Fak.*, **73**, 129.

Petty, T. L., Reiss, O. K., Paul, G. W., Silvers, G. W., and Elkins, N. D. (1977), *Am. Rev. Respir. Dis.*, **115**, 531.

Petty, T. L., Silvers, G. W., Paul, G. W., and Stanford, R. E. (1979), *Chest*, **75**, 571.

Peyser, P., Tutas, D. J., and Stromberg, R. R. (1967), *J. Polym. Sci. A-1*, **5**, 651.

Phillips, J. N., and Rideal, E. K. (1955), *Proc. R. Soc. London Ser. A*, **232**, 149.

Phillips, M. C., and Chapman, D. (1968), *Biochim. Biophys. Acta*, **163**, 301.

Phillips, M. C., and Hauser, H. (1974), *J. Colloid Interface Sci.*, **49**, 31.

Phillips, M. C., Williams, R. M., and Chapmann, D. (1969), *Chem. Phys. Lipid*, **3**, 234.

Phillips, M. C., Evans, M. T. A., Graham, D. E., and Oldani, D. (1975), *Colloid Polym. Sci.*, **253**, 424.

Pilpel, N., and Hunter, B. F. J. (1970), *J. Colloid Interface Sci.*, **33**, 615.

Pinkerton, M., Steinrauf, L. K., and Dawkins, P. (1969), *Biochim. Biophys. Acta*, **35**, 512.

Pippard, A. B. (1957), *The Elements of Classical Thermodynamics*, Cambridge University Press, Cambridge.

Pluckthun, A., DeBony, J., Fanni, T., and Dennis, E. A. (1986), *Biochim. Biophys. Acta*, **856**, 144.

Pommier, H. P., Baril, J., Giuda, I., and le Blanc, R. M. (1979), *Can. J. Chem.*, **57**, 1377.

Porter, E. F. (1937), *J. Am. Chem. Soc.*, **59**, 1883.

Porter, E. F., and Wyman, J., Jr. (1938), *J. Am. Chem. Soc.*, **60**, 1083.

Poskanzer, A., Goodrich, F. C., and Birdi, K. S. (1974), American Chem. Soc. Meeting, Los Angeles.

Poste, G., and Allison, A. C. (1973), *Biochim. Biophys. Acta*, **300**, 421.

Prabhakaran, M., and Ponnuswamy, P. K. (1979), *J. Theor. Biol.*, **80**, 485.

Prakash, M., Dutta, P., Ketterson, J. B., and Abraham, B. M. (1984), *Chem. Phys. Lett.*, **111**, 395.

Prats, M., Tocanne, J.-F., and Teissie, J. (1985), *Eur. J. Biochem.*, **149**, 663.

Pressman, B. C. (1976), *Annu. Rev. Biochem.*, **45**, 501.

Puddington, I. E. (1946), *J. Colloid Interface Sci.*, **1**, 505.

Puskin, J. S., and Gunter, T. E. (1975), *Biochemistry*, **14**, 187.

Putney, J. W., Weiss, S. J., van de Walle, C. M., and Haddas, R. A. (1980), *Nature (London)*, **284**, 345.

Quinn, P. J., and Dawson, R. M. C. (1969), *Biochem. J.*, **113**, 791; **115**, 165.

Quinn, P. J., and Dawson, R. M. C. (1970), *Biochem. J.*, **116**, 671.

Quinn, P. J., and Dawson, R. M. C. (1972), *Chem. Phys. Lipids*, **8**, 1.

Rabinovitch, W., Robertson, R. F., and Mason, S. G. (1960), *Can. J. Chem.*, **38**, 1881.

Rabolt, J. F., Burns, F. C., Schlotter, N. E., and Swalen, J. D. (1983), *J. Chem. Phys.*, **78**, 946.

Ralston, E., and de Coen, J. L. (1974a), *J. Mol. Biol.*, **83**, 393.

Ralston, E., and de Coen, J. L. (1974b), *Proc. Natl. Acad. Sci. U.S.A.*, **71**, 1142.

Ralston, E., and Healy, T. W. (1973), *Adv. Colloid Interface Sci.*, **9**, 303.

Ramachandran, G. N., and Sasisekharan, V. (1968), *Adv. Protein Chem.*, **23**, 283.

Ramachandran, L. K. (1963), *Biochemistry*, **2**, 1138.

Raman, C. V., and Ramdas, L. A. (1927), *Phil. Mag.*, **3**, 220.

Ramsay, W., and Shields, J. (1893), *J. Chem. Soc.*, 1089.

Rancell, C., and Ziritt, J. L. (1983), Informe Tecnico INTEVEP, SA-INT-00-691-83, February.

Rao, A. A., Wasan, D. T., and Manev, E. D. (1982), *Chem. Eng. Commun.*, **15**, 63.

Ree, F. H., and Hoover, W. G. (1964), *J. Chem. Phys.*, **40**, 939.

Reifenrath, R., and Zimmermann, I. (1976), *Respiration*, **33**, 303.

Reifenrath, R., Vatter, A., and Lin, C. (1981), *Prog. Resp. Res.*, **15**, 49.

Ries, H. E., and Kimball, W. A. (1955), *J. Phys. Chem.*, **59**, 992.

Ries, H. E., and Kimball, W. A. (1957), in *Proc. 2nd Inter. Congr. Surface Active Agents*, Vol. 1, p. 75.

Ries, H. E., and Walker, D. C. (1961), *J. Colloid Sci.*, **16**, 361.

Reiss-Husson, F. (1967), *J. Molec. Biol.*, **25**, 363.

Reiter, P. (1978), *Ann. Tropical Med. Parasitol.*, **72**, 169.

Reiter, P., and McMullen, A. I. (1978), *Ann. Tropical Med. Parasitol.*, **72**, 163.

Rideal, E. K. (1925), *J. Phys. Chem.*, **29**, 1585.

Rideal, E. K. (1930), *Surface Chemistry*, 2nd ed., Butterworths, London.

Rideal, E. K., and Mitchell, J. S. (1937), *Proc. R. Soc. London Ser. A*, **159**, 206.

Ringer, S. (1883), *J. Physiol. London*, **4**, 29.

Robbins, M. L., and La Mer, V. K. (1960), *J. Colloid Sci.*, **15**, 123.

Roberts, G. G. (1983), *Sens. Actuators*, **4**, 131.

Robeson, C. D., Blum, W. P., Dietrele, J. M., Crawley, J. D., and Baxter, J. G. (1955), *J. Am. Chem. Soc.*, **77**, 4120.

Robin, M. (1985), Proc. 3rd Europ. Meet. Improved Oil Recov., Rome, April.

Robinson, M. (1937), *Nature*, **139**, 626.

Robson, B., and Pain, R. H. (1971), *J. Mol. Biol.*, **58**, 237.

Rooney, S. A., and Gobran, L. I. (1977), *Lipids*, **12**, 1050.

Rooney, S. A., Canavan, P. M., and Motoyama, E. K. (1974), *Biochim. Biophys. Acta*, **360**, 56.

Rosano, H. L., and La Mer, V. K. (1956), *J. Phys. Chem.*, **60**, 348.

Rosmalen, R. J.-V., de Boer, R. B., and Keyzer, P. P. M. (1985), Proc. 3rd Europ. Meet. Improved Oil Recov., Rome, April.

Rothen, A. (1945), *Rev. Sci. Instr.*, **16**, 1626.

Rudol, V. M., and Ogarev, V. A. (1978), *Kolloidn. Zh.*, **40**, 270.

Ruttkay-Nedecky, G., and Bezuch, B. (1971), *Exp. Suppl.*, **18**, 553.

Sackman, E., and Trauble, H. (1972), *J. Amer. Chem. Soc.*, **94**, 4499.

Saffman, P. G., and Delbruck, M. (1975), *Proc. Natl. Acad. Sci. U.S.A.*, **72**, 3111.

Sagiv, J. (1979), *Israel J. Chem.*, **18**, 339.

Salem, L. (1960), *Molec. Phys.*, **3**, 441.

Salem, L. (1962a), *Can. J. Biochem. Physiol.*, **40**, 1287.

Salem, L. (1962b), *J. Chem. Phys.*, **37**, 2100.

Salem, L. (1962c), *Can. J. Biochem. Physiol.*, **40**, 1287.

Salzman, E. W. (1971), *Blood*, **38**, 509.

Sandu, C., and Lund, D. (1979), *Proc. Int. Conf. Fouling and Heat Exch. Surf.*, August, Rensselaer Polytechnic Institute, Troy, New York.

Saraga, L., T.-M. (1949), *C.R. Acad. Sci. Paris*, **228**, 548.

Saraga, L., T.-M. (1955), *J. Chim. Phys.*, **52**, 181.

Saraga, L., T.-M. (1957), in *Proc. 2nd Inter. Congr. Surf. Active Agents*, Vol. 1, Butterworths, London.

Saraga, L., T.-M. (1975), in *Progress in Surface and Membrane Science* (Cadenhead, D. A., Danielli, J. F., and Rosenberg, M. D., eds.), Vol. 9, Academic Press, New York.

Sarges, R., and Witkop, B. (1965), *J. Am. Chem. Soc.*, **87**, 2011.

Sasaki, K., and Matuura, R. (1951), *Bull. Chem. Soc. Jpn.*, **24**, 274.

Sato, S., and Kishimoto, H. (1982), *J. Colloid Interface Sci.*, **88**, 574.

Scarpa, A., Baldassare, J., and Inesi, G. (1972), *J. Gen. Physiol.*, **60**, 735.

Scatchard, G. (1976), *Equilibrium in Solutions*, Harvard University Press, Cambridge.

Scheibe, G., and Schuller, H. (1955), *Z. Elektrochem.*, **59**, 861.

Scheller, F., Jauchen, M., and Prumke, H. J. (1975), *Biopolymers*, **14**, 1553.

Schick, M. J., and Fowkes, F. M. (1967), *J. Phys. Chem.*, **61**, 1062.

Schirer, H. U., and Overath, P. (1969), *J. Mol. Biol.*, **44**, 209.

Schmidt, A., Varoqui, R., Uniyal, S., Brash, J. L., and Pusineri, C. (1983), *J. Colloid Interface Sci.*, **92**, 25.

Schoch, P., and Sargent, D. F. (1980), *Biochim. Biophys. Acta*, **402**, 234.

Schuerholz, T., and Schindler, H. (1983), *FEBS Lett.*, **152**, 187.

Schulman, J. H., and Dogan, M. E. (1954), *Faraday Soc. Disc.*, **16**, 158.

Schulman, J. H., and Hughes, A. H. (1932), *Proc. R. Soc. London Ser. A*, **138**, 430.

Schulman, J. H., and Hughes, A. H. (1935), *Biochem. J.*, **29**, 1236, 1243.

Schulman, J. H., and Rideal, E. K. (1931), *Proc. R. Soc. London Ser. A*, **130**, 259.

Schulman, J. H., and Rideal, E. K. (1937), *Proc. R. Soc. London Ser. B*, **122**, 29.

Schulman, J. H., and Teorell, T. (1938), *Trans. Faraday Soc.*, **34**, 1337.

Schultz, R. D., and Asunmaa, S. K. (1971), *Progress in Surface Science*, Vol. 3, Academic Press, New York.

Schultz, R. D., and Asumaa, S. K. (1971), *J. Biol. Chem.*, **246**, 2211.

Schutenberger, P., Bertani, R., and Kanzig, W. (1986), *J. Colloid Interface Sci.*, **114**, 82.

Schwartz, E. G., and Reid, W. J. (1964), *Industr. Eng. Chem.*, **56**, 26.

Schwarz, J. (1982), *Ber. Bunsenges. Phys. Chem.*, **86**, 613.

Scott, H. L. (1975), *J. Chem. Phys.*, **62**, 1347.

Sears, G. (1969), *Kolloid Z. Z. Polym.*, **234**, 1030.

Sebba, F. (1972a), *J. Colloid Interface Sci.*, **40**, 479.

Sebba, F. (1972b), *J. Colloid Interface Sci.*, **40**, 468.

Seimiya, T., and Ohki, S. (1972), *Biochim. Biophys. Acta*, **274**, 15.

Seimiya, T., Ashida, M., Heki, Y., Muramatsu, T., Hara, I., and Hayashi, M. (1976), *J. Colloid Interface Sci.*, **55**, 388.

Sekita, K., Nakamura, M., Motomura, K., and Matuura, R. (1976), *Mem. Fac. Sci. Kyushu Univ.*, *Ser. C*, **10**, 51.

Senior, W. (1943), *Indian Medical Gaz.*, **78**, 342.

Sessa, G., Freer, J. H., Colacicco, G., and Weissman, G. (1969), *J. Biol. Chem.*, **244**, 3575.

Shafer, P. T. (1974), *Biochim. Biophys. Acta*, **373**, 425.

Shah, D. O. (1969), *Biochim. Biophys. Acta*, **193**, 217.

Shah, D. O. (1974), *Am. Chem. Soc. Symp. Ser.*, **8**, 170.

Shah, D. O., and Schulman, J. H. (1967), *J. Colloid Interface Sci.*, **25**, 107.

Shah, D. O., Ellis, S. M., Aderangi, N., Chan, M. S., and McNamara, J. J. (1978), *Soc. Pet. Eng. J.*, 409.

Shapiro, E. (1975), Ph.D. thesis, University of Michigan.

Sheppard, E., Bronson, R. P., and Tcheurekdjian, N. (1964), *J. Colloid Sci.*, **19**, 833.

Sheppard, E., Bronson, R. P., and Tcheurekdjian, N. (1965), *J. Colloid Sci.*, **20**, 755.

Shirahama, H., and Shuzawa, T. (1985), *Colloid Polym. Sci.*, **263**, 141.

Siketa, K., Nakamura, N., Motomura, K., and Matuura, R. (1976), *Mem. Fac. Sci., Kyushu Univ.*, **C10**, 51.

Silman, I. H., and Katchalski, E. (1966), *Ann. Rev. Biochem.*, **35**, 873.

Silvius, J. R., and McElhaney, R. N. (1980), *Chem. Phys. Lipids*, **26**, 67.

Silvius, J. R., Mak, N., and McElhaney, R. N. (1980), *Biochim. Biophys. Acta*, **597**, 199.

Simovic, M., and Dobrilovic, L. J. (1978), *J. Radio Chem.*, **44**, 345.

Sims, B., and Zografi, G. (1972), *J. Colloid Interface Sci.*, **41**, 35.

Singer, S. J. (1948), *J. Chem. Phys.*, **16**, 872.

Singer, S. J. (1971), in *Membrane Structure and Function* (Rothfield, L. I., ed.), Academic Press, New York.

Singer, S. J. (1974), in *Annual Review of Biochemistry* (Snell, E. E., Boyer, P. D., Meister, A., and Richardson, C. C., eds.), Vol. 43, Annual Reviews, Palo Alto, California.

Singer, S. J., and Nicolson, G. L. (1972), *Science*, **175**, 720.

Singh, K. R. P., and Micks, D. W. (1957), *Mosq. News*, **17**, 70.

Sirhatta, R., and Scott, G. D. (1971), *Appl. Opt.*, **10**, 2192.

Skulachev, V. P. (1980), *Biochim. Biophys. Acta*, **604**, 297.

Small, D. M. (1967), *J. Am. Oil Chem. Soc.*, **45**, 108.

Smith, J. L., and Budenstein, P. P. (1969), *J. Appl. Phys.*, **40**, 3491.

Snow, A. W., Barger, W. R., Klusty, M., Wohltjen, H., and Jarvis, N. L. (1986), *Langmuir*, **2**, 513.

Snow, E. H., Grove, A. S., Deal, B. E., and Sah, G. T. (1965), *J. Appl. Phys.*, **36**, 1664.

Soberquist, M. E., and Walton, A. G. (1980), *J. Colloid Interface Sci.*, **75**, 386.

Sodergren, A. (1978), *Mitt. Internat. Verein. Limnol.*, **21**, 248.

Somasundaran, P., Dnaitz, M., and Mysels, K. J. (1974), *J. Colloid Interface Sci.*, **48**, 410.

Spink, J. A., and Sanders, J. V. (1955), *Proc. Roy. Soc., London*, **A227**, 537.

Srivastava, V. K., and Verma, A. R. (1962), *Proc. Phys. Soc. (London)*, **80**, 222.

Srivastava, V. K., and Verma, A. R. (1966), *Solid State Commun.*, **4**, 367.

Stefan, J. (1873), *Wien. Akad. Ber.*, **68**, 383.

Stefan, J. (1886), *Wien. Akad. Ber.*, **94**, IIA, 4; *Ann. Phys.*, **29**, 255.

Stein, W. D. (1967), *The Movement of Molecules across Cell Membranes*, Academic Press, New York.

Sterling, C. V., and Scriven, L. E. (1959), *Am. Ind. Chem. J.*, **5**, 514.

Steven, J. H., Hann, R. A., Barlow, A., and Laird, T. (1983), *Thin Solid Films*, **99**, 71.

Stoeckenius, W., and Engelman, D. M. (1969), *J. Cell Biol.*, **42**, 613.

Stoner, G., and Srinivasan, S. (1970), *J. Phys. Chem.*, **74**, 1088.

Stroeve, P., and Miller, I. (1975), *Biochim. Biophys. Acta*, **401**, 157.

Struck, D. K., and Pagano, R. E. (1980), *J. Biol. Chem.*, **225**, 5404.

Stuart, F. C. (1961), *Austr. J. Chem.*, **14**, 57.

Sturtevant, J. M. (1982), *Proc. Natl. Acad. Sci. U.S.A.*, **79**, 3963.

Suciu, D. G., Smigelschi, O., and Ruckenstein, E. (1970), *J. Colloid Interface Sci.*, **33**, 520.

Suggett, A. (1977), *Pharmaco. Chem. Libr.*, **1**, 95.

Sumper, M., and Trauble, H. (1973), *FEBS Lett.*, **30**, 29.

Szabo, G. (1974), *Nature*, **252**, 47.

Szabo, G., Eisenman, G., Laprade, R., and Ciani, S. M. (1973), in *Membranes*, Vol. 2 (Eisenmann, G., ed.), Marcel Dekker, New York.

Szoka, F., and Papahadjopoulos, D. (1978), *Proc. Natl. Acad. Sci. U.S.A.*, **75**, 4194.

Szundi, I. (1978), *Chem. Phys. Lipids*, **22**, 153.

Tajima, K., and Gershfeld, N. L. (1978), *Biophys. J.*, **22**, 489.

Tallon, J. L., and Cotterill, R. M. J. (1985), *Aust. J. Phys.*, **38**, 1.

Tancrede, P., Munger, G., and Leblanc, R. M. (1982), *Biochim. Biophys. Acta*, **689**, 45.

Tanford, C. (1962), *J. Am. Chem. Soc.*, **84**, 4240.

Tanford, C. (1980), *The Hydrophobic Effect*, John Wiley and Sons, New York.

Tanguy, J. (1972), *Thin Solid Films*, **13**, 33.

Taylor, J. A. G., Mingins, J., Pethica, B. A., Tan, B. Y. J., and Jackson, C. M. (1973), *Biochim. Biophys. Acta*, **323**, 157.

Teissie, J. (1979), *Chem. Phys. Lipids*, **25**, 357.

Teissie, J. (1981), *Biochemistry*, **20**, 1554.

Teissie, J., Tocanne, J. F., and Baudras, A. (1976), *FEBS Lett.*, **70**, 123.

Teissie, J., Tocanne, J. F., and Baudras, A. (1978), *Eur. J. Biochem.*, **83**, 77.

Teissie, J., Soucaille, P., and Tocanne, J. F. (1987), *Proc. Natl. Acad. Sci. U.S.A.*, **84**, 341.

Theies, C. (1966), *J. Phys. Chem.*, **70**, 3783.

Theorell, T. (1938), *Biochem. Z.*, **298**, 258.

Thiessen, J., de Wael, J. and Havinga, E. (1940), *Rec. Trav. Chim.*, **59**, 770.

Thin Solid Films (1980), **68** (entire issue).

Thorne, R. S. W. (1964), *Brew. Digest*, **39**, 50.

Thuman, W. C., Brown, A. G., and McBain, J. W. (1949), *J. Am. Chem. Soc.*, **71**, 3129.

Tien, H. T. (1970), *Adv. Expt. Med. Biol.*, **7**, 135.

Tien, H. T. (1974), *Bilayer Lipid Membranes [BLM]. Theory and Practice*, Marcel Dekker, New York.

Tien, H. T., and Verma, S. P. (1970), *Nature*, **227**, 1232.

Ting, L., Wasan, D. T., Miyano, K., and Xu, S. Q. (1984), *J. Colloid Interface Sci.*, **102**, 248.

Tornberg, E. (1978), *J. Sci. Food. Agr.*, **29**, 762.

Tosteson, M. T., and Tosteson, D. C. (1981), *Biophys. J.*, **36**, 109.

Tosteson, M. T., and Tosteson, D. C. (1984), *Biophys. J.*, **45**, 112.

Traube, J. (1891), *Annalen*, **265**, 27.

Trapeznikov, A. A. (1939), *Acta Physicochim.*, **10**, 65.

Trapeznikov, A. A. (1941), *Dokl. Akad. Nauk. SSSR*, **30**, 321.

Trapeznikov, A. A. (1945), *Acta Physicochim. URSS*, **20**, 589.

Trapeznikov, A. A. (1964), in *Proc. 4th Int. Congr. Surface Active Substances*, Vol. 2, p. 857, Brussels.

Trapeznikov, A. A., and Avetisyan, R. A. (1970), *Russ. J. Phys. Chem.*, **44**, 76.

Trapeznikov, A. A., and Lonomosova, T. A. (1967), *Russ., J. Phys. Chem.*, **41**, 139.

Triggle, D. J. (1972), in *Prog. in Surface Membrane Science* (Danielli, J. F., Rosenberg, M. D., and Cadenhead, D. A., eds.), Vol. 7, Academic Press, New York.

Tromstad, L., and Fichum, C. G. P. (1934), *Proc. R. Soc. London Ser. A*, **145**, 115, 127.

Tscharner, V., and McConnell, H. M. (1981), *Biophys. J.*, **36**, 409.

Tschoegl, N. W. (1962), *Kolloid-Z.*, **181**, 19.

Tschoegl, N. W., and Alexander, A. E. (1960), *J. Colloid Sci.*, **15**, 168.

Trurnit, H. J. (1960), *J. Colloid Sci.*, **15**, 1.

Trurnit, H. J., and Lauer, W. E. (1959), *Rev. Sci. Instrum.*, **30**, 975.

Trurnit, H. J., and Schidlovsky, G. (1961), in *Proc. Europ. Region Conf. Electron Microscopy*, **2**, 721.

Tvaroha, B. (1954), *Chem. Listy*, **48**, 183.

Tweet, A. G. (1963), *Rev. Sci. Instrum.*, **34**, 1412.

Uhlig, A. (1937), *Kolloid-Z.*, **75**, 81.

Urry, D. W. (1971), *Proc. Natl. Acad. Sci. U.S.A.*, **68**, 672.

Uzgiris, E. E., and Fromageot, H. P. M. (1976), *Biopolymers*, **15**, 257.

van den Tempel, M. (1977), *J. Non-Newtonian Flow Mech.*, **2**, 205.

Vanderkooi, J. M., and Callis, J. B. (1974), *Biochemistry*, **13**, 4000.

van Emdenkroon, C. C. M., Schoeman, E. N., and van Senter, H. A. (1974), *Environmental Pollution*, **6**, 296.

Van Mau, N., and Amblard, J. (1983), *J. Colloid Interface Sci.*, **91**, 138.

Van Mau, N., Tenebre, L., and Gavach, C. (1980), *J. Electroan. Chem. Int. Electrochem.*, **114**, 225.

Verkley, A. J., Zwaal, R. F. A., Roelofsen, B., Kastelijn, D., and van Deenen, L. L. M. (1973), *Biochim. Biophys. Acta*, **323**, 178.

Verger, R., and Haas, G. H. de (1973), *Chem. Phys. Lipids*, **10**, 127.

Verger, R., Rietsch, J., van Dam-Mieras, M. C. E., and De Haas, G. H. (1976), *J. Biol. Chem.*, **251**, 3128.

Vincent, P. S., and Roberts, G. G. (1980), *Thin Solid Films*, **68**, 135.

Vollhardt, D., and Wuestneck, R. (1974), *Kolloid Z., Moskva*, **36**, 1116.

Vollhardt, D., Wuestneck, R., and Zastrow, I. (1978), *Colloid Polym. Sci.*, **256**, 983.

Volta, A. (1801), *Ann. Xhim. Phys.*, **40**, 225.

von Hippel, P. H., and Scheich, T. (1959), *Structure and Stability of Macromolecules* (Timasheff, S. N., and Fasman, G. D., eds.), Marcel Dekker, New York.

von Neergaard, K. (1929), *Z. Gesamte Exp. Med.*, **66**, 373.

Von Wichert, P., and Kohl, F. V. (1977), *Intens. Care Med.*, **3**, 27.

Vroman, L. (1969), *J. Biom. Mat. Res.*, **3**, 669.

Vroman, L., Kanor, S., and Adams, A. L. (1968), *Rev. Sci. Instrum.*, **39**, 278.

Waggoner, A. S., Wang, C. H., and Tolles, R. L. (1977), *J. Membr. Biol.*, **33**, 109.

Wald, G., Brown, P. K., and Gibbons, I. R. (1963), *J. Opt. Soc. Am.*, **53**, 20.

Ward, A. F. H., and Tordai, L. (1946), *J. Chem. Phys.*, **14**, 453.

Wasan, D. J., Gupta, L., and Vora, M. K. (1971), *Am. Ind. Chem. J.*, **17**, 1287.

Watterson, J. C., Schaub, N. C., and Waser, P. G. (1974), *Biochim. Biophys. Acta*, **356**, 133.

Weetall, H. H. (1971), *Res. Dev.*, **22**, 18.

Wegener, P. P., and Parlange, J. Y. (1964), *Z. Phys. Chem.*, *Frankfurt am Main*, **43**, 245.

Weidmann, S. (1955), *J. Physiol.*, *London*, **129**, 568.

Weitzel, G., Fretzdorff, A. M., and Heller, S. (1952), *Hoppe-Seylers Z. Physiol. Chem.*, **290**, 32.

Weitzel, G., Fretzdorff, A. M., and Heller, S. (1956), *Hoppe-Seylers Z. Physiol. Chem.*, **303**, 14.

Weitzman, P., Kennedy, J., and Caldwell, R. (1971), *FEBS Lett.*, **17**, 241.

Wen, W.-Y., and Muccitelli, J. A. (1979), *J. Solution Chem.*, **8**, 225.

Wigglesworth, V. B. (1972), *The Principles of Insect Physiology*, Methuen, London.

Wilhelmy, L. (1863), *Ann. Phys. (Leipzig)*, **119**, 177.

Winterton, L. C., Andrade, J. D., Feijen, J., and Kim, S. W. (1986), *J. Colloid Interface Sci.*, **111**, 314.

Wojciak, J. F., Notter, R. H., and Oberdorster, G. (1985), *J. Colloid Interface Sci.*, **106**, 547.

Wolf, K. L. (1959), *Physik und Chemie Grenflachen*, 2 vols., Springer-Verlag, Berlin.

Wolstenholme, T., and Schulman, J. H. (1950), *Trans. Faraday Soc.*, **46**, 475.

Wolstenholme, T., and Schulman, J. H. (1951), *Trans. Faraday Soc.*, **47**, 788.

Wu, S. H. W., and McConnell, H. M. (1973), *Biochem. Biophys. Res. Commun.*, **55**, 484.

Wuestneck, R., and Zastrow, L. (1985), *Colloid Polym. Sci.*, **263**, 778.

Wulf, J., and Pohl, W. G. (1977), *Biochim. Biophys. Acta*, **465**, 471.

Wulf, J., Benze, R., and Pohl, W. G. (1977), *Biochim. Biophys. Acta*, **465**, 429.

Yamashita, S., and Yamashita, T. (1978), *Maku*, **4**, 295.

Yamashita, T. (1977), *Maku*, **4**, 283.

Yamashita, T., and Bull, H. B. (1967), *J. Colloid Interface Sci.*, **24**, 310.

Yamashita, T., Shibata, A., and Yamashita, S. (1978a), *Bull. Chem. Soc. Jpn.*, **51**, 2751.

Yamashita, T., Shibata, A., and Yamashita, S. (1978b), *Chem. Lett.*, *Chem. Soc. Jpn.*, **51**, 11.

Yamins, H. G., and Zisman, W. A. (1933), *J. Chem. Phys.*, **1**, 656.

Yasuaki, M., and Toshizo, I. (1967), *Nature*, **215**, 765.

Yatsyuk, M. D. (1972), *Kollod. Zh.*, **34**, 135.

Yin, T. P., and Wu, S. (1971), *J. Polym. Sci.*, **C34**, 265.

Young, T. (1855), *Miscellaneous Works* (Peacock, G., ed.), Vol. I, J. Murray, London.

Yue, B. Y., Jackson, C. M., Taylor, J. A. G., Mingins, J., and Pethica, B. A. (1976), *J. Chem. Soc., Faraday Trans.*, *I*, **72**, 2658.

Zaborsky, O. (1973), *Immobilized Enzymes*, CRC Press, Cleveland, Ohio.

Zastrow, L., Wuestneck, R., and Kretzschmar, G. (1985), *Colloid Polym. Sci.*, **263**, 749.

Zilversmit, D. V. (1963), *J. Colloid Sci.*, **794**, 18.

Ziritt, J. L., Paz, F., and Rancel, C. (1985), Proc. 3rd Europ. Meet. Improved Oil Recov., Rome, April.

Zisman, W. A. (1932), *Rev. Sci. Instrum.*, **3**, 369.

Zittle, C. A., Della Monica, E. S., Rudd, R. K., and Custer, J. H. (1957), *J. Am. Chem. Soc.*, **79**, 4661.

Zografi, G., Verger, R., and Haas, G. H. de (1971), *Chem. Phys. Lipids*, **7**, 185.

Zwaal, R. F. A., Roelofsen, B., and Colley, C. M. (1973), *Biochim. Biophys. Acta*, **300**, 159.

Zwanzig, R. (1963), *J. Colloid Sci.*, **39**, 2251.

INDEX

317

Caseinate–Na, 202, 206
Cell
 adsorption at an interface, 219
 surface hydrophobicity, 219
Cell adhesion, 219
Charged lipid monolayers, 77
Charged monolayer films, 222
 biopolymers, 170, 183, 222
 lipids, 77, 210, 217
Charges on surfaces, 21
Chlorophyll monolayers, 24, 119
Chloroplasts, 254
Cholesterol
 mixed monolayers of, 109, 119–120, 123,
  130, 140, 144, 177, 210, 216, 220,
  222, 224
 monolayers, 31, 54, 101, 103, 107, 147,
  152, 226
Chromatophore membrane, lipid monolayers,
  255
Cis/trans
 fatty acid, 61
 retinal, 61
Clapeyron equation, 96
Clausius–Clapeyron equation, 96
Cleaning of troughs: *see* Langmuir balance
Cohesive forces in monolayers (*see also* van
  der Waals forces), estimation of, 71
Collapse states, 44, 75, 108–109, 128, 248
Comparison of data on phase transition by
  different monolayer spreading methods,
  93
Complexes, in films, 107
Compressibility of monolayer, 164
 definition, 23, 88
Compression barrier, 31
Compression rates, 128, 248
Concentration at surface: *see* Surface
  concentration
Condensed monolayers, polymers, 178
Condensing effect in mixtures: *see* Mixed
  monolayers
Conformation of steroid compounds, 144
Contact angle, 13
Controlled-pore glass (CPG), 290
 binding sites of, 290
Coordination number(z), 64
Copper distearate monolayers, 122
Corresponding states theory, 8
Critical pressure, 10
Critical temperature, 9
Cytochrome-C, 210

Davies equation, 61, 82, 184
DDT monolayers, 119
Debye–Huckel theory, 80
Debye length, 80, 236
Degree of unfolding of proteins, 157
Desorption kinetics of lipid monolayers, 206
Dialkyl chain lipids, monolayers, 102
Dialkyl-phosphatidylcholines
 phase-transition
  enthalpy, 95
  entropy, 95
  temperatures, 95
Diffusion
 coefficient, 198
 constants, 203
 of gases through monolayers, 25
 in monolayer, 197
 translational, 147
Dilaurin, 102
Dioleoyl lecithin, 103
Dipalmitoyl-lecithin (DPL or DPPC), mono-
  layers of, 42, 49, 75
Dipalmitoylphosphatidylethanolamine, transi-
  tion temperature, 140
Dipole moment, 47
Discrete charges in protein monolayers, 235
Dispersion forces: *see* van der Waals
Distearoyl-lecithin (DSL), monolayers of, 91
Docosanoic acid monolayers, 128
Docosyl sulfate, mixed films of, 258
Dodecanol, 138
Double bonds, effect on monolayer properties,
  59
Double layer, electrical, 77
Drop weight method, 36
Dynamic isotherms, of mixed lipids, 130
Dynamic lipid monolayer measurements, 128–
  131, 225

Effect of light on monolayers, 132
Effect of monolayers on water evaporation
  rates, 124
Effect of pH
 on biopolymer monolayers, 183
 on lipid monolayers, 82
Effect of solvents on monolayers, 37
Eicosanoic acid monolayers, 128
Elasticity
 of films, 281
 Gibbs, 141, 281
 of monolayers, 281